KARL CHANG

495

D0849270

Fundamentals
and Modeling
of
Separation Processes

PRENTICE-HALL INTERNATIONAL SERIES
IN THE PHYSICAL AND CHEMICAL ENGINEERING SCIENCES

Neal R. Amundson, editor, *University of Minnesota*

ADVISORY EDITORS

Andreas Acrivos, *Stanford University*
John Dahler, *University of Minnesota*
Thomas J. Hanratty, *University of Illinois*
John M. Prausnitz, *University of California*
L. E. Scriven, *University of Minnesota*

HOLLAND *Fundamentals and Modeling of Separation Processes:
Absorption, Distillation, Evaporation, and Extraction*

HOLLAND *Multicomponent Distillation*

HOLLAND *Unsteady State Processes with Applications in
Multicomponent Distillation*

KOPPEL *Introduction to Control Theory with Applications to Process Control*

LEVICH *Physicochemical Hydrodynamics*

MEISSNER *Processes and Systems in Industrial Chemistry*

MODELL AND REID *Thermodynamics and its Applications*

NEWMAN *Electrochemical Systems*

OHARA AND REID *Modeling Crystal Growth Rates from Solution*

PERLMUTTER *Stability of Chemical Reactors*

PETERSEN *Chemical Reaction Analysis*

PRAUSNITZ *Molecular Thermodynamics of Fluid-Phase Equilibria*

PRAUSNITZ, ECKERT, ORYE, AND O'CONNELL *Computer Calculations for
Multicomponent Vapor-Liquid Equilibria*

RUDD et al. *Process Synthesis*

SCHULTZ *Polymer Materials Science*

SEINFELD AND LAPIDUS *Mathematical Methods in Chemical Engineering:
Vol. III, Process Modeling, Estimation, and Identification*

WHITAKER *Introduction to Fluid Mechanics*

WILDE *Optimum Seeking Methods*

WILLIAMS *Polymer Science and Engineering*

PRENTICE-HALL, INC.
PRENTICE-HALL INTERNATIONAL, INC.,
UNITED KINGDOM AND EIRE
PRENTICE-HALL OF CANADA, LTD., CANADA

Fundamentals and Modeling of Separation Processes

Absorption, Distillation, Evaporation, and Extraction

CHARLES D. HOLLAND

Professor and Head
Department of Chemical Engineering
Texas A&M University

PRENTICE-HALL, INC.

Englewood Cliffs, New Jersey

Library of Congress Cataloging in Publication Data

HOLLAND, CHARLES DONALD.
   Fundamentals and modeling of separation processes.

   (Prentice-Hall international series in the physical
and chemical engineering sciences)
   Includes bibliographies.
   1. Separation (Technology)   2. Separation (Technology)—
Mathematical models.   I. Title.
TP156.S45H65   660.2′842   74–8872
ISBN   0–13–344390–6

©1975 by Prentice-Hall, Inc.
Englewood Cliffs, New Jersey

All rights reserved. No part of this book
may be reproduced in any form or by any means
without permission in writing from the publisher.

10   9   8   7   6   5   4   3   2

Printed in the United States of America

PRENTICE-HALL INTERNATIONAL, INC., *London*
PRENTICE-HALL OF AUSTRALIA, PTY. LTD., *Sydney*
PRENTICE-HALL OF CANADA, LTD., *Toronto*
PRENTICE-HALL OF INDIA PRIVATE LIMITED, *New Delhi*
PRENTICE-HALL OF JAPAN, INC., *Tokyo*

To

*Nancy
Charlotte
and Thomas*

Contents

Section I. Fundamentals and Modeling of Equilibrium Stage Processes

Appendices

Preface

Although this book is oriented toward the modeling of processes, a casual examination of the contents shows that this orientation is not at the expense of the presentation of the fundamental principles that are involved in separation processes. The treatment of each of the separation processes, evaporation, distillation, absorption and stripping, and extraction begins with first principles in Section 1. The sequel then advances to the modeling of processes from the results of field tests.

The term *model* used here means the complete set of assumptions pertaining to the behavior of a process and the corresponding set of equations required to predict the observable behavior. Model is simply a new word for the description of physical phenomena such as heat transfer, mass transfer, and chemical reactions. Models that have been found useful such as Fourier's law, Fick's law, Newton's law of cooling, and the rate expressions for convective heat and mass transfer are generally classified as fundamentals. Nonetheless, these relationships seldom constitute exact descriptions of the given processes.

Well-accepted models such as the equilibrium stage at steady-state operation are presented in Section 1. Rate processes and processes at unsteady-state operation are then treated in Section 2. The second section is initiated by the presentation of some common rate expressions as well as basic techniques required to set up the equations. This is first done to describe some of the well-known classical models. The section is concluded by the modeling of absorption, distillation, and extraction columns.

Since many of the models proposed in Section 2 are not commonplace, they are presented with the supporting results of field tests. The models presented make use of only those data that are commonly available at the plant site. Thus, the practicing engineer may use the techniques presented to assist him in improving the design and operation of his equipment.

In the modeling of packed columns, field tests run on a packed distillation column and a packed absorber at the Zoller Gas Plant (owned jointly by Hunt Oil Company and Exxon Company) in Refugio, Texas, were used. To test a new vaporization plate efficiency model, a field test made by the Monsanto Company was employed. In the modeling of liquid-liquid extraction columns, field tests made on a column used in the Edeleanu process at the Baytown Refinery of the Exxon Company were employed.

This book was written with the objective of covering each of the separation processes enumerated, and to do so from first principles to the frontiers of the present state of the art. As such, this book is arranged with chapters that are suitable for both undergraduate and graduate students. The presentation of material in each chapter is in the order of increasing difficulty. On the other hand, if an in-depth treatment of a given separation process is desired, only those chapters pertaining to the given separation process need be covered. That is, each set of chapters that pertains to a particular process is almost independent of the other chapters. For example, for an in-depth treatment of distillation, Chapters 2, 3, 4, and 8–11 should be covered in the order listed. Absorption and stripping are treated in Chapter 4 and Chapters 8 through 10. Extraction is presented in Chapters 5 and 12.

The development of the modeling techniques presented in this book required the combined efforts of many people to whom the author is deeply indebted. In particular, the author appreciates the support, assistance, and encouragement given by J. H. Galloway, M. F. Clegg, W. B. Franklin, W. M. Harp, R. B. Bennett, E. H. Oliver, and others of Humble Oil and Refining Company; R. V. Randall, W. E. Muzacz, and C. E. Cooke, Jr., of Esso Production Research Company; W. E. Vaughn, J. W. Thompson, J. D. Dyal, and J. P. Smith of Hunt Oil Company; B. J. Claybourn, L. G. Sharp, and C. L. Humphries of Mobil Oil Company; D. I. Dystra and Charles Grua of the Office of Saline Water, U.S. Department of Interior; J. P. Lennox, K. S. Campbell, and D. L. Williams of Stearns-Roger Corporation; L. H. Ballard, W. H. Lane, J. A. Glass, J. W. Fulton, J. P. Graham, and K. S. McMahon of the Monsanto Company; D. L. Rooke, Everett Jacob, E. A. Rozas, B. A. Weaver, and N. J. Tetlow of Dow Chemical Company. Direct and indirect financial support provided by the aforementioned companies and organizations is gratefully acknowledged. The author acknowledges the many contributions made by graduate students, particularly those by A. A. Bassyoni,

J. W. Burdett, A. J. Gonzalez, S. E. Gallun, A. E. Hutton, M. S. Kuk, Ronald McDaniel, G. P. Pendon, and R. E. Rubac. For the helpful advice and assistance given by Professors E. C. Klipple, H. A. Luther, and B. C. Moore of the Department of Mathematics, and Professors P. T. Eubank and D. T. Hanson of the Department of Chemical Engineering, the author is thankful. The author also appreciates the financial assistance provided by Alcoa, Dow Chemical Company, E. I. du Pont de Nemours & Company, Goodyear Tire and Rubber Company, NASA, National Science Foundation, Petro-Tex Chemical Corporation, Phillips Petroleum Company, Sun Oil Company, Texas Eastman, The Texas Engineering Experiment Station, Texaco, Inc., and Union Carbide Corporation for the support of graduate students who worked on subjects treated in this book. Finally, the author wishes to express his appreciation to Dean Fred J. Benson for his continued support of the research projects upon which this book is based.

CHARLES D. HOLLAND

Section I:
Fundamentals and Modeling
of Equilibrium Stage
Processes

Introduction 1

In the first section of this book, selected separation processes are modeled by use of the concept of the *equilibrium stage* (also called a *theoretical plate* and a *perfect plate*). In the second section, the rates of simultaneous mass and heat transfer are taken into consideration in the modeling of separation processes.

The word *model* as used herein means the complete set of assumptions and corresponding equations required to describe either a process or part of a process. The formulation of the model for a process (or system) involves (1) the application of the laws of conservation of mass and energy and (2) appropriate expressions to account for the transfer of mass and energy across the boundaries of the system.

The equilibrium stage processes considered in Section I are taken to be at steady-state operation; the rate processes considered in Section II are taken to be steady state for some models and at unsteady state for others. By *unsteady-state operation* it is meant that the variables at all points within the system do not change with time. Steady-state operation means that at least one of the variables at some point within the system varies with time.

The concept of the equilibrium stage, which is used in the models of Section I, is also used as one of the building blocks of the more general models presented in Section II. Separation processes with actual rather than theoretical (or perfect) plates are modeled in Section II. Assumptions used to model the processes presented in Section 2 are believed to be compatible with both the availability of data for physical properties and the present instrumentation available to measure process variables.

Models for the separation processes presented in Chapters 2 through 5 are based in part on the laws of conservation of mass and energy. For a process at steady-state operation, the law of conservation of mass may be stated as follows:

Input of mass to the system — output of mass from the system = 0   (1-1)

This law is the basis of the component and total material balances used in the description of the separation processes in Chapters 2 through 5.

The energy balances for these separation processes are based on the first law of thermodynamics which takes the following form for one pound mass of fluid flowing through the system

$$\Delta H + \Delta KE + \Delta PE = Q - W_s \qquad (1\text{-}2)$$

where ΔH, ΔKE, and ΔPE denote the values of the quantities possessed by one pound of material leaving the system minus the respective values per pound of material entering the system. Usually, the changes in kinetic and potential energy (ΔKE and ΔPE) are negligible, as is the shaftwork W_s done by the system on the surroundings. Under these conditions, Equation (1-2) reduces to

$$\Delta H = Q \qquad (1\text{-}3)$$

which says that the change in enthalpy ($\Delta H = H_{out} - H_{in}$) of the one pound mass of fluid is equal to the heat adsorbed by the system per pound mass of fluid flowing. Equation (1-3) serves as the basis of the enthalpy balances in the description of the separation processes in Chapters 2 through 5.

In addition to the laws of conservation of mass and energy, the concept of *equilibrium stages* is used to describe the mass transfer that occurs in these separation processes. The names *equilibrium stage, ideal stage, theoretical plate,* and *perfect plate* are all used to describe a contacting device, such as a plate, in which the vapor leaving the device is in equilibrium with the liquid leaving. A state of equilibrium is said to exist between a vapor and a liquid phase provided (1) the temperature of the vapor phase is equal to the temperature of the liquid phase, (2) the pressures throughout the vapor phase are equal to each other and to the pressure at each point throughout the liquid phase, and (3) the tendency of each component to escape from the vapor to the liquid phase is exactly equal to its tendency to escape from the liquid to the vapor phase. Various expressions are used to describe the escaping tendency, for example, Raoult's law,

$$Py_i = P_i x_i \qquad (1\text{-}4)$$

and Henry's law,

$$y_i = K_i x_i \qquad (1\text{-}5)$$

where P is the total pressure, P_i is the vapor pressure of pure component i, K_i is a Henry law constant that depends on temperature and pressure alone, and y_i and x_i are the mole fractions of component i in the vapor and liquid phases, respectively. For the case of a nonideal solution, P_i and K_i in the above expressions are preceded by an appropriate activity coefficient that depends on the composition of the liquid phase as well as on the temperature and pressure of the mixture.

The separation process, distillation, is based on the fact that the vapor phase of an equilibrium mixture is always richer in the more volatile components ($K_i > 1$ or $P_i/P > 1$) than is the liquid phase, and the liquid phase is richer in the components for which $K_i < 1$ and $P_i/P < 1$ than is the vapor phase. For example, consider a binary mixture composed of components 1 and 2. By the definition of a mole fraction, $y_1 + y_2 = 1$ and $x_1 + x_2 = 1$, and Equation (1-5) may be stated for each of the components 1 and 2 and the results added to give

$$1 = K_1 x_1 + K_2 x_2 \tag{1-6}$$

Since $x_1 + x_2 = 1$, it is evident that one K must be greater than unity and the other one must be less than unity. Suppose component 1 is the more volatile component; then at the equilibrium temperature and pressure $K_1 > 1$ and $K_2 < 1$. Then by Equation (1-5) it follows that

$$y_1 > x_1, y_2 < x_2 \tag{1-7}$$

Thus, at the state of equilibrium, the vapor phase is richer in the more volatile component than the liquid phase. Conversely, the liquid phase is richer in the least volatile component than the vapor phase. By connecting equilibrium stages in series, as is demonstrated in Chapter 3 for distillation columns, a terminal vapor stream may be obtained which is almost pure component 1 and a terminal liquid stream may be obtained which is almost pure component 2.

Evaporation 2

The separation of a solvent from a solution of a solvent and a nonvolatile
solute is commonly effected by use of the *unit operation* known as *evaporation*. Since energy is transferred in an evaporator from a condensing vapor
to a boiling liquid, evaporation may be regarded as a special case of the unit
operation called *heat transfer*. Also, evaporation may be regarded as that
special case of the unit operation *distillation* in which a solvent is separated
from a mixture of the solvent and a nonvolatile solute. When the solution to
be separated contains more than one volatile component and the separation
is effected by use of the evaporator equipment described herein, the equations
describing this process are analogous to those presented in Chapters 3 and 4
as well as those presented previously for multicomponent separations (7, 8,
11).* Consequently, only the special case of the separation of a solution of
a solvent and a nonvolatile solute is considered in the following treatment
of evaporation. Evaporators are commonly found in the inorganic, organic,
paper, and sugar industries. Typical applications include the concentration
of sodium hydroxide, brine, organic colloids, and fruit juices. Generally,
the solvent is water.

In this section the following topics are considered: the design of evaporator systems and the determination of the separations that may be effected
by an existing system of evaporators at a specified set of steady-state operating
conditions.

*1 Numbers in parentheses refer to References at the end of the chapter.

Part 1. Design of Evaporator Systems

The design of evaporators is generally taken to mean the determination of the heat transfer area and the steam consumption required to effect a specified separation at a specified set of steady-state operating conditions. Typical sets of specifications are presented in subsequent sections. Prior to the development of the design equations, certain terms, notations, and characteristics of the evaporation process are presented.

A typical forced-circulation evaporator is shown in Figure 2-1. The feed mixture to be separated or concentrated is introduced to the circulating system. The energy required to evaporate the solvent is supplied by the latent heat of vaporization given up by the steam upon condensing. The steam is introduced to the space outside the tubes and is removed as condensate or "drips." The circulating pump forces the liquid up through the tubes at a relatively high velocity (say from 6 to 18 ft/sec), and some evaporation occurs each pass. After the vapor-liquid mixture leaves the tubes, it strikes a deflector that separates a major portion of the entrained liquid. The liquid is returned to the holdup section and the vapor is withdrawn and either condensed or used as the heating medium in the chest of the next effect of a multiple effect system. Other kinds of evaporators in common use are shown in Figures 2-2 and 2-3. Discussions of the advantages enjoyed by the various kinds of evaporators in each of several applications have been presented (2, 3).

In *single-effect operation*, as the name implies, only one evaporator is employed. The feed upon entering this effect must be heated to the boiling point temperature of the effect at the operating pressure. Then the solvent, generally water, is evaporated and removed as a vapor. (Since water is the most common solvent, it is for definiteness regarded as the solvent in the development of the equations. The final equations and solutions apply, however, for any solvent.) To evaporate one pound of water from, say, a sodium hydroxide solution, about 1200 Btu are needed, and this requires better than one pound of steam. The concentrated solution withdrawn from the evaporator is known as the *thick liquor* or *process liquid*.

In *multiple-effect operation*, several evaporators are connected in series. The vapor or steam produced in the first effect is introduced to the steam chest of the second effect and thus becomes the heating medium for the second effect. Similarly, the vapor from the second effect becomes the steam for the third effect. In the case of series operation with *forward feed*, depicted in Figure 2-4, the thick liquor leaving the first effect becomes the feed for the second effect. For each effect added to the system, approximately one additional pound of solvent is evaporated per pound of steam fed to the first effect. This increase in the pounds of solvent evaporated per pound of steam

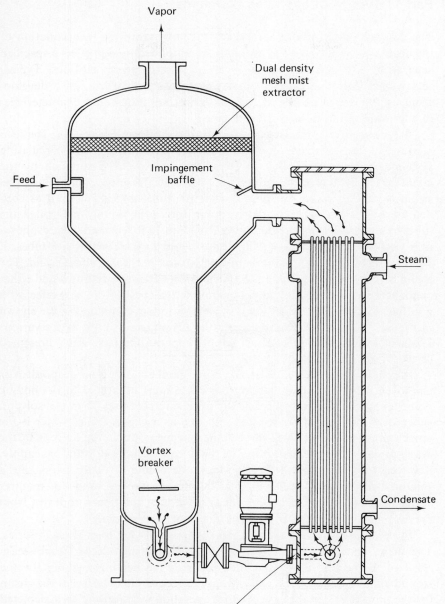

Figure 2-1. Forced-circulation vertical evaporator. (*Courtesy Trentham Corporation, Houston, Texas.*)

7

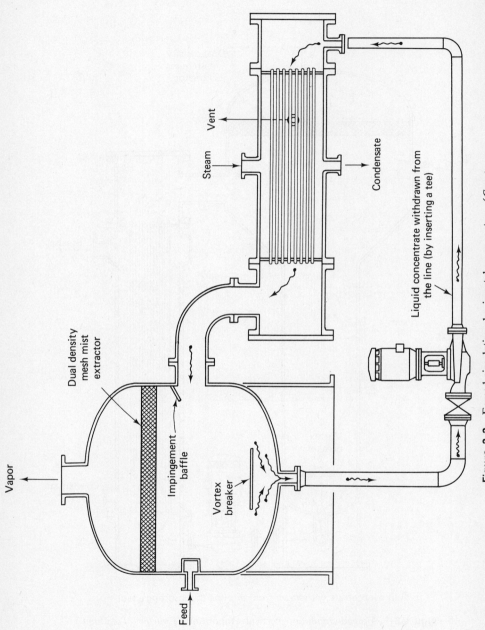

Figure 2-2. Forced-circulation horizontal evaporator. (*Courtesy Trentham Corporation, Houston, Texas.*)

Vapor

Dual density mesh mist extractor

Impingement baffle

Vortex breaker

Feed

Steam

Vent

Condensate

Liquid concentrate withdrawn from the line (by inserting a tee)

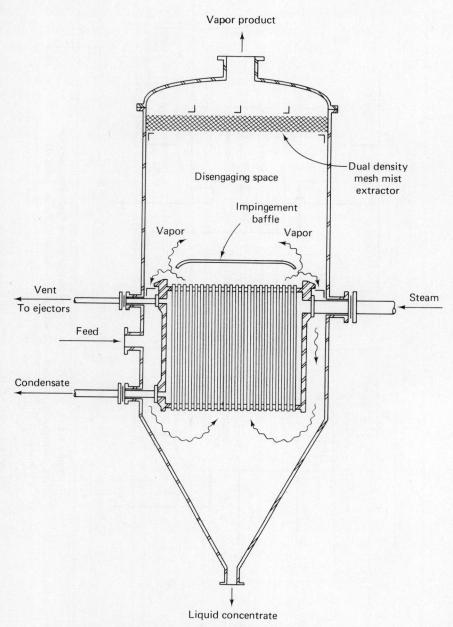

Figure 2-3. Cross-sectional view of a basket-type evaporator. (*Courtesy Trentham Corporation, Houston, Texas.*)

9

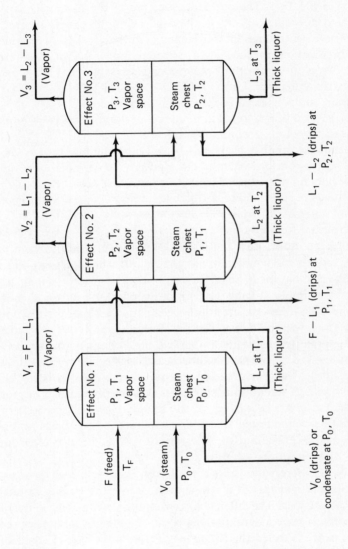

Figure 2-4. A triple-effect evaporator with forward feed. The temperature distribution shown is for a system with negligible boiling point elevations.

fed is achieved at the expense of the additional capital outlay required for the additional effects.

In order to provide the temperature potential required for heat transfer to occur in each effect, it is necessary that each effect be operated at a successively lower pressure. The operating pressure of the last effect is determined by the condensing capacity of the condenser following this effect. The pressure distribution throughout the remainder of the system is determined by the design specifications for the system. The term *evaporator system* is used to mean either one evaporator or any number of evaporators that are connected in some prescribed manner. Unless otherwise noted, it will be supposed that the evaporators are connected in series with forward feed.

To describe evaporator operation, the three terms, *capacity, economy*, and *steam consumption* are commonly used. By *capacity* of an evaporator system is meant the number of pounds of solvent evaporated per hour. The *economy* of an evaporator system is the total number of pounds of solvent vaporized per pound of steam fed to the evaporator system. *Steam consumption* is the pounds of steam fed to the system per hour. Note that the *economy* is the ratio of *capacity* to *steam consumption*.

If a true state of equilibrium existed between the vapor and the liquid phases in an evaporator, then the temperature and pressure in each phase would be equal and the temperature would be called the boiling point temperature of the evaporator. In an actual evaporator, however, the temperature of the vapor and liquid streams leaving an evaporator may be measurably different from each other and from other temperatures measured within the evaporator. Thus, the *boiling point* of an evaporator is commonly taken to be the boiling point temperature of the thick liquor (leaving the evaporator) at the pressure in the vapor space within the evaporator. Because of the effect of hydrostatic head, the pressure and consequently the corresponding boiling point of the liquid at the bottom of the liquid holdup within an evaporator is greater than it is at the surface of the liquid. However, because of the turbulent motion of the liquid within an evaporator, there exists no precise quantitative method for taking the effect of hydrostatic head into account in the analysis of evaporator operation.

Generally, the pure vapor above a solution is superheated because at a given pressure it condenses at a temperature below the boiling point temperature of the solution. The difference between the boiling point temperature of the solution and the condensation temperature of the vapor at the pressure of the vapor space is called the *boiling point elevation* of the effect. That an elevation of boiling point should be expected follows immediately by consideration of the equilibrium relationship between the two phases. If the fugacities (5, 7) are taken to be equal to their corresponding pressures, then

$$p_{\text{solvent}} = \gamma_{\text{solvent}} P_{\text{solvent}} x_{\text{solvent}} \tag{2-1}$$

where

$p_{solvent}$ = partial pressure of the solvent in the vapor phase;

$P_{solvent}$ = vapor pressure of the pure solvent at the boiling point temperature of the liquid solution of solvent and solute;

$x_{solvent}$ = mole fraction of the solvent in the liquid solution;

$\gamma_{solvent}$ = the thermodynamic activity coefficient of the solvent in the solution; the activity coefficient is a function of temperature, pressure, and composition of the solution.

Since the vapor is pure solvent, the partial pressure is, of course, equal to the total pressure P and Equation (2-1) reduces to

$$P = \gamma_{solvent} P_{solvent} x_{solvent} \qquad (2\text{-}2)$$

In view of the fact that the mole fraction of the solvent in the solution decreases as the mole fraction of the solute is increased,

$$x_{solvent} = 1 - x_{solute} \qquad (2\text{-}3)$$

it follows that at a given pressure P, the vapor pressure $P_{solvent}$ (or more precisely the product $\gamma_{solvent} P_{solvent}$) must be increased as the concentration of the solute in the solution is increased. Since the product $\gamma_{solvent} P_{solvent}$ is generally an increasing function of temperature, the total pressure P may be maintained constant as the concentration of the solute is increased by increasing the temperature of the solution. This property of solutions containing dissolved nonvolatile solutes gives rise to the term *boiling point elevation*. The boiling point temperatures of many aqueous solutions containing dissolved solids follow the Dühring rule in that the boiling point temperature of the solution is a linear function of the boiling point temperature of pure water. A typical Dühring plot for sodium hydroxide is shown in Figure 2-5. These data were taken from the work of Gerlack (6). Each concentration of dissolved solute yields a separate Dühring line.

Design of a single-effect evaporator

The equations describing a single-effect evaporator are developed in the following manner. Component-material balances on the solute and solvent are

$$FX = Lx \qquad (2\text{-}4)$$

and

$$F(1 - X) = V + L(1 - x) \qquad (2\text{-}5)$$

respectively,
where

F = feed rate, lb/hr;

L = thick liquor rate, lb/hr;

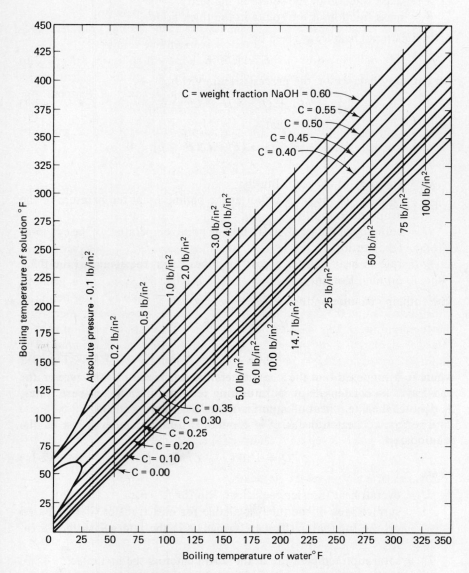

Figure 2-5. Dühring lines for solutions of sodium hydroxide in water [Taken from W. L. McCabe, "The Enthalpy Concentration Chart—A Useful Device for Chemical Engineering Calculations," *Trans. Am. Inst. Chem. Engrs.*, **31**, (1935), 129.]

V = vapor rate, lb/hr;
X = mass fraction of the solute in the feed;
x = mass fraction of the solute in the thick liquor.

A total material balance is given by

$$F = V + L \tag{2-6}$$

An enthalpy balance on the process stream yields,

$$Fh_F + Q - VH - Lh = 0 \tag{2-7}$$

Since $V = F - L$, it follows that

$$F(h_F - h) + Q - (F - L)(H - h) = 0 \tag{2-8}$$

where

h_F = enthalpy of the feed, Btu/lb;
h = enthalpy of the thick liquor at the boiling point temperature of the evaporator, Btu/lb;
H = enthalpy of the vapor at the boiling point temperature of the evaporator, Btu/lb;
Q = rate of heat transfer across the tubes (from the steam to the thick liquor), Btu/hr.

The enthalpy balance on the steam is given by

$$V_0 H_0 - Q - V_0 h_0 = 0 \tag{2-9}$$

or

$$Q = V_0(H_0 - h_0) = V_0 \lambda_0 \tag{2-10}$$

where it is supposed that the steam enters the steam chest of the evaporator and leaves as condensate at its saturation temperature and pressure. Thus, λ_0 denotes the latent heat of vaporization of the entering steam.

The rate of heat transfer Q is commonly approximated by use of the relationship

$$Q = UA(T_0 - T) \tag{2-11}$$

where

U = overall heat transfer coefficient, Btu/(hr ft^2 °F);
A = surface area of the tubes available for heat transfer (if U is based on the internal surface area, then A is the internal surface area, ft^2);
T_0 = saturation temperature of the steam entering the first effect;
T = boiling point temperature of the thick liquor at the pressure of the vapor space.

Actually, the temperature of the liquid varies along the length of the tubes, and the rate of heat transfer should be computed more precisely by use of the

$$Q = \int_0^{z_T} Ua(T_0 - T)\, dz \tag{2-12}$$

where the distance z is measured along the length of the tube and T becomes the temperature of the liquid at each z in the interval $0 \leq z \leq z_T$. However, the overall heat transfer coefficients that are commonly available for design purposes have been computed from the results of field tests by use of Equation (2-11). Thus, if the U's available were determined by use of Equation (2-11), this same expression should be used to compute Q in evaporator design. The use of Equations (2-4) through (2-11) are best demonstrated by the following numerical example.

ILLUSTRATIVE EXAMPLE 2-1

A single-effect evaporator is to be designed to concentrate a 20% (by weight) solution of sodium hydroxide to a 50% solution (see Figure 2-6). The dilute solution (the feed) at 200°F is to be fed to the evaporator at the rate of 40,000 lb/hr. For heating purposes, saturated steam at 350°F is used. Sufficient condenser area is available to maintain a pressure of 0.9492 lb/in.² (absolute) in the vapor space of the evaporator. On the basis of an overall heat transfer coefficient of 300 Btu/(hr ft² °F), compute (a) the heating area required and (b) the steam consumption and the steam economy.

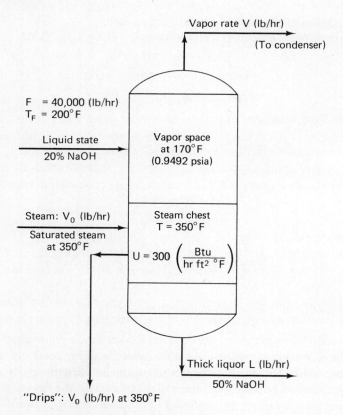

Figure 2-6. Design specifications for Illustrative Example 2-1.

Solution: The rate L at which the thick liquor leaves the evaporator is computed by use of Equation (2-4), the component-material balance on the solute NaOH,

$$L = \frac{FX}{x} = \frac{(40,000)(0.2)}{(0.5)} = 16,000 \text{ lb/hr}$$

The evaporation rate (vapor rate V) follows from Equation (2-6)

$$V = F - L = 40,000 - 16,000 = 24,000 \text{ lb/hr}$$

The boiling point of water at 0.9492 psia is 100°F; see for example Keenan and Keyes (9). Use of this temperature and Figure 2-7 gives a boiling point temperature of 170°F for a 50% NaOH solution.

The following enthalpies were taken from Figure 2-7.

$$h_F(@ \ 200°F \text{ and } 20\% \text{ NaOH}) = 145 \text{ Btu/lb}$$

$$h(@ \ 170°F \text{ and } 50\% \text{ NaOH}) = 200 \text{ Btu/lb}$$

From Keenan and Keyes (9),

$$H(@ \ 170°F \text{ and } 0.9492 \text{ psia}) = 1,136.94 \text{ Btu/lb}$$

$$\lambda_0(\text{saturated at } 134.63 \text{ psia}) = 870.7 \text{ Btu/lb } @$$

$$T_0 = 350°F$$

(a) Calculation of the heat transfer area A required:
   The rate of heat transfer Q is computed by use of Equation (2-8).

$$Q = (F - L)(H - h) - F(h_F - h)$$

$$Q = (24,000)(1,136.94 - 200) - 40,000(145 - 200)$$

$$= 24.686 \times 10^6 \text{ Btu/hr}$$

Then, from Equation (2-11),

$$A = \frac{Q}{U(T_0 - T)} = \frac{24.686 \times 10^6}{(300)(350 - 170)} = 457.15 \text{ ft}^2$$

(b) Calculation of the steam economy:
   From Equation (2-10), it follows that the steam consumption is given by

$$V_0 = \frac{Q}{\lambda_0} = \frac{24.686 \times 10^6}{870.7} = 28,353 \text{ lb/hr}$$

Then,

$$\text{Steam economy} = \frac{V}{V_0} = \frac{24,000}{28,353} = 0.84647$$

Design of multiple-effect evaporator systems

For illustrative purposes, the equations describing a triple-effect evaporator system are developed for the case where boiling point elevations are negligible. Also, the effect of composition on liquid enthalpy is neglected. The equations so obtained are generalized to include the case where boiling point elevations cannot be neglected. For definiteness, forward feeds are employed.

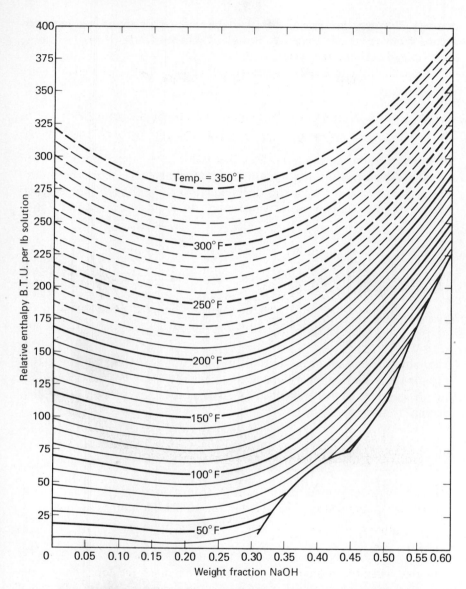

Figure 2-7. Enthalpy concentration chart for solutions of sodium hydroxide in water. (Taken from McCabe, "The Enthalpy Concentration Chart," p. 129.)

1. SPECIFICATIONS: F, X, T_F, T_0, P_0, P_3 (or T_3), x_3 (or L_3), U_1, U_2, U_3, Equal Areas, Forward Feed, and Negligible Boiling Point Elevations
   To Find: V_0, T_1, L_1, T_2, L_2, and A

Actually, four additional dependent variables exist; namely, V_1, V_2, x_1, and x_2. However, as shown, these variables are either eliminated or determined from equations that may be solved independently of the remaining equations describing the system.

A sketch of the three-effect system under consideration is shown in Figure 2-4.

EFFECT NO. 1. An enthalpy balance on the process fluid stream yields

$$Fh_F + Q_1 - (F - L_1)H_1 - L_1h_1 = 0 \qquad (2\text{-}13)$$

In this expression, as well as in those that follow, the vapor rates have been eliminated by use of the appropriate total material balance. Also, the enthalpies are approximated by taking them equal to those for the pure solvent. Thus, the enthalpies depend on temperature alone. For the first effect

$$V_1 = F - L_1 \qquad (2\text{-}14)$$

and Equation (2-13) is readily restated in the following form:

$$F(h_F - h_1) + Q_1 - (F - L_1)\lambda_1 = 0 \qquad (2\text{-}15)$$

where

$\lambda_j = H_j - h_j$, the latent heat of vaporization of the solvent from the thick liquor at temperature T_j and pressure P_j ($j = 1, 2, 3$, the effect number).

Alternately, the result given by Equation (2-15) may be obtained by taking the enthalpy reference for the first effect to be the enthalpy of the thick liquor leaving the effect at the temperature T_1. The enthalpy balance on the heating medium is given by

$$Q_1 = V_0(H_0 - h_0) = V_0\lambda_0 \qquad (2\text{-}16)$$

and the rate of heat transfer by

$$Q_1 = U_1 A(T_0 - T_1) \qquad (2\text{-}17)$$

If Equation (2-16) is used to eliminate Q_1 from Equations (2-15) and (2-17), the results so obtained may be stated in functional notation as follows:

Enthalpy balance: $f_1 = F(h_F - h_1) + V_0\lambda_0 - (F - L_1)\lambda_1 \qquad (2\text{-}18)$

Heat transfer rate: $f_2 = U_1 A(T_0 - T_1) - V_0\lambda_0 \qquad (2\text{-}19)$

Note that when the desired solution has been obtained, the equations for the first effect as well as those that follow for the second and third effects will be satisfied simultaneously.

EFFECT NO. 2 In a manner analogous to that demonstrated for the first effect, the following equations are obtained for the second effect.

$$L_1(h_1 - h_2) + Q_2 - (L_1 - L_2)\lambda_2 = 0$$

$$Q_2 = (F - L_1)\lambda_1$$

$$Q_2 = U_2 A(T_1 - T_2)$$

Thus,

Enthalpy balance: $f_3 = L_1(h_1 - h_2) + (F - L_1)\lambda_1 - (L_1 - L_2)\lambda_2$ (2-20)

Heat transfer rate: $f_4 = U_2 A(T_1 - T_2) - (F - L_1)\lambda_1$ (2-21)

EFFECT NO. 3 As before

$$L_2(h_2 - h_3) + Q_3 - (L_2 - L_3)\lambda_3 = 0$$

$$Q_3 = (L_1 - L_2)\lambda_2$$

$$Q_3 = U_3 A(T_2 - T_3)$$

and

Enthalpy balance: $f_5 = L_2(h_2 - h_3) + (L_1 - L_2)\lambda_2 - (L_2 - L_3)\lambda_3$ (2-22)

Heat transfer rate: $f_6 = U_3 A(T_2 - T_3) - (L_1 - L_2)\lambda_2$ (2-23)

The six independent equations [Equations (2-18) through (2-23)] may be solved for the six unknowns $V_0, L_1, L_2, T_1, T_2,$ and A. In addition to these six independent equations, three additional equations that contain three additional independent variables, $x_1, x_2,$ and x_3 (or L_3) exist; namely, the component-material balances which may be stated in the form

$$FX - L_j x_j = 0, \qquad (j = 1, 2, 3) \tag{2-24}$$

Since either x_3 or L_3 is specified, the value for the unspecified variable follows immediately from Equation (2-24). After Equations (2-18) through (2-23) have been solved, the values of L_1 and L_2 so obtained may be used to compute the corresponding values of x_1 and x_2.

The equations describing the triple evaporator system constitute a set of nonlinear algebraic equations that may be solved in a variety of ways. Two methods are presented after some of the fundamental principles involved in evaporation have been demonstrated by solving several relatively simple problems.

ILLUSTRATIVE EXAMPLE 2-2

(a) For the triple-effect evaporator system described by the above equations (which neglect boiling point elevations), show that if the sensible heat effects are negligible, then (1) the evaporation rates for each effect are equal and (2) the rates of heat transfer for each effect are equal.

(b) If in addition to the suppositions of Part (a), it is specified that the areas are to be equal, then show that

$$\frac{U_2}{U_1} = \frac{\Delta T_1}{\Delta T_2}, \qquad \frac{U_3}{U_1} = \frac{\Delta T_1}{\Delta T_3}$$

where

$$\Delta T_1 = (T_0 - T_1), \qquad \Delta T_2 = (T_1 - T_2), \qquad \Delta T_3 = (T_2 - T_3)$$

(c) If in addition to the conditions given by Parts (a) and (b), it is given that the overall heat transfer coefficients are equal, then show that

$$\Delta T_1 = \Delta T_2 = \Delta T_3$$

Solution:

Part (a). Since it is given that the sensible heats effects are negligible, then it follows that

$$h_F = h_0 = h_1 = h_2 = h_3 \quad \text{and} \quad H_0 = H_1 = H_2 = H_3$$

Then,

$$H_0 - h_0 = H_1 - h_1 = H_2 - h_2 = H_3 - h_3$$

or

$$\lambda_0 = \lambda_1 = \lambda_2 = \lambda_3 = \lambda$$

For this simple case, Equations (2-18) through (2-23) may be solved directly by first setting $f_0 = f_1 = \cdots = f_5 = f_6 = 0$. First observe that Equations (2-18) and (2-14) give

$$0 = V_0 \lambda_0 - (F - L_1)\lambda_1 = V_0 \lambda - V_1 \lambda$$

and, therefore,

$$V_0 = V_1$$

Similarly, since $L_1 - L_2 = V_2$ and $L_2 - L_3 = V_3$, it follows from Equations (2-20) and (2-22), respectively, that

$$V_1 = V_2 \quad \text{and} \quad V_2 = V_3$$

Consequently,

$$V_0 = V_1 = V_2 = V_3 \tag{2-25}$$

By using the results for Part (a) and the above expressions for Q_1, Q_2, Q_3, one finds

$$Q_1 = V_0 \lambda_0 = V_0 \lambda, \qquad Q_2 = (F - L_1)\lambda_1 = V_1 \lambda_1 = V_0 \lambda,$$

$$Q_3 = (L_1 - L_2)\lambda_2 = V_2 \lambda_2 = V_0 \lambda$$

and, therefore,

$$Q_1 = Q_2 = Q_3 \tag{2-26}$$

Part (b). Since

$$Q_1 = Q_2 = Q_3$$

it follows that

$$Q_1 = U_1 A_1 \, \Delta T_1 = U_2 A_2 \, \Delta T_2 = U_3 A_3 \, \Delta T_3$$

and when the areas are equal, it is evident that

$$\frac{U_2}{U_1} = \frac{\Delta T_1}{\Delta T_2}, \qquad \frac{U_3}{U_1} = \frac{\Delta T_1}{\Delta T_3} \tag{2-27}$$

Part (c). If $U_1 = U_2 = U_3$, it is evident from Equation (2-27) that

$$\Delta T_1 = \Delta T_2 = \Delta T_3 \tag{2-28}$$

Thus, when the sensible heats are negligible for a triple-effect evaporator system for which the boiling point elevations are also negligible [the conditions of Part (a) of Illustrative Example 2-2], the system has the following characteristics:

$$\text{Steam consumption} = V_0$$

$$\text{Capacity} = V_1 + V_2 + V_3 = 3V_0$$

$$\text{Economy} = \frac{V_1 + V_2 + V_3}{V_0} = \frac{3V_0}{V_0} = 3$$

ILLUSTRATIVE EXAMPLE 2-3

Suppose that the evaporators in Part (c) of Illustrative Example 2-2 are disconnected and operated separately and at the same terminal conditions as the system, that is, the ΔT for each evaporation is given by

$$\Delta T = \Delta T_1 + \Delta T_2 + \Delta T_3 = 3 \, \Delta T_1$$

Compute the steam consumption, capacity, and economy of the three evaporators operated separately, and express the results in terms of Q_1, V_0, and λ for the system in Illustrative Example 2-2.

Solution: Let the rates of heat transfer for the single effects be denoted by $\bar{Q}_1$, $\bar{Q}_2$, and $\bar{Q}_3$. Then,

$$\bar{Q}_1 = \bar{Q}_2 = \bar{Q}_3 = UA \, \Delta T = 3UA \, \Delta T_1 = 3V_0\lambda = 3Q_1$$

Thus,

Steam consumption of the three-single effects $= 3V_0 + 3V_0 + 3V_0 = 9V_0$

Capacity of the three single effects $= 9V_0$

Economy of the three single effects $= \dfrac{9V_0}{9V_0} = 1$

Comparison of these results with those obtained for triple-effect operation shows that triple-effect operation has one-third the capacity and one-ninth the steam consumption of a single-effect operation.

Thus, multiple-effect operation versus single-effect operation is seen to be a trade off between capital costs and operating costs. That is, the capital costs per pound of product for triple-effect operation are three times those for single-effect operation while the operating costs per pound of product (cost of steam consumption per pound of product) is only one-third of those for single-effect operation. As shown in a subsequent section, however the effect of boiling point elevation reduces the steam economy for multiple-effect operation relative to single-effect operation.

Next, two calculational procedures (the Badger and McCabe method and the Newton-Raphson method) for solving design problems of the kind formulated above for a triple-effect evaporator system [see Equations (2-13) through (2-23)] are presented.

Use of the Badger and McCabe method for solving evaporator design problems (2)

This method consists of the following steps for the triple-effect system:

1. Assume values for the temperatures for the first and second effects.
2. Determine evaporation rates V_1, V_2, and V_3 by means of the enthalpy balances [Equations (2-18), (2-20), and 2-22)].

3. Use the heat transfer rate equations to compute the heating surfaces A_1, A_2, and A_3.

4. If the areas the unequal, redistribute the temperature drops and repeat Steps 2 and 3.

This procedure is repeated until the areas differ from one another by no more than a preassigned amount.

The redistribution of temperature drops may be effected by using the following procedure: First, the areas are computed in Step 3 by using equations $Q_1 = U_1 A_1 (\Delta T_1)_a$, $Q_2 = U_2 A_2 (\Delta T_2)_a$, and $Q_3 = U_3 A_3 (\Delta T_3)_a$, where the subscript a denotes the ΔT's corresponding to the temperatures assumed in Step 1. If the correct ΔT's had been assumed, then one would have obtained the same A for each evaporator; that is, one would have obtained: $Q_1 = U_1 A (\Delta T_1)_{co}$, $Q_2 = U_2 A (\Delta T_2)_{co}$, $Q_3 = U_3 A (\Delta T_3)_{co}$. The latter equations imply that the same rates Q_1, Q_2, and Q_3 would have been obtained for the corrected ΔT's as were obtained for the assumed ΔT's, and this approximation is made in the convergence procedure. It should also be noted that the sum of the corrected ΔT's is equal to the total temperature drop across the system. The two sets of equations enumerated above may be restated as follows.

Basis: Assumed ΔT's	Basis: Corrected ΔT's
$\dfrac{Q_1}{U_1 A} = \dfrac{A_1}{A}(\Delta T_1)_a$	$\dfrac{Q_1}{U_1 A} = (\Delta T_1)_{co}$
$\dfrac{Q_2}{U_2 A} = \dfrac{A_2}{A}(\Delta T_2)_a$	$\dfrac{Q_2}{U_2 A} = (\Delta T_2)_{co}$
$\dfrac{Q_3}{U_3 A} = \dfrac{A_3}{A}(\Delta T_3)_a$	$\dfrac{Q_3}{U_3 A} = (\Delta T_3)_{co}$

Since the left-hand sides of the respective equations of these sets are seen to be equal, it follows that

$$\frac{A_1}{A}(\Delta T_1)_a + \frac{A_2}{A}(\Delta T_2)_a + \frac{A_3}{A}(\Delta T_3)_a = (\Delta T_1)_{co} + (\Delta T_2)_{co} + (\Delta T_3)_{co} = \Delta T$$

where

$\Delta T = T_0 - T_3$, the total temperature drop across the evaporator system. Thus,

$$A = \frac{A_1 (\Delta T_1)_a + A_2 (\Delta T_2)_a + A_3 (\Delta T_3)_a}{\Delta T} \tag{2-29}$$

After A has been computed, the corrected temperature drops are computed as follows:

$$(\Delta T_1)_{co} = \frac{A_1}{A}(\Delta T_1)_a; \quad (\Delta T_2)_{co} = \frac{A_2}{A}(\Delta T_2)_a; \quad (\Delta T_3)_{co} = \frac{A_3}{A}(\Delta T_3)_a$$

$$\tag{2-30}$$

Instead of computing A by use of Equation (2-29), Badger and McCabe (2) took A equal to the arithmetic average of A_1, A_2, and A_3. Consequently, the corresponding set of corrected ΔT's computed by Equation (2-30) did not necessarily have a sum equal to the total temperature drop across the system. To achieve the latter equality, Badger and McCabe adjusted the corrected ΔT's by inspection.

The use of the Badger and McCabe method for solving evaporator design problems is demonstrated by solving the following example.

ILLUSTRATIVE EXAMPLE 2-4

It is desired to design a triple-effect evaporator system to concentrate the solute from a 10% solution (feed) to 50% by weight. The feed rate is 50,000 lb/hr, and it enters the first effect as a liquid at 100°F. Forward feed is to be used. Saturated vapor of the solvent at 250°F is available for satisfying the heating requirement for the first effect. The third effect is to be operated at an absolute pressure corresponding to a boiling point of 125°F for the pure solvent. Neglect boiling point elevation as well as the variation of heat capacities and latent heats of vaporization with temperature and composition. Determine the area A per effect (equal areas are to be employed), the temperatures T_1 and T_2, and the flow rates L_1, L_2, and L_3, the compositions x_1 and x_2, and the rate V_0.

Given:

$C_p = 1.0$ Btu/(lb °F) for the feed and all other liquid streams;

$\lambda_0 = \lambda_1 = \lambda_3 = 1,000$ Btu/lb;

$U_1 = 500$, $U_2 = 300$, $U_3 = 200$.

Solution:

Step 1. To initiate the calculational procedure, the total temperature drop across the system $(250 - 125 = 125°F)$ is distributed by use of Equation (2-27). The relationships given by this equation are, of course, approximate for this problem because the sensible heat effects are not negligible. Then,

$$\frac{\Delta T_2}{\Delta T_1} = \frac{U_1}{U_2} = \frac{500}{300}; \qquad \frac{\Delta T_3}{\Delta T_1} = \frac{U_1}{U_3} = \frac{500}{200}$$

Since

$$\Delta T = 125 = \Delta T_1 + \left(\frac{\Delta T_2}{\Delta T_1}\right)\Delta T_1 + \left(\frac{\Delta T_3}{\Delta T_1}\right)\Delta T_1$$

it follows that

$$\Delta T_1 = \frac{125}{1 + \dfrac{5}{3} + \dfrac{5}{2}} = 24.1936$$

and

$$\Delta T_2 = \left(\frac{5}{3}\right)(24.1936) = 40.3226; \qquad \Delta T_3 = \left(\frac{5}{2}\right)(24.1936) = 60.4839$$

Step 2. Next, L_3 may be computed directly by use of Equation (2-24), namely,

$$L_3 = \frac{FX}{x_3} = \frac{(50,000)(0.1)}{(0.5)} = 10,000 \text{ lb/hr}$$

Since $C_p = 1$, and $\lambda_0 = \lambda_1 = \lambda_2 = \lambda_3 = \lambda$, Equations (2-18), (2-20), and (2-22) may be restated as the following simultaneous equations

$$0 = F(T_F - T_1) + V_0\lambda - (F - L_1)\lambda$$

$$0 = L_1(T_1 - T_2) + (F - L_1)\lambda - (L_1 - L_2)\lambda$$

$$0 = L_2(T_2 - T_3) + (L_1 - L_2)\lambda - (L_2 - L_3)\lambda$$

These equations contain the three unknowns V_0, L_1, and L_2. Since the second and third equations contain only L_1 and L_2, they may be solved simultaneously for these flow rates to give

$$L_1 = 38,194.4 \text{ lb/hr}$$

$$L_2 = 24,848.7 \text{ lb/hr}$$

When these results are substituted into the first equation, it is found that

$$V_0 = 18,095.9 \text{ lb/hr}$$

Step 3. The areas are computed as follows:

$$A_1 = \frac{Q_1}{U_1 \, \Delta T_1} = \frac{V_0\lambda}{U_1 \, \Delta T_1} = \frac{(18,095.9)(10^3)}{(500)(24.1936)} = 1,495.93 \text{ ft}^2$$

$$A_2 = \frac{Q_2}{U_2 \, \Delta T_2} = \frac{(F - L_1)\lambda}{U_2 \, \Delta T_2} = \frac{(11,805.6)(10^3)}{(200)(60.4839)} = 975.932 \text{ ft}^2$$

$$A_3 = \frac{Q_3}{U_3 \, \Delta T_3} = \frac{(L_1 - L_2)\lambda}{U_2 \, \Delta T_2} = \frac{(13,345.7)(10^3)}{(200)(60.4839)} = 1,103.24 \text{ ft}^2$$

Step 4. A new set of ΔT's is obtained by use of Equations (2-29) and (2-30). By Equation (2-29),

$$A = \frac{(1,495.93)(24.1936) + (975.932)(40.3226) + (1,103.24)(60.4839)}{125}$$

$$= 1,138.18 \text{ ft}^2$$

and Equation (2-30) gives

$$(\Delta T_1)_{co} = \left(\frac{1,495.93}{1,138.18}\right)(24.1936) = 31.798°\text{F}$$

$$(\Delta T_2)_{co} = \left(\frac{975.932}{1,138.18}\right)(40.3226) = 34.575°\text{F}$$

$$(\Delta T_3)_{co} = \left(\frac{1,103.24}{1,138.18}\right)(60.4839) = 58.627°\text{F}$$

TABLE 2-1
SOLUTION OF ILLUSTRATIVE EXAMPLE 2-4 BY USE OF THE
BADGER-McCABE METHOD

Trial No.	L_1	L_2	V_0	A	$(\Delta T_1)_{co}$	$(\Delta T_2)_{co}$	$(\Delta T_3)_{co}$
1	38,194.4	24,848.7	18,095.9	1,138.1	31.798	34.575	58.627
2	38,026.6	24,738.5	17,883.5	1,137.0	31.459	35.104	58.438
3	38,038.8	24,742.4	17,888.2	1,137.0	31.465	35.065	58.470
4	38,038.1	24,742.4	17,888.7	1,137.0	31.466	35.067	58.467
5	38,038.1	24,742.4	17,888.5	1,137.0	31.465	35.067	58.467

On the basis of each set of corrected ΔT's obtained, the calculational procedure described was repeated to give the results shown in Table 2-1.

The nonlinear equations for the design of the triple-effect evaporator system may be solved by use of the Newton-Raphson method which is described in detail in Appendix A.

Use of the Newton-Raphson method for solving evaporator design problems

As described in Appendix A, the Newton-Raphson method consists of the repeated use of the linear terms of the Taylor series expansions of the functions f_1, f_2, f_3, f_4, f_5 and f_6 [Equations (2-18) through (2-23)]; namely,

$$0 = f_j + \frac{\partial f_j}{\partial V_0} \Delta V_0 + \frac{\partial f_j}{\partial T_1} \Delta T_1 + \frac{\partial f_j}{\partial L_1} \Delta L_1 + \frac{\partial f_j}{\partial T_2} \Delta T_2 + \frac{\partial f_j}{\partial L_2} \Delta L_2$$
$$+ \frac{\partial f_j}{\partial A} \Delta A, \quad (j = 1, 2, \ldots, 5, 6) \tag{2-31}$$

where

$$\Delta V_0 = V_{0,k+1} - V_{0,k};$$
$$\Delta T_1 = T_{1,k+1} - T_{1,k};$$
$$\Delta T_2 = T_{2,k+1} - T_{2,k};$$
$$\Delta L_2 = L_{2,k+1} - L_{2,k};$$
$$\Delta A = A_{k+1} - A_k;$$

and where the subscripts k and $k + 1$ denote the kth and $k + 1$st trials.

These six equations may be stated in compact form by means of the following matrix equation

$$J_k \Delta X_k = -f_k \tag{2-32}$$

where J_k is called the Jacobian matrix and,

$$\Delta X_k = X_{k+1} - X_k = [\Delta V_0\, \Delta T_1\, \Delta T_2\, \Delta L_2\, \Delta A]^T$$

The subscripts k and $k + 1$ denote that elements of the matrices carrying these subscripts are those given by the kth and $k + 1$st trials, respectively. In the interest of simplicity, the subscript k is omitted from the elements of X_k, J_k, and f_k. On the basis of an assumed set of values for the elements of the column vector (or column matrix) X_k, which may be stated as the transpose of the corresponding row vector (or row matrix)

$$X_k = [V_0 T_1 L_1 T_2 L_2 A]^T \tag{2-33}$$

the corresponding values of the elements of J_k and f_k are computed. A display of the elements of J_k and f_k follows:

$$J_k = \begin{bmatrix} \dfrac{\partial f_1}{\partial V_0} & \dfrac{\partial f_1}{\partial T_1} & \dfrac{\partial f_1}{\partial L_1} & \dfrac{\partial f_1}{\partial T_2} & \dfrac{\partial f_1}{\partial L_2} & \dfrac{\partial f_1}{\partial A} \\ \dfrac{\partial f_2}{\partial V_0} & \dfrac{\partial f_2}{\partial T_1} & \dfrac{\partial f_2}{\partial L_1} & \dfrac{\partial f_2}{\partial T_2} & \dfrac{\partial f_2}{\partial L_2} & \dfrac{\partial f_2}{\partial A} \\ \cdot & \cdot & \cdot & \cdot & \cdot & \cdot \\ \cdot & \cdot & \cdot & \cdot & \cdot & \cdot \\ \cdot & \cdot & \cdot & \cdot & \cdot & \cdot \\ \dfrac{\partial f_6}{\partial V_0} & \dfrac{\partial f_6}{\partial T_1} & \dfrac{\partial f_6}{\partial L_1} & \dfrac{\partial f_6}{\partial T_2} & \dfrac{\partial f_6}{\partial L_2} & \dfrac{\partial f_6}{\partial A} \end{bmatrix} ; \quad f_k = \begin{bmatrix} f_1 \\ f_2 \\ \cdot \\ \cdot \\ \cdot \\ f_6 \end{bmatrix} \qquad (2\text{-}34)$$

The convergence of the Newton-Raphson method is considered in detail in Appendix A. However, suffice it to say here that if the functions $f_1, f_2, \ldots,$ f_6 and their partial derivatives which appear in J_k are continuous and the determinant of J_k is not equal to zero, then the Newton-Raphson method will converge, provided a set of assumed values of the variables which are close enough to the solution set can be found.

If the changes in the specific heats with temperature in the neighborhood of the solution to Equations (2-18) through (2-23) are negligible, then the sensible heat terms $(h_F - h_1)$, $(h_1 - h_2)$, and $(h_2 - h_3)$ may be replaced by their respective equivalents: $C_p(T_F - T_1)$, $C_p(T_1 - T_2)$, and $C_p(T_2 - T_3)$. If the variation of the latent heats with temperature are also regarded as negligible in the neighborhood of the solution, then

$$J_k = \begin{bmatrix} \lambda_0 & -FC_p & \lambda_1 & 0 & 0 & 0 \\ -\lambda_0 & -U_1A & 0 & 0 & 0 & U_1(T_0 - T_1) \\ 0 & L_1C_p & b_{33} & -L_1C_p & \lambda_2 & 0 \\ 0 & U_2A & \lambda_1 & -U_2A & 0 & U_2(T_1 - T_2) \\ 0 & 0 & \lambda_2 & L_2C_p & b_{55} & 0 \\ 0 & 0 & -\lambda_2 & U_3A & \lambda_2 & U_3(T_2 - T_3) \end{bmatrix} \qquad (2\text{-}35)$$

where

$$b_{33} = C_p(T_1 - T_2) - (\lambda_1 + \lambda_2)$$
$$b_{55} = C_p(T_2 - T_3) - (\lambda_2 + \lambda_3)$$

To demonstrate the use of the Newton-Raphson method, the previous example, Illustrative Example 2-4, is solved.

Solution of illustrative example 2–4 by the Newton-Raphson method

The calculational procedure may be initiated on the basis of any reasonable set of assumptions, say

1. $(T_0 - T_1) = 42°F$, $(T_1 - T_2) = 42°F$, and $(T_2 - T_3) = 41°F$.

2. Solvent evaporated in first effect $= 14{,}000$ lb/hr.
   Solvent evaporated in second effect $= 14{,}000$ lb/hr.
   Solvent evaporated in third effect $= 12{,}000$ lb/hr.
   (Note: These assumptions are consistent with the fact that the total amount of solvent evaporated follows immediately from the statement of the problem and a component-material balance, namely,

$$L_3 = \frac{FX}{x_3} = 10{,}000 \text{ lb/hr}$$

3. $A = 1000$ ft^2 for each effect.
4. $V_0 = 15{,}000$ lb/hr.

A scaling procedure is used to reduce the magnitude of the terms appearing in the functional equations and matrices. For computational purposes, it is desirable to have terms with magnitudes near unity. The choice of factors for reducing the terms is arbitrary, but more meaning may be associated with the reduced terms if values associated with the parameters of the problem are selected such as feed rate, steam temperature, and the latent heat of vaporization of steam. The following scaling procedure was used.

1. Each functional equation was divided by the product $F\lambda_0$, and the new functional expression so obtained was denoted by g_j ($1 \leq j \leq 6$); where

$$g_j = \frac{f_j}{F\lambda_0}$$

2. All flow rates were expressed as a fraction of the feed rate F, that is, $L_j = l_j F$ and $V_j = v_j F$.
3. All temperatures were expressed as a fraction of the steam temperature as follows: $T_j = u_j T_0$, which defines the fractional temperature u_j.
4. The area of each effect was expressed as a fraction of a term proportional to the feed rate in the following manner: $A_j = a_j(F/50)$, which defines the fractional area a_j.

After this scaling procedure has been applied to the functional expressions, the matrices J_k, ΔX_k, and f_k take the following forms.

$$J_k = \begin{bmatrix} 1 & b_{12} & \lambda_1/\lambda_0 & 0 & 0 & 0 \\ -1 & b_{22} & 0 & 0 & 0 & b_{26} \\ 0 & b_{32} & b_{33} & b_{34} & \lambda_2/\lambda_0 & 0 \\ 0 & b_{42} & \lambda_1/\lambda_0 & b_{44} & 0 & b_{46} \\ 0 & 0 & \lambda_2/\lambda_0 & b_{54} & b_{55} & 0 \\ 0 & 0 & -\lambda_2/\lambda_0 & b_{64} & \lambda_2/\lambda_0 & b_{66} \end{bmatrix}$$

$$\Delta X_k = [\Delta v_0 \; \Delta u_1 \; \Delta l_1 \; \Delta u_2 \; \Delta l_2 \; \Delta a]^T$$

$$f_k = [g_1 g_2 g_3 g_4 g_5 g_6]^T$$

where the elements of J_k consist of the partial derivatives of the g_j's with respect to the new set of variables (v_0, u_1, l_1, u_2, l_2, and a)

$$b_{12} = \frac{-C_pT_0}{\lambda_0}; \qquad b_{34} = \frac{-l_1C_pT_0}{\lambda_0}; \qquad b_{22} = \frac{-U_1aT_0}{50\lambda_0};$$

$$b_{44} = \frac{-U_2aT_0}{50\lambda_0}; \qquad b_{32} = \frac{l_1C_pT_0}{\lambda_0}; \qquad b_{54} = \frac{l_2C_pT_0}{\lambda_0};$$

$$b_{42} = \frac{U_2aT_0}{50\lambda_0} \qquad b_{64} = \frac{U_3aT_0}{50\lambda_0}; \qquad b_{26} = \frac{[U_1(1-u_1)T_0]}{50\lambda_0};$$

$$b_{33} = \frac{[C_p(u_1-u_2)T_0 - (\lambda_1+\lambda_2)]}{\lambda_0}; \qquad b_{46} = \frac{[U_2(u_1-u_2)T_0]}{50\lambda_0};$$

$$b_{55} = \frac{[C_p(u_2-u_3)T_0 - (\lambda_2+\lambda_3)]}{\lambda_0}; \qquad b_{66} = \frac{[U_3(u_2-u_3)T_0]}{50\lambda_0}$$

Then, on the basis of the assumed values of the variables, the matrices J_0, f_0, and X_0 used in the first trial follow:

$$J_0 = \begin{bmatrix} 1 & -0.25 & 1 & 0 & 0 & 0 \\ -1 & -2.5 & 0 & 0 & 0 & 0.42 \\ 0 & 0.18 & -1.958 & -0.18 & 1 & 0 \\ 0 & 1.5 & 1 & -1.5 & 0 & 0.252 \\ 0 & 0 & 1 & 0.11 & -1.959 & 0 \\ 0 & 0 & -1 & 1.0 & 1 & 0.164 \end{bmatrix}$$

$$f_0 = \begin{bmatrix} -0.088 \\ 0.120 \\ 0.030 \\ -0.028 \\ 0.058 \\ -0.116 \end{bmatrix}, \qquad X_0 = \begin{bmatrix} 0.300 \\ 0.832 \\ 0.720 \\ 0.664 \\ 0.440 \\ 1.000 \end{bmatrix}$$

Covergence to within about six significant numbers was achieved in three trials. The values of the elements of X computed at the end of the first four trials are displayed in Table 2-2.

TABLE 2-2
SOLUTION OF EXAMPLE 2-4 BY USE OF THE NEWTON-RAPHSON METHOD

Trial No.	Value of the Variables (Scaled)					
	v_0	u_1	l_1	u_2	l_2	a
1	0.359374	0.879493	0.760498	0.743069	0.494740	1.13835
2	0.357773	0.874144	0.760762	0.733878	0.494848	1 13703
3	0.357771	0.874138	0.760763	0.733868	0.494848	1.13703
4	0.357771	0.874138	0.760763	0.733868	0.494848	1.13703

Thus, the desired solution is:

$$V_0 = v_0 F = 17,888.5 \text{ lb/hr}; \qquad T_1 = u_1 T_0 = 218.534°\text{F};$$

$$L_1 = l_1 F = 38,038.1 \text{ lb/hr}; \qquad T_2 = u_2 T_0 = 183.467°\text{F};$$

$$L_2 = l_2 F = 24,742.4 \text{ lb/hr}; \qquad A = \frac{aF}{50} = 1,137.03 \text{ ft}^2;$$

$$L_3 = \frac{FX}{x_3} = 10,000 \text{ lb/hr}; \qquad x_1 = \frac{FX}{L_1} = 0.131447;$$

$$x_2 = \frac{FX}{L_2} = 0.202082$$

A generalized scaling procedure which is very similar to the scaling procedure used in the solution of this example is presented in Appendix A. Such a generalized procedure is particularily useful for the solution of problems in which a large number of variables are involved such as those considered by Burdett (4). For the remaining illustrative examples presented in this chapter, however, scaling procedures similar to the one used in the solution of Illustrative Example 2-4 are employed. Also, the problems are formulated in terms of the Newton-Raphson method. The formulation of the solution of problems in terms of the Newton-Raphson method is helpful because it forces one to display the independent equations and the independent variables. However, in the solution of relatively simple problems "by hand" such as those involving two-effect evaporator systems, trial procedures such as the Badger and McCabe method and appropriate variations of it may be easier to apply than the Newton-Raphson method (see Problem 2-3 for a comparison).

Part 2. Analysis of Existing Evaporator Systems

Many of the problems involving evaporators consist of case studies whose objective is the improvement of the economics of an evaporator system or of some process involving an evaporator system. The procedure that follows may be used also in the design of evaporator systems in which the restriction of equal areas is not imposed.

2. SPECIFICATIONS: F, X, T_F, T_0, P_0, P_3 (or T_3), U_1, U_2, U_3, A (or A_1, A_2, and A_3), and Forward Feed
   To Find: V_0, T_1, L_1, T_2, L_2, L_3, x_3

This set of specifications represents an existing triple-effect evaporator system for which the operating conditions have been specified. The primary objective of problems of this kind is the determination of the separation

(x_3, L_3) that can be achieved by an evaporator system at the specified set of operating conditions. This problem differs from the design problem in that the role of the area and x_3 (or L_3) has been reversed. In the present problem the area (or areas) is specified and both x_3 and L_3 are unknowns. However, since x_3 does not appear in any of the other equations, it may be regarded as a dependent variable and computed by use of Equation (2-24) after a solution has been obtained. The functions f_1 through f_6 are applicable provided the rate expressions are altered in the following manner: Replace A in f_2 by A_1, A in f_4 by A_2, and A in f_6 by A_3.

Let the ordering of the Newton-Raphson equations be the same as that used to identify the functions. Also, let the elements of the column vectors X_k and ΔX_k be ordered in the same manner as shown in Equation (2-33) except that the elements A and ΔA are to be replaced by L_3 and ΔL_3, respectively. When the resulting equations are scaled by use of the same procedure described, the elements of the resulting matrix are as follows:

$$
J_k = \begin{bmatrix}
1 & b_{12} & \lambda_1/\lambda_0 & 0 & 0 & 0 \\
-1 & b_{22} & 0 & 0 & 0 & 0 \\
0 & b_{32} & b_{33} & b_{34} & \lambda_2/\lambda_0 & 0 \\
0 & b_{42} & \lambda_1/\lambda_0 & b_{44} & 0 & 0 \\
0 & 0 & \lambda_2/\lambda_0 & b_{54} & b_{55} & \lambda_3/\lambda_0 \\
0 & 0 & -\lambda_2/\lambda_0 & b_{64} & \lambda_2/\lambda_0 & 0
\end{bmatrix}
\tag{2-36}
$$

where the particular b's that appear in this expression have the same definition as those stated in Illustrative Example 2-4 for these respective b's.

The following illustrative example is presented to demonstrate the use of the Newton-Raphson method for solving separation problems involving an existing evaporator system.

ILLUSTRATIVE EXAMPLE 2-5

The statement of this example is the same as Example 2-4 except that L_3 and x_3 are unknown and the specified values of the areas are as follows: $A_1 = 1,000$ ft², $A_2 = 1,050$ ft², and $A_3 = 1,340$ ft².

Solution: The calculational procedure may be initiated on the basis of the following assumptions:

1. $(T_0 - T_1) = 42°F$, $(T_1 - T_2) = 42°F$, and $(T_2 - T_3) = 41°F$.
2. Solvent evaporated in first effect = 14,000 lb/hr.
   Solvent evaporated in second effect = 14,000 lb/hr.
   Solvent evaporated in third effect = 12,000 lb/hr.
3. $V_0 = 15,000$ lb/hr.

On the basis of these assumptions, the elements appearing in matrices J_0, X_0, and f_0

are readily computed to give

$$
J_0 = \begin{bmatrix}
1 & -0.25 & 1 & 0 & 0 & 0 \\
-1 & -2.5 & 0 & 0 & 0 & 0 \\
0 & 0.18 & -1.958 & -0.18 & 1 & 0 \\
0 & 1.575 & 1 & -1.575 & 0 & 0 \\
0 & 0 & 1 & 0.11 & -1.958 & 1 \\
0 & 0 & -1 & 1.34 & 1 & 0
\end{bmatrix}
$$

$$
f_0 = \begin{bmatrix} g_1 \\ g_2 \\ g_3 \\ g_4 \\ g_5 \\ g_6 \end{bmatrix} = \begin{bmatrix} -0.088 \\ 0.120 \\ 0.03024 \\ -0.0154 \\ 0.05804 \\ -0.06024 \end{bmatrix}; \quad
X_0 = \begin{bmatrix} v_0 \\ u_1 \\ l_1 \\ u_2 \\ l_2 \\ l_3 \end{bmatrix} = \begin{bmatrix} 0.30 \\ 0.832 \\ 0.720 \\ 0.664 \\ 0.440 \\ 0.20 \end{bmatrix}
$$

Again, convergence to the solution set of values of the variables was achieved in three trials as shown in Table 2-3.

TABLE 2-3
SOLUTION OF EXAMPLE 2-3 BY USE OF THE NEWTON-RAPHSON METHOD

Trial No.	Value of the Variables (Scaled)					
	v_0	u_1	l_1	u_2	l_2	l_3
1	0.357090	0.857176	0.757235	0.703040	0.485162	0.188902
2	0.357108	0.857157	0.757182	0.702986	0.485179	0.188556
3	0.357107	0.857157	0.757182	0.702987	0.485179	0.188556
4	0.357107	0.857157	0.757182	0.702987	0.485179	0.188556

The solution is:

$$V_0 = v_0 F = 17{,}855.2 \text{ lb/hr}; \qquad T_1 = u_1 T_0 = 219.289^\circ\text{F};$$

$$L_1 = l_1 F = 37{,}859.1 \text{ lb/hr}; \qquad T_2 = u_2 T_0 = 175.747^\circ\text{F};$$

$$L_2 = l_2 F = 24{,}859.1 \text{ lb/hr}; \qquad L_3 = l_3 F = 9{,}427.8 \text{ lb/hr};$$

$$x_1 = \frac{FX}{L_1} = 0.132069; \qquad x_2 = \frac{FX}{L_2} = 0.206109;$$

$$x_3 = \frac{FX}{L_3} = 0.530346$$

It should be observed that, in the solution of Illustrative Example 2-3, the variables x_1, x_2, and x_3 were taken to be dependent. These variables could have been regarded as independent and corresponding functions formulated by setting the expressions given by Equation (2-24) for $j = 1, 2$, and 3, equal to three new functions, say f_7, f_8, and f_9. In general, the number of equations and the number of variables in the Newton-Raphson method may be reduced as desired by setting one function equal to zero for each variable taken to be dependent. However, the policy employed in the solution of the illustrative examples was to take as dependent only those variables that did not appear in any other equations of the set. This policy eliminated the necessity of taking into account the dependency of one variable on another in the calculation of the partial derivatives appearing in the Newton-Raphson equations.

Design of evaporator systems with unequal areas

The procedure demonstrated in the preceding section is seen to be applicable for the case where any given set of areas is specified. Thus, for the design of a system of evaporators with unequal areas, some additional condition must be imposed upon the system. Problems of this kind generally fall into the category of optimization problems, and the condition to be optimized is formulated in terms of an objective function.

Effect of boiling point elevations

Systems that exhibit boiling point elevations generally possess liquid enthalpies that depend on both temperature and composition. In the following analysis, the general case is considered wherein both variations are involved. Additional symbols are required to describe evaporation processes of this type. The symbols used are displayed in Figure 2-8 for a triple-effect evaporator system and defined as follows:

$T_j =$ saturation temperature of the pure solvent at the pressure P_j of effect j;

$\mathfrak{I}_j =$ boiling point temperature of evaporator effect j at the pressure P_j and composition x_j;

$h(T_j), H(T_j) =$ enthalpies of the pure solvent in the liquid and vapor states, respectively, evaluated at the saturation temperature T_j corresponding to the pressure P_j of effect j;

$H(\mathfrak{I}_j) =$ enthalpy of the pure solvent in the vapor state, evaluated at pressure P_j and temperature $\mathfrak{I}_j$;

$h(\mathfrak{I}_j, x_j) =$ enthalpy of the liquid solution leaving the jth effect; evaluated at P_j, $\mathfrak{I}_j$, and x_j;

$h(T_F, X) =$ enthalpy of the liquid feed solution, evaluated at its entering temperature T_F, composition X, and pressure.

Since the enthalpy of the thick liquor depends on composition, the three variables x_1, x_2 and x_3 appear in the equations describing a triple-effect system. Corresponding to each x_j, a component-material balance given by Equation (2-24) may be stated. At a given composition or mass fraction x_j,

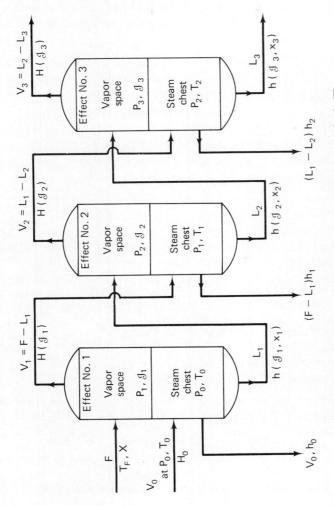

Figure 2-8. A triple-effect evaporator system with forward feed. The temperature distribution is shown for a system with boiling point elevations.

33

the temperatures $\mathfrak{I}_j$ and T_j for evaporator effect j are related by the Dühring line,

$$\mathfrak{I}_j = m(x_j)T_j + b(x_j) \tag{2-37}$$

where

$m(x_j), b(x_j) = $ slope and intercept of the Dühring lines; evaluated at the mass fraction x_j of the solute in the liquor leaving the jth effect.

The following functions are developed in a manner analogous to that demonstrated for the case where boiling point elevations are negligible.

Effect 1 $\begin{cases} \text{Enthalpy balance: } f_1 = F[h(T_F, X) - h(\mathfrak{I}_1, x_1)] + V_0\lambda_0 \\ \qquad\qquad\qquad\quad - (F - L_1)[(H(\mathfrak{I}_1) - h(\mathfrak{I}_1, x_1)] \\ \text{Heat transfer rate: } f_2 = U_1A(T_0 - \mathfrak{I}_1) - V_0\lambda_0 \\ \text{Phase equilibrium: } f_3 = m(x_1)T_1 + b(x_1) - \mathfrak{I}_1 \\ \text{Mass balance: } \qquad f_4 = FX - L_1x_1 \end{cases}$

$\tag{2-38}$

Effect 2 $\begin{cases} \text{Enthalpy balance: } f_5 = L_1[H(\mathfrak{I}_1, x_1) - h(\mathfrak{I}_2, x_2)] + (F_1 - L_1) \\ \qquad\qquad\qquad\quad \times [H(\mathfrak{I}_1) - h(T_1)] - (L_1 - L_2) \\ \qquad\qquad\qquad\quad \times [H(\mathfrak{I}_2) - h(\mathfrak{I}_2, x_2)] \\ \text{Heat transfer rate: } f_6 = U_2A[T_1 - \mathfrak{I}_2] - (F - L_1)[H(\mathfrak{I}_1) - h(T_1)] \\ \text{Phase equilibrium: } f_7 = m(x_2)T_2 + b(x_2) - \mathfrak{I}_2 \\ \text{Mass balance: } \qquad f_8 = FX - L_2x_2 \end{cases}$

Effect 3 $\begin{cases} \text{Enthalpy Balance: } f_9 = L_2[h(\mathfrak{I}_2, x_2) - h(\mathfrak{I}_3, x_3)] + (L_1 \\ \qquad\qquad\qquad\quad - L_2)[H(\mathfrak{I}_2) - h(T_2)] - (L_2 - L_3)[H(\mathfrak{I}_3) \\ \qquad\qquad\qquad\quad - h(\mathfrak{I}_3, x_3)] \\ \text{Heat transfer rate: } f_{10} = U_3A(T_2 - \mathfrak{I}_3) - (L_1 - L_2)[H(\mathfrak{I}_2) \\ \qquad\qquad\qquad\quad - h(T_2)] \\ \text{Phase equilibrium: } f_{11} = m(x_3)T_3 + b(x_3) - \mathfrak{I}_3 \\ \text{Mass balance: } \qquad f_{12} = FX - L_3x_3 \end{cases}$

Before demonstrating the solution of these equations by use of the Newton-Raphson method, it is informative to demonstrate the effect of boiling point elevation on evaporator operation by reconsidering Illustrative Examples 2-2 and 2-3.

Effect of boiling point elevation on multiple-effect evaporator operation

To demonstrate the effect of boiling point elevation, suppose for convenience that the boiling point elevation for each effect in multiple-effect operation is equal to one-sixth of the total ΔT across the system, that is, suppose in Illustrative Example 2-2 that

$$(\mathfrak{I}_1 - T_1) = (\mathfrak{I}_2 - T_2) = (\mathfrak{I}_3 - T_3) = \tfrac{1}{6}(T_0 - \mathfrak{I}_3)$$

Then Parts (a) through (c) of this example yield the following results:

$$\lambda_1 = \lambda_2 = \lambda_3 = \lambda; \qquad V_0 = V_1 = V_2 = V_3$$

$$Q_1 = Q_2 = Q_3 = Q; \qquad (T_0 - \mathfrak{I}_1) = (T_1 - \mathfrak{I}_2) = (T_2 - \mathfrak{I}_3)$$

and

$$Q = UA(T_0 - \mathfrak{I}_1) = UA(T_1 - \mathfrak{I}_2) = UA(T_2 - \mathfrak{I}_3)$$

In this case, however, the sum of the ΔT's is no longer equal to the total drop $(T_0 - \mathfrak{I}_3)$ across the system, but instead

$$\sum_{j=1}^{3} \Delta T_j = (T_0 - \mathfrak{I}_1) + (T_1 - \mathfrak{I}_2) + (T_2 - \mathfrak{I}_3) = (T_0 - \mathfrak{I}_3)$$

$$- (\mathfrak{I}_1 - T_1) - (\mathfrak{I}_2 - T_2) = \tfrac{2}{3}(T_0 - \mathfrak{I}_3)$$

(As the number of effects in the system is increased, the value of the sum decreases until operation becomes impossible.)

Then for triple-effect operation,

$$\text{Steam consumption} = V_0 = \frac{UA}{\lambda}(T_0 - \mathfrak{I}_1) = \frac{UA}{\lambda}\left(\frac{\sum\limits_{j=1}^{3} \Delta T_j}{3}\right) = \frac{2UA}{9\lambda}(T_0 - \mathfrak{I}_3)$$

$$\text{Capacity} = V_1 + V_2 + V_3 = \frac{3UA}{\lambda}(T_0 - \mathfrak{I}_1)$$

$$= \frac{3UA}{\lambda}\left(\frac{\sum\limits_{j=1}^{3} \Delta T_j}{3}\right) = \frac{2UA}{3\lambda}(T_0 - \mathfrak{I}_3)$$

$$\text{Economy} = \frac{V_1 + V_2 + V_3}{V_0} = 3$$

When the evaporators of the triple-effect system described above are operated as three single effects with the same terminal conditions as used for the triple-effect system, then the following results are obtained instead of those found in Illustrative Example 2-3.

$$\text{Steam consumption for the 3 single effects} = \frac{3UA(T_0 - \mathfrak{I}_3)}{\lambda}$$

$$\text{Capacity for the 3 single effects} = \frac{3UA(T_0 - \mathfrak{I}_3)}{\lambda}$$

$$\text{Economy for the 3 single effects} = 1$$

Thus, in this case

$$\frac{\text{Steam consumption for triple effect}}{\text{Steam consumption for three single effects}} = \left(\frac{2}{3}\right)\left(\frac{1}{9}\right) = \frac{2}{27}$$

and

$$\frac{\text{Capacity for triple effect}}{\text{Capacity for three single effects}} = \left(\frac{2}{3}\right)\left(\frac{1}{3}\right) = \frac{2}{9}$$

Recall that without boiling point elevation, the ratio of the capacities was equal to $\frac{1}{3}$. Thus, it is evident that the effect of boiling point elevations is to further reduce the capacity ratio, which eventually limits the number of effects that can be employed in a multiple-effect system.

Now consider the use of the equations stated above [Equation (2-38)] for the purpose of solving design problems as well as problems involving existing systems of evaporators. Consider first the following design problem.

1. SPECIFICATIONS: F, X, T_F, T_0, P_0, P_3 (or T_3), x_3 (or L_3), U_1, U_2, U_3, Equal Areas, and Forward Feed.

   To Find: V_0, $\mathfrak{I}_1$, T_1, x_1, L_1, $\mathfrak{I}_2$, T_2, x_2, L_2, A, $\mathfrak{I}_3$, and L_3 (or x_3).

The specifications of this problem permit a reduction of the number of equations to be solved simultaneously because some of them (namely f_{11} and f_{12}) may be solved independently of the remaining equations at the outset. Suppose x_3 is specified rather than L_3. Then, by setting $f_{12} = 0$, it follows that

$$L_3 = \frac{FX}{x_3}$$

Next, if P_3 is specified, then the temperature T_3 at which the pure solvent has the vapor pressure P_3 follows immediately from the physical data for the pure solvent. Since x_3 is also fixed, the temperature $\mathfrak{I}_3$ follows immediately upon setting $f_{11} = 0$ and solving for $\mathfrak{I}_3$ to obtain

$$\mathfrak{I}_3 = m(x_3)T_3 + b(x_3)$$

Thus, for the particular set of specifications stated, Equation (2-38) may be reduced to ten equations (f_1 through f_{10}) in ten unknowns (V_0, $\mathfrak{I}_1$, T_1, x_1, L_1, $\mathfrak{I}_2$, T_2, x_2, L_2, and A). These equations may be solved by using the Newton-Raphson technique as demonstrated by the following example.

ILLUSTRATIVE EXAMPLE 2-6

Repeat Illustrative Example 2-1 for the case where three effects (with equal areas and forward feed) are to be employed rather than a single effect. The overall heat transfer coefficients may be considered constant at the values: $U_1 = 300$, $U_2 = 250$, and $U_3 = 200$ Btu/(hr ft^2 °F).

Solution: The variables were scaled in a manner similar to that used in Example 2-5. Flow rates were again expressed as a fraction of the total feed rate, and the temperatures as a fraction of the steam temperature. The area was scaled by use of the relationship: $A = a(F/40)$. (When the mass fraction X of the feed is relatively small, all mass fractions should be stated relative to X.) Although the precise choice of the assumed set of variables needed to make the first trial is fairly arbitrary, the relatively simple scheme employed for picking the initial set gave very satisfactory results. This scheme assumed equal vapor rates, equal rates of heat transfer, and boiling point rises proportional to the concentrations in the respective effects.

The particular set of values of the variables assumed for the first trial as well as the subsequent sets of values of the variables obtained by use of the Newton-Raphson method for each of the first six trials are tabulated in Table 2-4.

TABLE 2-4
SOLUTION OF EXAMPLE 2-6 BY USE OF THE NEWTON-RAPHSON METHOD

Scaled Variable	Assumed Set	Trial No.					Solution Set
		1	2	3	4	5	
v_0	0.324	0.30014	0.304701	0.304660	0.304680	0.304679	$V_0 = 12{,}187.2$ lb/hr
$u_1{}^*$	0.926	0.88643	0.880640	0.881008	0.881002	0.881002	$\mathfrak{I}_1 = 308.351°F$
u_1	0.824	0.81997	0.814208	0.814574	0.814568	0.814568	$T_1 = 285.099°F$
x_1	0.250	0.24343	0.243718	0.243694	0.243694	0.243694	$x_1 = 0.243694$
l_1	0.800	0.82102	0.820619	0.820702	0.820700	0.820700	$L_1 = 32{,}828.0$ lb/hr
$u_2{}^*$	0.735	0.72579	0.724361	0.724476	0.724473	0.724473	$\mathfrak{I}_2 = 253.566°F$
u_2	0.600	0.62603	0.624404	0.624528	0.624525	0.624525	$T_2 = 218.584°F$
x_2	0.333	0.32470	0.325106	0.325079	0.325080	0.325080	$x_2 = 0.325080$
l_2	0.600	0.61553	0.615183	0.615235	0.615233	0.615233	$L_2 = 24{,}609.3$ lb/hr
a	1.026	0.79741	0.849149	0.849149	0.849164	0.849163	$A = 849.163$ ft²
							Steam Economy $= 1.97$

*Note $u_1 = \mathfrak{I}_1/T_0$, $u_2 = \mathfrak{I}_2/T_0$.

In the solution of Example 2-4, the Dühring lines shown in Figure 2-6 were represented by Equation (2-37) by taking

$$m(x) = 1.0 + 0.1419526x$$

and

$$b(x) = 271.3627x^2 - 9.419608x$$

When the areas are specified and the problem is to determine the separation that may be achieved, the problem is formulated as follows:

2. SPECIFICATIONS: F, X, T_F, T_0, P_0, P_3 (or T_3), U_1, U_2, U_3, A (or A_1, A_2, and A_3), and Forward Feed
   To Find: V_0, $\mathfrak{I}_1$ T_1, x_1, L_1, $\mathfrak{I}_2$, T_2, x_2 L_2, $\mathfrak{I}_3$, x_3, and L_3

To solve problems of this kind, the 12 equations given by Equation (2-38) may be solved simultaneously by using the Newton-Raphson procedure.

In addition to the procedures presented in this chapter for applying the Newton-Raphson method, other procedures are demonstrated in Chapters 3 and 4. For the case where the variation of the overall heat transfer coefficient with temperature is known, this dependency may be taken into account in the partial differentiation of the functions f_j. Also, if known, the effect of scale formation on the heat transfer surface may be taken into account.

PROBLEMS

2-1. In order to apply the Badger and McCabe calculational procedure described in the text to problems in which the boiling point elevations cannot be neglected, how must the equations involving the sum of the ΔT_j's such as Equation (2-29) be modified?

Hint: Note that the sum of the ΔT_j's may be expressed in terms of $(T_0 - \mathfrak{I}_3)$ and T_1 and T_2 by use of the boiling elevation expression $\mathfrak{I}_j = m_j T_j + b_j$, $(j = 1, 2)$.

2-2. Set up the energy balances and rate expressions for the steady-state operation of the double-effect evaporator system shown in Figure P2-2. Boiling point elevations are negligible as well as the variation of the heat capacities and latent heats with temperature. For a given design problem, F, X, T_F, T_0, P_0,, P_2 (or T_2), x_2 (or L_2), U_1, U_2, and equal areas A are specified. The unknowns are V_0, T_1, L_1, and A. Formulate the corresponding matrix equations for the Newton-Raphson method.

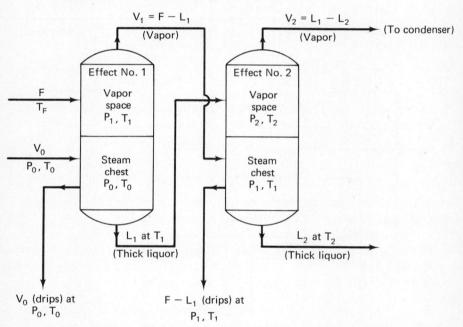

Figure P2-2. Double-effect evaporator system with forward feed. Boiling point elevations are negligible.

2-3(a). Evaluate the elements in the Newton-Raphson equations obtained in Problem 2-2 for the following design problem at the set of assumed values of the variables stated below.

$$F = 50,000 \text{ lb/hr};$$
$$L_2 = 10,000 \text{ lb/hr};$$
$$U_1 = 500, \ U_2 = 300 \text{ Btu/(hr ft}^2 \text{ °F)};$$
$$C_p = 1 \text{ Btu/(1b°F)} \text{ for all effects};$$
$$\lambda_0 = \lambda_1 = \lambda_2 = 1,000 \text{ Btu/lb};$$
$$T_0 = 250°F; \ T_F = 100°F; \ T_2 = 125°F.$$

For the first trial, assume

1. $(T_0 - T_1) = \dfrac{250 - 125}{2}$ °F;

2. Amount evaporated in each effect $= \dfrac{40,000}{2}$ lb/hr;

3. $A = 1,000$ ft^2;

4. $V_0 = 20,000$ lb/hr.

2-3(b). On the basis of the same assumed value of T_1 stated in Part (a), make one complete trial by the Badger and McCabe method.

2-4. Repeat Problem 2-2 for the case where backward feed is employed as indicated in Figure P2-4; it is supposed that L_1 is specified rather than L_2.

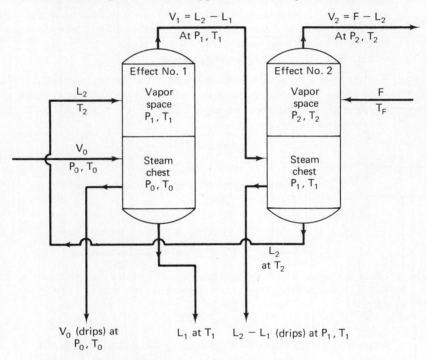

Figure P2-4. Double-effect evaporator system with backward feed. Boiling point elevations are negligible.

2-5. Evaluate the elements in the Newton-Raphson equations obtained in Problem 2-4. Use the same assumptions as those stated in Problems 2-3(a) and (b) except in this case take L_1 to be specified at 10,000 lb/hr instead of L_2.

2-6. Using the results obtained in Illustrative Examples 2-1 and 2-6, construct bar graphs that demonstrate the effect of boiling point elevation on evaporator systems.

2-7. A solution of an organic compound is to be concentrated from 20% to 60% in a double-effect evaporator system. Steam is available at 10 psig, and a vacuum of 2 psia is to be maintained in the vapor space of the second effect. The feed at a rate of 40,000 lb/hr enters the first effect at 200°F (forward feed is to be used). The heat capacity of the feed may be taken to be equal to 0.9 (Btu/lb °F), and that of the liquid streams L_1 and L_2 to be 0.8 Btu/lb °F. Boiling point elevation effects are negligible. The overall heat transfer coefficients are as follows: $U_1 = 700$, $U_2 = 500$ Btu/(hr ft^2 °F). Find the steam rate required as well as the areas (A_1 and A_2), where the evaporators are to have equal areas.

2-8. A single-effect evaporator is used to concentrate a sodium hydroxide solution from a 30% to 60% sodium hydroxide by weight. Steam is available at 40 psig and the condenser is of sufficient size to permit an operating pressure of 4.7 psia. The feed, 15,000 lb/hr, enters the evaporator at a temperature of 75°F. The overall heat transfer coefficient is 400 Btu/(hr ft^2 °F). Find the steam rate and the evaporator area required.

2-9(a). For the case in which sensible heat effects can be neglected and boiling point elevations are negligible in an N-effect evaporator system, show that an overall heat transfer coefficient U for the system exists and that it satisfies the rate equation for the system,

$$Q = UA\,\Delta T$$

where

A = an arbitrarily selected area upon which U is based;
$Q = Q_1 = Q_2 = \cdots = Q_N$;
$\Delta T = T_0 - T_N$

$$U = \frac{1}{\dfrac{A}{U_1 A_1} + \dfrac{A}{U_2 A_2} + \cdots + \dfrac{A}{U_N A_N}}.$$

2-9(b). For the case in which the sensible heat effects can be neglected but the boiling point elevations cannot be neglected, show that

$$Q = UA \sum_{j=1}^{N} \Delta T_j$$

where A, Q, and U have the same definitions stated in Part (a) and

$$\sum_{j=1}^{N} \Delta T_j = \sum_{j=1}^{N} (T_{j-1} - \mathfrak{I}_j) = (T_0 - \mathfrak{I}_N) - \sum_{j=1}^{N-1} (\mathfrak{I}_j - T_j)$$

2-10. In the brief description of evaporator operation given at the beginning of this chapter, the statement is made that "the operating pressure in the last effect is determined by the condensing capacity of the condenser following this effect." In most unit operation laboratories, the student will find a vacuum pump connected in some manner to the condenser or the accumulator tank for the condensed vapor produced by the last effect. If the first statement is true, then what purpose does the vacuum pump serve?

NOTATION

$a =$ sq ft of surface area per ft of length of pipe (a denotes the particular surface area upon which the overall heat transfer coefficient U is based).

$A_j =$ total heat transfer surface for evaporator effect j.

$b(x_j) =$ intercept of that Dühring line having as its composition x_j.

$C_p =$ specific heat, Btu/(lb °F).

$f_k =$ column vector of the N functions $f_1, f_2, \ldots, f_N$.

$F =$ feed rate to the evaporator system, lb/hr.

$h(T_j),$ $h(\mathfrak{I}_j)$ = enthalpy of the pure solvent in the liquid state at the temperatures T_j and $\mathfrak{I}_j$, respectively, and pressure P_j, Btu/lb. Where boiling point elevations are negligible, the notation h_j, which is equal to $h(T_j)$, is used.

$H(T_j),$ $H(\mathfrak{I}_j)$ = same as above except that the capital H denotes the vapor state.

$h(\mathfrak{I}_j, x_j) =$ enthalpy of the liquid at temperature $\mathfrak{I}_j$, composition x_j, and pressure P_j, Btu/lb.

$h(T_F, X) =$ enthalpy of the feed at its entering temperature, pressure, and composition, Btu/lb. Where boiling point elevations are negligible, the enthalpy of the feed is denoted by h_F.

$J =$ used in the analysis to denote the Jacobian matrix.

$L_j =$ mass flow rate of liquid from evaporator effect j, lb/hr.

$N =$ total number of effects in the evaporator system.

$p_i =$ partial pressure of component i.

$P_i =$ vapor pressure of pure component i.

$P_j =$ total pressure in evaporator j.

$Q_j =$ rate of heat transfer for evaporator effect j, Btu/hr.

$T_F, T_0 =$ temperature of the feed and steam, respectively, to an evaporator.

$T_j =$ saturation temperature at the pressure P_j of the vapor leaving the jth effect of a multiple-effect evaporator system.

$\mathfrak{I}_j =$ temperature of the liquid leaving the jth effect.

$V_j =$ mass flow rate of the vapor from the jth effect of a multiple-effect evaporator system.

$x_j =$ mass fraction of the solute in the liquid leaving effect j. [Also x is used to denote the mole fraction of the solvent in Equation (2-1).]

$X_k =$ column vector of the values of the variables used to make the kth trial.

$\Delta X_k =$ column vector; $\Delta X_k = X_{k+1} - X_k$.

$X^T =$ the transpose of a column matrix X is equal to a row matrix and conversely (1).

$z =$ the distance measured from a specified reference point; $z_T =$ total length.

SUBSCRIPTS

a = assumed value of the variable.

co = corrected value of the variable.

k, n = counting integers.

GREEK LETTERS

γ = thermodynamic activity coefficient.

$\lambda_j = H_j - h_j$, latent heat of vaporization of the pure solvent at its saturation temperature T_j and pressure P_j.

MATHEMATICAL SYMBOLS

$$\sum_{j=1}^{n} x_j = x_1 + x_2 + \cdots + x_n.$$

REFERENCES

1. Amundson, N. R., *Methematical Methods in Chemical Engineering, Matrices, and Their Applications.* Englewood Cliffs, N.J.: Prentice-Hall, Inc., 1966.

2. Badger, W. L., and W. L. McCabe, *Elements of Chemical Engineering.* New York: McGraw-Hill Book Company, Inc., 1936.

3. Brown, G. G., and Associates, *Unit Operations.* New York: John Wiley & Sons, Inc., 1950.

4. Burdett, J. W., "Prediction of Steady State and Unsteady State Response Behavior of a Multiple Effect Evaporator System," Ph.D. dissertation, Texas A&M University, College Station, Texas, 1969. See also, J. W. Burdett and C. D. Holland, "Dynamics of a Multiple-Effect Evaporator System," *A.I.Ch.E. Journal,* **17**, (1971), 1080.

5. Carnahan, Brice, H. A. Luther, and J. O. Wilkes, *Applied Numerical Methods.* New York: John Wiley & Sons, Inc., 1969.

6. Denbigh, Kenneth, *The Principles of Chemical Equilibrium.* New York: Cambridge University Press, 1955.

7. Gerlack, A., "Ueber Siedetemperaturen der Salzosungen and Vergleiche der Eihohung der Siedetemperaturen Mit der Ubrigen Eigenschaften der Salzlosungen," *Z. Analytical Chemistry,* **26**, (1887), 412.

8. Holland, C. D., *Multicomponent Distillation.* Englewood Cliffs, N.J.: Prentice-Hall, Inc., 1963.

9. ———, *Unsteady State Processes with Applications in Multicomponent Distillation.* Englewood Cliffs, N.J.: Prentice-Hall, Inc., 1966.

10. Keenan, J. H., and F. G. Keyes, *Thermodyanamic Properties of Steam.* New York: John Wiley & Sons, Inc., 1936.

11. McCabe, W. L., "The Enthalpy Concentration Chart—A Useful Device for Chemical Engineering Calculations," *Trans. Am. Inst. Chem. Engrs.,* **31**, (1935), 129.

12. Nartker, T. A., J. M. Srygley, and C. D. Holland, "Solution of Problems Involving Systems of Distillation Columns," *Can. J. Chem. Engr.,* **44**, (1966), 217.

Introduction to the Fundamentals of Distillation 3

In this chapter, the fundamental principles and relationships involved in making multicomponent distillation calculations are developed from first principles. To enhance the visualization of the fundamental relationships involved in this separation process, the development of the calculational procedures are presented first for the special case of binary mixtures and then for the general case of multicomponent mixtures. All mixtures that contain three or more components are referred to as *multicomponent mixtures*. Proposed calculational procedures are illustrated by numerical examples.

Description of the separation of mixtures by distillation

The general objective of distillation is the separation of substances that have different vapor pressures at any given temperature. The word *distillation* as used here refers to the physical separation of a mixture into two or more fractions that have different boiling points.

If a liquid mixture of two volatile materials is heated, the vapor that comes off will have a higher concentration of the lower boiling material than the liquid from which it was evolved. Conversely, if a warm vapor is cooled,

the higher boiling material has a tendency to condense in a greater proportion than the lower boiling material. The early distillers of alcohol for bever-
ages applied these fundamental ideas. Although distillation was known and
practiced in antiquity and a commercial still had been developed by Coffey
in 1832, the theory of distillation was not studied until the work of Sorel
(19) in 1893. Other early workers were Lord Rayleigh (16) and Lewis (11).
Current technology has permitted the large-scale separation by distillation
of ethylbenzene and *p*-xylene, which have only a 3.9°F difference in boiling
points (1).

A distillation column consists of a series of plates (or trays). In normal
operation, there is a certain amount of liquid on each plate, and some arrangement is made for ascending vapors to pass through the liquid and make
contact with it. The descending liquid flows down from the plate above
through a downcomer, across the next plate, and then over a *weir* and into
another downcomer to the next lower plate as shown in Figure 3-1. For
many years, *bubble caps* were used (of which a variety of designs are shown
in Figure 3-2) for contacting the vapor with the liquid. These contacting devices promote the production of small bubbles of vapor with relatively large
surface areas.

Recent developments of devices for contacting the vapor and liquid
streams have tended to displace bubble caps. New columns are usually
equipped with either *ballast trays* (see Figure 3-3), sometimes called *valve*

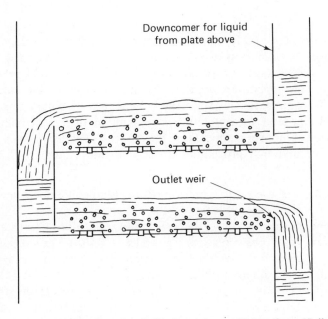

Figure 3-1. The interior of a distillation column. (Taken from Holland
and Lindsay, *Encyclopedia Chemical Technology*, Vol. 7, 2nd ed., p. 206.)

Figure 3-2. Various types of bubble caps used in distillation columns. (*Courtesy of Fritz W. Glitsch & Sons, Inc.*)

Figure 3-3. Portion of a Glitsch V–1 ballast tray. (*Courtesy of Fritz W. Glitsch & Sons, Inc.*)

Figure 3-4. A perforated or sieve tray, 9 feet in diameter. (*Courtesy of Fritz W. Glitsch & Sons, Inc.*)

trays, or *perforated trays* (see Figure 3-4), sometimes called *sieve trays*. In the valve trays, the valve opens wider as the vapor velocity increases and closes as the vapor velocity decreases. This opening and closing allows the valve to remain immersed in liquid and thereby preserve a liquid seal over wide ranges of liquid and vapor flow rates.

Distillation columns have been built as high as 200 ft. Diameters as large as 44 ft have been used. Construction of a plate for a column with a 40 ft diameter is shown in Figure 3-5. Operating pressures for distillation columns have been reported which range from 15 mm to 500 psia.

As indicated in Figure 3-6, the overhead vapor V_1, upon leaving the top plate enters the condenser where it is either partially or totally condensed. The liquid formed is collected in an accumulator from which the liquid stream L_0 (called *reflux*) and the top product stream D (called the *distillate*) are withdrawn. When the overhead vapor V_1 is totally condensed to the liquid state and the distillate D is withdrawn as a liquid, the condenser is called a *total condenser*. If V_1 is partially condensed to the liquid state to produce the reflux L_0 and the distillate D is withdrawn as a vapor, the condenser is called a *partial condenser*. The amount of liquid reflux is commonly expressed

Figure 3-5. A Glitsch A–2 ballast tray with mist eliminator in the process of construction, 40 feet in diameter. (*Courtesy of Fritz W. Glitsch & Sons, Inc.*)

in terms of the *reflux ratio*, L_0/D. Although the internal liquid to vapor ratio, L/V, is sometimes referred to as the internal reflux ratio, the term *reflux ratio* will be reserved herein to mean L_0/D.

The liquid that leaves the bottom plate of the column enters the reboiler, where it is partially vaporized. The vapor produced is allowed to flow back up through the column, and the liquid is withdrawn from the reboiler and called the *bottoms* or *bottom product B*. In practice, the reboiler is generally located externally from the column. A typical commercial installation is shown in Figure 3-7.

Part 1. Fundamental Principles Involved in Distillation

To compute the composition of the top product D and the bottom product B which may be expected by use of a given distillation column operated at a given set of conditions, it is necessary to obtain a solution to the following

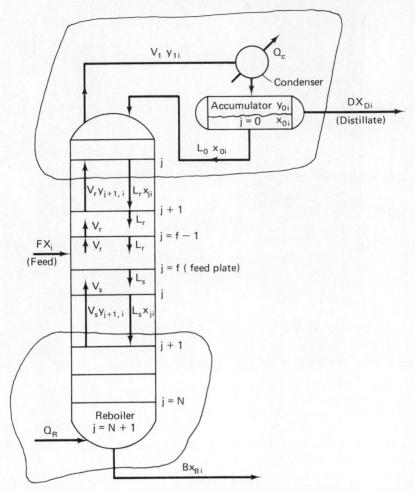

Figure 3-6. Sketch of a conventional column in which the total flow rates are constant within the rectifying and stripping sections.

equations:

1. Equilibrium relationships
2. Component-material balances
3. Total-material balances
4. Energy balances

Consider first equilibrium relationships.

Physical equilibrium

A two-phase multicomponent mixture is said to be in equilibrium if the following necessary conditions are satisfied (5).

Figure 3-7. Typical view of distillation columns at the Mobil Refinery at Beaumont, Texas. (*Courtesy Mobil Oil Corporation.*)

1. The temperature T^V of the vapor phase is equal to the temperature T^L of the liquid phase.
2. The total pressure P^V throughout the vapor phase is equal to the total pressure P^L throughout the liquid phase. $\qquad$ (3-1)
3. The tendency of each component to escape from the liquid phase to the vapor phase is exactly equal to its tendency to escape from the vapor phase to the liquid phase.

In the following analysis it is supposed that a state of equilibrium exists, $T^V = T^L = T$, $P^V = P^L = P$, and the escaping tendencies are equal.

Now consider the special case in which the third condition may be represented by Raoult's law.

$$Py_i = P_i x_i \qquad (3-2)$$

where x_i and y_i are the mole fractions of component i in the liquid and vapor phases, respectively, and P_i is the vapor pressure of pure component i at the temperature T of the system.

The separation of a binary mixture by distillation may be represented in two-dimensional space, but n-dimensional space is required to represent the separation of a multicomponent mixture. The graphical method proposed by McCabe and Thiele (13) for the solution of problems involving binary mixtures is presented in a subsequent section. The McCabe-Thiele method makes use of an equilibrium curve that may be obtained from the "boiling point diagram."

49

Construction and interpretation of the boiling point diagram for
binary mixtures

When a state of equilibrium exists between a vapor and liquid phase
composed of two components A and B, the system is described by the following set of independent equations,

$$\text{Equilibrium relationships}\begin{cases} Py_A = P_A x_A \\ Py_B = P_B x_B \\ y_A + y_B = 1 \\ x_A + x_B = 1 \end{cases} \tag{3-3}$$

where it is understood that Raoult's law is obeyed. Since the vapor pressures
P_A and P_B depend on T alone, Equation (3-3) consists of four equations in
six unknowns. Thus, to obtain a solution to this set of equations, two variables must be fixed. (Observe that this result is in agreement with the Gibbs
phase rule: $\mathcal{P} + \mathcal{V} = c + 2$. For the above case, the number of phases
$\mathcal{P} = 2$, the number of components $c = 2$, and thus the number of degrees
of freedom $\mathcal{V} = 2$, that is, the number of variables that must be fixed $= 2$.)
In the construction of the boiling point diagram for a binary mixture, the
total pressure P is fixed and a solution is obtained for each of several temperatures lying between the temperatures at which the respective vapor
pressures P_A and P_B are equal to the total pressure P.

The solution of the expressions given by Equation (3-3) for x_A in terms
of P_A, P_B, and P is effected as follows. Addition of the first two expressions
followed by the elimination of the sum of the y's by use of the third expression yields

$$P = P_A x_A + P_B x_B \tag{3-4}$$

Elimination of x_B by use of the fourth expression given by Equation (3-3)
followed by rearrangement of the result so obtained yields

$$x_A = \frac{P - P_B}{P_A - P_B} \tag{3-5}$$

From the definition of a mole fraction ($0 \leq x_A \leq 1$), Equation (3-5) has a
meaningful solution at a given P for any T lying between the boiling point
temperatures T_A and T_B of pure A and pure B, respectively. (At T_A, $P_A = P$,
and at T_B, $P_B = P$.) After x_A has been computed by use of Equation (3-5)
at the specified P and T, the corresponding value of y_A which is in equilibrium
with the value of x_A so obtained is computed by use of the first expression of
Equation (3-3), namely,

$$y_A = \left(\frac{P_A}{P}\right) x_A \tag{3-6}$$

By plotting T versus x_A and T versus y_A, the lower and upper curves, respec-

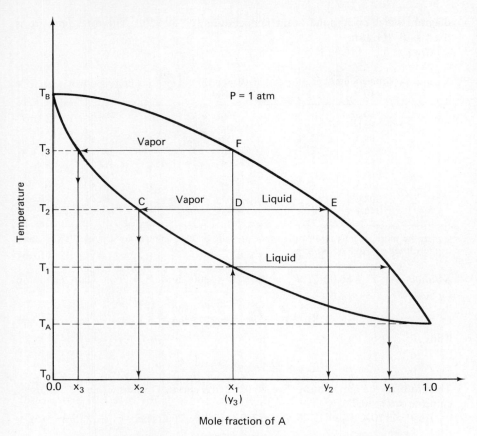

Figure 3-8. The boiling point diagram. (Taken from Holland and Lindsay, *Encyclopedia Chemical Technology*, Vol. 7, 2nd ed., p. 216.)

tively, of Figure 3-8 are typical of those obtained when component A is more
volatile than B. Component A is said to be more volatile than component
B, if for all T in the closed interval $T_A \leqq T \leqq T_B$, the vapor pressure of A
is greater than the vapor pressure of B, that is, $P_A > P_B$. The parallel lines
such as $\overline{CE}$ that join equilibrium pairs (x, y), computed at a given T and P
by use of Equations (3-5) and (3-6), are commonly called *Tie lines*.

ILLUSTRATIVE EXAMPLE 3-1*

By use of the following vapor pressures for benzene and toluene (taken from
The Chemical Engineer's Handbook, 2nd ed., J. H. Perry, editor, McGraw-Hill,
New York, 1941), compute the three equilibrium pairs (x, y) on a boiling point

*Taken from C. D. Holland, *Introduction to the Fundamentals of Distillation*. Proceedings to the Fourth Annual Education Symposium of the ISA, April 5–7, 1972, Wilmington,
Del. Courtesy Instrument Society of America.

diagram which correspond to the temperature $T = 80.02°C$. The total pressure is fixed at $P = 1$ atm.

   Given:

Temperature (°C)	P_A(Benzene) (mm Hg)	P_B(Toluene) (mm Hg)
80.02	760	300.0
84.0	852	333.0
88.0	957	379.5
92.0	1,078	432.0
96.0	1,204	492.5
100.0	1,344	559.0
104.0	1,495	625.5
108.0	1,659	704.5†
110.4	1,748	760.0

   †In the more recent editions, the vapor pressure of 704.5 mm for toluene as 108°C is inaccurately listed as 740.5 or 741 mm.

Solution: At $T = 80.02°C$, $P_A = 760$, $P_B = 300$, and $P = 760$. Then Equation (3-5) gives

$$x_A = \frac{P - P_B}{P_A - P_B} = \frac{760 - 300}{760 - 300} = 1$$

Thus,

$$y_A = \frac{P_A}{P} x_A = 1$$

Therefore, at the temperature $T = 80.02°C$, the curves T versus x_A and T versus y_A coincide at (1, 80.02).

   At $T = 110.4$, $P_A = 1,748$, $P_B = 760$, and $P = 760$. Then, by Equation (3-5),

$$x_A = \frac{760 - 760}{1,748 - 760} = 0$$

and thus,

$$y_A = \left(\frac{1,748}{760}\right)(0) = 0$$

Hence, the curves T versus x_A and T versus y_A again coincide at the point (0, 110.4).

   At any temperature between T_A and T_B, say $T = 100°C$, the calculations are carried out as follows:

$$x_A = \frac{760 - 559}{1,344 - 559} = \frac{201}{785} = 0.256$$

and

$$y_A = \left(\frac{1,344}{760}\right)(0.256) = 0.453$$

These results give the point (0.256, 100) on the T versus x_A curve and the point (0.453, 100) on the T versus y_A curve. Other points on these curves for temperatures lying between T_A and T_B are located in the same manner.

A boiling point diagram is a most convenient aid in the visualization of phase behavior. For definiteness, suppose P is fixed at 1 atm. Consider first the case of the liquid mixture of A and B at a temperature T_0, at a pressure of 1 atm, and with the composition $x_A = x_1$, $x_B = 1 - x_1$. As indicated by Figure 3-8, such a mixture is in the single-phase region. Suppose the pressure is held fixed at 1 atm throughout the course of the following changes. First, suppose the mixture is heated to the temperature T_1. At this temperature, the first evidence of a vapor phase, a "bubble of vapor," may be observed. The temperature T_1 is called the *bubble point* temperature of a liquid with the composition x_1. The mole fraction of A in the vapor in equilibrium with this liquid is seen to be y_1. As the mixture is heated from T_1 to T_2, vaporization continues. Since A has a greater escaping tendency than B, the liquid becomes leaner in A ($x_2 < x_1$). The relative amounts of A and B vaporized also depend on their relative amounts in the liquid phase. As the liquid phase becomes richer in B, the vapor phase also becomes richer in B ($y_2 < y_1$). Point D (the intersection of the horizontal line passing through T_2 and the vertical line passing through x_1) is seen to lie in the two-phase region. It is readily shown that the ratio of the moles of vapor to the moles of liquid at T_2 is equal to the ratio of $\overline{CD}/\overline{DE}$. Also, note that all initial liquid mixtures (at the temperature T_0) with the mole fraction of A lying between x_2 and y_2 will have the same equilibrium composition (x_2, y_2) at T_2. This property permits the equilibrium state (x_2, y_2, T_2, 1 atm) to be approached from different directions. If the particular mixture $x_A = x_1$ at T_0 is heated until point F is reached, the equilibrium mixture (x_3, y_3) at T_3 is obtained. The temperature T_3 is called the *dew point* temperature. At F, the last point in the two-phase region, all of the liquid is vaporized with the exception of, say, one drop. Thus, the dew point temperature is seen to be that temperature at which the first drop of liquid is formed when a vapor with the composition $y_3 = x_1$ is cooled from a temperature greater than its dew point to its dew point temperature, T_3.

Generalized equilibrium relationships

Unfortunately, the phase behavior of many mixtures is not adequately described by Raoult's law. A more precise statement of the third condition of Equation (3-1) is that the partial molar free energies are equal (5) from which the following alternate but equivalent statement may be deduced,

$$\bar{f}_i^V = \bar{f}_i^L \tag{3-7}$$

where $\bar{f}_i^V$ and $\bar{f}_i^L$ are the fugacities of component i in the vapor and liquid phases, evaluated at the P and T of the system. Equation (3-7) may be restated in the following equivalent form:

$$\gamma_i^V f_i^V y_i = \gamma_i^L f_i^L x_i \tag{3-8}$$

where

f_i^L, f_i^V = fugacities of pure component i in the liquid and vapor states, respectively, evaluated at the total pressure P and temperature T of the system;

x_i, y_i = mole fractions of component i in the liquid and vapor phases, respectively;

γ_i^L, γ_i^V = activity coefficients of component i in the liquid and vapor phases, respectively. $\gamma_i^L = \gamma_i^L(P, T, x_1, \ldots, x_c)$; $\gamma_i^V = \gamma_i^V(P, T, y_1, \ldots, y_c)$.

If, as is usually the case, the vapor may be assumed to form an ideal solution, then $\gamma_i^V = 1$ for each i, and Equation (3-8) may be restated as follows:

$$y_i = \gamma_i^L K_i x_i \tag{3-9}$$

where

$K_i = f_i^L / f_i^V$, the ideal solution K value.

The expression given by Equation (3-9) is recognized as one form of Henry's law. If the liquid phase also forms an ideal solution ($\gamma_i^L = 1$ for all i), then Equation (3-9) reduces to

$$y_i = K_i x_i \tag{3-10}$$

In some of the literature, the activity coefficient γ_i^L is absorbed in K_i, that is, the product $\gamma_i^L K_i$ is called K_i and an equation of the form of Equation (3-10) is obtained which is applicable to systems described by Equation (3-9).

If the effect of total pressure on the liquid fugacity is negligible in the neighborhood of the vapor pressure of pure component i, then

$$f_i^L |_{P,T} \cong f_i^L |_{P_i,T} = f_i^V |_{P_i,T} \tag{3-11}$$

where P_i is the vapor pressure of pure component i. If in addition to the assumptions required to obtain Equations (3-10) and (3-11), one also assumes that the vapor phase obeys the perfect gas law ($PV = RT$), then Equation (3-10) reduces to Raoult's law, Equation (3-2).

Determination of the bubble point and dew point temperatures of multicomponent mixtures

In the interest of simplicity, the equilibrium relationship given by Equation (3-10) is used in the following developments. The state of equilibrium for a two-phase (vapor and liquid) system is described by the following equations in which any number of components c are distributed between the two phases

$$\text{Equilibrium relationships} \begin{cases} y_i = K_i x_i & (1 \leq i \leq c)^* \\ \sum_{i=1}^{c} y_i = 1 \\ \sum_{i=1}^{c} x_i = 1 \end{cases} \tag{3-12}$$

Since K_i is a function of the total pressure P and the temperature T [$K_i = K_i(P, T)$], it is evident that the expressions represented by Equation (3-12) consist of $c + 2$ equations in $2c + 2$ unknowns. Thus, to obtain a solution to these equations, c variables must be fixed.

When $c - 1$ values of x_i and the total pressure P are fixed, the temperature T required to satisfy these equations is called the *bubble point temperature*. The cth mole fraction may be found by use of the $(c - 1)$ fixed values of x_i and the last expression given by Equation (3-12). When the first expression is summed over all components and the sum of the y_i's eliminated by use of the second expression given by Equation (3-12), the following result is obtained,

$$1 = \sum_{i=1}^{c} K_i x_i \tag{3-13}$$

Equation (3-13) consists of one equation in one unknown, the temperature. Since K_i is generally an implicit function of T, the solution of Equation (3-13) for the bubble point temperature becomes a trial-and-error problem. Of the many numerical methods for solving such a problem, only Newton's method (4, 8) is presented. In the application of this method, it is convenient to restate Equation (3-13) in function notation as follows:

$$f(T) = \sum_{i=1}^{c} K_i x_i - 1 \tag{3-14}$$

Thus, the bubble point temperature becomes that T which makes $f(T) = 0$. In the application of Newton's method, the following expression for the first derivative of $f(T)$ is needed.

$$f'(T) = \sum_{i=1}^{c} x_i \frac{dK_i}{dT} \tag{3-15}$$

Newton's method is initiated by the selection of an assumed value for T, say T_n. Then the values of $f(T_n)$ and $f'(T_n)$ are determined. The improved value of T, denoted by T_{n+1}, is found by application of Newton's formula [see Equation (A-9)]

$$T_{n+1} = T_n - \frac{f(T_n)}{f'(T_n)} \tag{3-16}$$

*The counting integer i for component number takes on only integral values, and the notation $(1 \leq i \leq c)$ is used herein to mean $i = 1, 2, \ldots, c - 1, c$.

The value so obtained for T_{n+1} becomes the assumed value for the next trial. This procedure is repeated until $|f(t)|$ is less than some small preassigned positive number ϵ. Observe that when the T has been found that makes $f(T) = 0$, each term $K_i x_i$ of the summation in Equation (3-14) is equal to y_i, the composition of the vapor. In Illustrative Example 3-2, as well as in those that follow, synthetic functions for the K-values and the enthalpies were selected in order to keep the arithmetic simple.

ILLUSTRATIVE EXAMPLE 3-2*

If for a three-component mixture, the following information is available, compute the bubble point temperature at the specified pressure of $P = 1$ atm by use of Newton's method. Take the first assumed value of T_n to be equal to 100°F.

Given:

Component No.	K_i	x_i
1	$K_1 = \dfrac{0.01T\dagger}{P\dagger}$	$\dfrac{1}{3}$
2	$K_2 = \dfrac{0.02T}{P}$	$\dfrac{1}{3}$
3	$K_3 = \dfrac{0.03T}{P}$	$\dfrac{1}{3}$

$\dagger T$ is in °F and P is in atm.

Solution: Assume $T_1 = 100$°F. The total pressure $P = 1$ atm. Then,

Component No.	x_i	K_i @ $P = 1$ atm $T = 100$°F	$K_i x_i$	$\dfrac{dK_i}{dT}\Big|_{T_n=100}$	$x_i \dfrac{dK_i}{dT}$
1	$\dfrac{1}{3}$	1	$\dfrac{1}{3}$	0.01	$\dfrac{0.01}{3}$
2	$\dfrac{1}{3}$	2	$\dfrac{2}{3}$	0.02	$\dfrac{0.02}{3}$
3	$\dfrac{1}{3}$	3	$\dfrac{3}{3}$	0.03	$\dfrac{0.03}{3}$
			$\dfrac{6}{3} = 2$		$\dfrac{0.06}{3} = 0.02$

From the above results, it follows that

$$f(100) = \sum_{i=1}^{c} K_i x_i - 1 = 2 - 1 = 1$$

$$f'(100) = \sum_{i=1}^{c} x_i \frac{dK_i}{dT} = 0.02$$

Then

$$T_2 = T_1 - \frac{f(T_1)}{f'(T_1)} = 100 - \frac{1}{0.02} = 50°F$$

*Taken from Holland, *Introduction to the Fundamentals of Distillation*. Courtesy Instrument Society of America.

Assume $T_2 = 50°F$

Component No.	K_i @ $P = 1$ atm $T = 50°F$	$K_i x_i = y_i$
1	$\frac{1}{2}$	$\frac{1}{6}$
2	$\frac{2}{2}$	$\frac{2}{6}$
3	$\frac{3}{2}$	$\frac{3}{6}$
		$\frac{6}{6} = 1$

Therefore, the bubble point temperature is 50°F.

When the y_i's and P are fixed rather than the x_i's and P, the solution temperature of the expressions given by Equation (3-12) is called the *dew point temperature*. By rearranging the first expression of Equation (3-12) to the form $x_i = y_i/K_i$ and carrying out steps analogous to those described above, the dew point function $F(T)$ is obtained.

$$F(T) = \sum_{i=1}^{c} \frac{y_i}{K_i} - 1 \tag{3-17}$$

The dew point temperature is that T that makes $F(T) = 0$. In this case

$$F'(T) = -\sum_{i=1}^{c} \frac{y_i}{K_i^2} \frac{dK_i}{dT} \tag{3-18}$$

Observe again that when the T is found that makes $F(T) = 0$, each term y_i/K_i of the summation in Equation (3-18) is equal to x_i, the composition of the liquid.

Now observe that if after the bubble point temperature has been determined for a given set of x_i's, the set of y_i's so obtained are used to determine the dew point temperature at the same pressure, it will be found that these two temperatures are equal. For a binary mixture, this result is displayed graphically in Figure 3-8. For example, a bubble point temperature calculation on the basis of the $\{x_{1i}\}$ yields the bubble point temperature T_1 and the composition of the vapor $\{y_{1i}\}$. Then a dew point temperature on the set $\{y_{1i}\}$ yields the dew point temperature T_1 and the original set of x_{1i}'s.

The K_b method for the determination of bubble point and dew point temperature

Robinson and Gilliland (17) pointed out that if the relative values of the K_i's are independent of temperature, the expressions given by Equation (3-12) may be rearranged in a manner such that trial-and-error calculations are avoided in the determination of the bubble point and dew point temperatures. The ratio K_i/K_b is called the relative volatility α_i of component i with respect to component b, that is,

$$\alpha_i = \frac{K_i}{K_b} \tag{3-19}$$

where K_i and K_b are evaluated at the same temperature and pressure. Component b may or may not be a member of the given mixture under consideration.

When the x_i's and the pressure P are given and it is desired to determine the bubble point temperature, the formula needed may be developed by first rewriting the first expression of Equation (3-12) as follows:

$$y_i = \left(\frac{K_i}{K_b}\right) K_b x_i = \alpha_i K_b x_i \tag{3-20}$$

Summation of the members of Equation (3-20) over all components i, followed by rearrangement yields

$$K_b = \frac{1}{\sum\limits_{i=1}^{c} \alpha_i x_i} \tag{3-21}$$

Since the α_i's are independent of temperature, they may be computed by use of the values of K_i and K_b evaluated at any arbitrary value of T and at the specified pressure. After K_b has been evaluated by use of Equation (3-21), the desired bubble point temperature is found from the known relationship between K_b and T.

If the y_i's are known instead of the x_i's, then the desired formula for the determination of the dew point temperature is found by first rearranging Equation (3-20) to the following form

$$K_b x_i = \frac{y_i}{\alpha_i}$$

and then summing over all components to obtain

$$K_b = \sum\limits_{i=1}^{c} \frac{y_i}{\alpha_i} \tag{3-22}$$

This equation is used to determine the dew point temperature in a manner analogous to that described for Equation (3-21).

Many families of compounds are characterized by the fact that their vapor pressures may be approximated by the Clausius-Clapeyron equation and by the fact that their latent heats of vaporization are approximately equal. The vapor pressures of the members of such families of compounds fall on parallel lines when plotted against the reciprocal of the absolute temperature. For any two members i and b of such a mixture, it is readily shown that α_i is independent of temperature.

Although there exists many systems whose α_i's are very nearly constant and Equations (3-21) and (3-22) are applicable for the determination of the bubble point and dew point temperatures, respectively, the greatest use of these relationships lies in their application in the iterative procedures for solving multicomponent distillation problems as described in a subsequent section.

ILLUSTRATIVE EXAMPLE 3-3

Repeat Illustrative Example 3-2 by use of the K_b method.

Solution: Since K_b may be selected arbitrarily, take $K_b = K_1$. Assume $T = 100°F$.

Component No.	x_i	$K_i @ P = 1$ atm $T = 100°F$	$\alpha_i = \dfrac{K_i}{K_b}$	$\alpha_i x_i$
1	$\dfrac{1}{3}$	1	1	$\dfrac{1}{3}$
2	$\dfrac{1}{3}$	2	2	$\dfrac{2}{3}$
3	$\dfrac{1}{3}$	3	3	$\dfrac{3}{3}$
				$\dfrac{6}{3} = 2$

Then

$$K_b = \frac{1}{\sum\limits_{i=1}^{c} \alpha_i x_i} = \frac{1}{2} = 0.5$$

and since $K_b = K_1$,

$$0.5 = 0.01T, \quad \text{or} \quad T = 50°F$$

Since the α_i's are independent of temperature, 50°F is the correct value for the bubble point temperature.

Part 2. Separation of Multicomponent Mixtures by Use of a Single Equilibrium Stage

Each of the separation processes considered in this and in Part 3 are special cases of the general separation problem in which a multicomponent mixture is to be separated into two or more parts through the use of any number of equilibrium stages.

Flash calculations

The boiling point diagram (Figure 3-8) is useful for visualizing the necessary conditions required for a flash to occur. Suppose the feed to be flashed has the composition $X_i = x_{1i}(x_{1,A}$ and $x_{1,B})$, and further suppose that this liquid mixture at the temperature T_0 and pressure $P = 1$ atm is to be flashed by raising the temperature to the specified flash temperature $T_F = T_2$ at the specified flash pressure $P = 1$ atm. First observe that the bubble point temperature of the feed $T_{B.P.}$ at $P = 1$ atm is T_1. The dew point temperature, $T_{D.P.}$, of the feed at the pressure $P = 1$ atm is seen to be T_3. Then it is obvious from Figure 3-8 that a necessary condition for a flash to occur at the specified pressure is that

$$T_{B.P.} < T_F < T_{D.P.} \qquad (3\text{-}23)$$

In practice, the flash process is generally carried out by reducing the pressure on the feed stream rather than heating the feed at constant pressure as described above.

To determine whether or not the feed will flash at a given T_F and P, the above inequality may be tested by determining the bubble point and dew point temperatures of the feed at the specified pressure P. In determination of the bubble point temperature of the feed at the specified P of the flash, the x_i's in Equation (3-14) are replaced by the X_i's of the feed, and in the determination of the dew point temperature at the specified pressure, the y_i's in Equation (3-17) are replaced by the X_i's. Alternately, the inequality given by Equation (3-23) is satisfied if at the specified T_F and P,

$$f(T_F) > 0, \quad \text{and} \quad F(T_F) > 0 \qquad (3\text{-}24)$$

where these functions are defined as follows:

$$f(T_F) = \sum_{i=1}^{c} K_{Fi}X_i - 1, \quad \text{and} \quad F(T_F) = \sum_{i=1}^{c} \frac{X_i}{K_{Fi}} - 1 \qquad (3\text{-}25)$$

The two kinds of flash calculations that are commonly made are generally referred to as *isothermal* and *adiabatic* flashes.

ISOTHERMAL FLASH. In the isothermal flash, the following specifications are made: $T_F, P, \{X_i\}$, and F. It is required to find $V_F, L_F, \{y_{Fi}\}$, and $\{x_{Fi}\}$. In addition to the $c + 2$ equations required to describe the state of equilibrium between the vapor and liquid phases [see Equation (3-12)], "c additional component-material balances are required to describe the isothermal flash process." Thus, the independent equations required to describe this flash process are as follows:

$$\text{Equilibrium relationships} \begin{cases} y_{Fi} = K_{Fi}x_{Fi} & (1 \leqq i \leqq c) \\ \sum\limits_{i=1}^{c} y_{Fi} = 1 \\ \sum\limits_{i=1}^{c} x_{Fi} = 1 \end{cases} \qquad (3\text{-}26)$$

Material balances $\{FX_i = V_F y_{Fi} + L_F x_{Fi} \quad (1 \leqq i \leqq c)$

Equation (3-26) is seen to represent $2c + 2$ equations in $2c + 2$ unknowns $[V_F, L_F, \{y_{Fi}\}, \{x_{Fi}\}]$.

This system of nonlinear equations is readily reduced to one equation in one unknown (say, V_F) in the following manner. First observe that the total material balance expression (a dependent equation) may be obtained by summing each member of the last expression of Equation (3-26) over all components to give

$$F \sum_{i=1}^{c} X_i = V_F \sum_{i=1}^{c} y_{Fi} + L_F \sum_{i=1}^{c} x_{Fi}, \quad \text{or} \quad F = V_F + L_F \qquad (3\text{-}27)$$

Elimination of the y_{Fi}'s from the last expression given by Equation (3-26) by use of the first expression, followed by rearrangement yields

$$x_{Fi} = \frac{X_i}{\dfrac{L_F}{F} + \dfrac{V_F K_{Fi}}{F}} \qquad (3\text{-}28)$$

Elimination of L_F from Equation (3-28) by use of Equation (3-27) yields

$$x_{Fi} = \frac{X_i}{1 - \Psi(1 - K_{Fi})} \qquad (3\text{-}29)$$

where

$$\Psi = \frac{V_F}{F}$$

When each member of Equation (3-29) is summed over all components i and the result so obtained is restated in functional notation, one obtains

$$P(\Psi) = \sum_{i=1}^{c} \frac{X_i}{[1 - \Psi(1 - K_{Fi})]} - 1 \qquad (3\text{-}30)$$

and

$$P'(\Psi) = \sum_{i=1}^{c} \frac{X_i(1 - K_{Fi})}{[1 - \Psi(1 - K_{Fi})]^2} \qquad (3\text{-}31)$$

From a graph of the branch of the function $P(\Psi)$ (see Figure 3-9) which contains the positive root, it is evident that Newton's method (8) always converges to the desired root when $\Psi = 1$ is taken to be the first assumed value for the root. After this root (the value of $\Psi > 0$ that makes $P(\Psi) = 0$) has been found, both V_F and L_F may be calculated by using the total material balance [Equation (3-27)] and the fact that $\Psi = V_F/F$. Also, it is evident from Equation (3-29) that when the solution value of Ψ has been found, each term in the summation of $P(\Psi) = 0$ is one of the solution values of $\{x_{Fi}\}$. Then the corresponding solution set of y_{Fi}'s is obtained by using the first expression of Equation (3-26), $y_{Fi} = K_{Fi}x_{Fi}$.

ILLUSTRATIVE EXAMPLE 3-4*

It is proposed to flash the following feed at a specified temperature $T_F = 100°F$ and a pressure $P = 1$ atm.

*Taken from Holland, *Introduction to the Fundamentals of Distillation.* Courtesy Instrument Society of America.

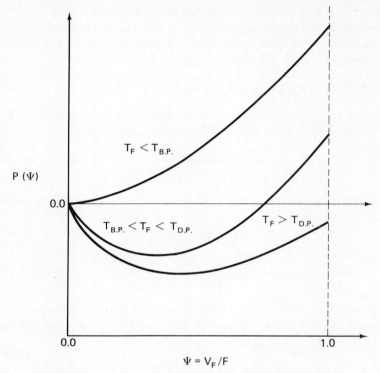

Figure 3-9. Graphical representation of the flash function $P(\Psi)$.

Component No.	K_i	X_i
1	$K_1 = \dfrac{10^{-2}T\dagger}{3P\dagger}$	$\dfrac{1}{3}$
2	$K_2 = \dfrac{2 \times 10^{-2}T}{P}$	$\dfrac{1}{3}$
3	$K_3 = \dfrac{7 \times 10^{-2}T}{2P}$	$\dfrac{1}{3}$

$\dagger T$ is in °F and P is in atm.

If the feed rate to the flash drum is $F = 100$ moles/hr, compute the vapor and liquid rates V_F and L_F leaving the flash as well as the respective mole fractions $\{y_{Fi}\}$ and $\{x_{Fi}\}$ of these streams.

Solution: First, the specified value of T_F will be checked to determine whether or not it lies between the bubble point and dew point temperatures of the feed.

Component No.	K_{Fi} @ $P = 1$ $T = 100$°F	X_i	$K_{Fi}X_i$	$\dfrac{X_i}{K_{Fi}}$
1	$\dfrac{1}{3}$	$\dfrac{1}{3}$	$\dfrac{1}{9}$	1
2	2	$\dfrac{1}{3}$	$\dfrac{2}{3}$	$\dfrac{1}{6}$
3	$\dfrac{7}{2}$	$\dfrac{1}{3}$	$\dfrac{7}{6}$	$\dfrac{2}{21}$
			1.94	1.2619

Thus,

$$f(T_F) = \sum_{i=1}^{c} K_{Fi}X_i - 1 = 1.94 - 1 = 0.94 > 0$$

$$F(T_F) = \sum_{i=1}^{c} \frac{X_i}{K_{Fi}} - 1 = 1.2619 - 1 = 0.2619 > 0$$

and thus,

$$T_{\text{B.P.}} < 100 < T_{\text{D.P.}}$$

Trial No. 1: Assume $\Psi = 1$.

Component No.	K_i	$1 - K_{Fi}$	$\Psi(1 - K_{Fi})$	$1 - \Psi(1 - K_{Fi})$
1	$\frac{1}{3}$	$\frac{2}{3}$	$\frac{2}{3}$	$\frac{1}{3}$
2	2	-1	-1	2
3	$\frac{7}{2}$	$-\frac{5}{2}$	$-\frac{5}{2}$	$\frac{7}{2}$

$\dfrac{X_i}{1 - \Psi(1 - K_{Fi})}$	$\dfrac{X_i}{[1 - \Psi(1 - K_{Fi})]^2}$	$\dfrac{X_i(1 - K_i)}{[1 - (1 - K_{Fi})]^2}$
1.0000	3.0000	2.0000
0.1667	0.0833	−0.0833
0.0952	0.0272	−0.0680
1.2619		1.8487

$$P(1) = 1.2619 - 1 = 0.2619$$

$$P'(1) = 1.8487$$

$$\Psi_2 = 1 - \left(\frac{0.2614}{1.8487}\right) = 1 - 0.1417 = 0.8583$$

Trial No. 2: Assume $\Psi = 0.8583$ and repeat the steps shown in the first trial. The results so obtained are as follows:

$$P(0.8583) = 1.0651 - 1 = 0.0651$$

$$P'(0.8583) = 1.0358$$

$$\Psi_3 = 0.8583 - \left(\frac{0.0651}{1.0358}\right) = 0.7955$$

Continuation of this procedure gives the solution value of $\Psi = 0.787$. Thus, $V_F = 78.7$, $L_F = 21.3$, and the solution sets $\{x_{Fi}\}$ and $\{y_{Fi}\}$ are as follows:

Component No.	$1 - K_{Fi}$	$\Psi(1 - K_{Fi})$	$1 - \Psi(1 - K_{Fi})$	$x_{Fi} = \dfrac{X_i}{1 - \Psi(1 - K_{Fi})}$	$y_{Fi} = K_{Fi}x_{Fi}$
1	0.667	0.525	0.475	0.701	0.234
2	−1.000	−0.787	1.787	0.187	0.374
3	−2.500	−1.968	2.968	0.112	0.392

Up to this point no mention has been made of the manner of satisfying the energy requirement of the flash. The specification of T_F implies that the feed either possesses precisely the correct amount of energy for the flash to occur at T_F at the specified P or that energy is to be added or withdrawn at the flash drum as required. It is common practice to adjust the heat content of the feed before it reaches the flash drum so that the flash occurs adiabatically, that is, the heat added Q at the flash drum is equal to zero.

After the solution $[V_F, L_F, \{y_{Fi}\}, \{x_{Fi}\}]$ has been found for a given isothermal flash problem, the heat content H that the feed must possess in order for the flash to occur adiabatically ($Q = 0$ at the flash drum) may be found by using the enthalpy balance that encloses the entire process

$$FH = V_F H_F + L_F h_F \qquad (3-32)$$

When the vapor V_F and liquid L_F form ideal solutions, the enthalpies H_F and h_F of the vapor and liquid streams, respectively, may be computed as follows:

$$H_F = \sum_{i=1}^{c} H_{Fi} y_{Fi} \quad \text{and} \quad h_F = \sum_{i=1}^{c} h_{Fi} x_{Fi} \qquad (3-33)$$

The above procedure may also be used to solve adiabatic flash problems as described below.

ILLUSTRATIVE EXAMPLE 3-5*

On the basis of the solution to Illustrative Example 3-4, compute the enthalpy H which the feed must possess in order for the flash to occur adiabatically.

Given:

Component No.	h_i (Btu/lb mole)	H_i (Btu/lb mole)
1	$h_1 = 10,000 + 30T$†	$H_1 = 17,000 + 30T$†
2	$h_2 = 8,000 + 20T$	$H_2 = 13,000 + 20T$
3	$h_3 = 500 + T$	$H_3 = 800 + T$

†T is in °F.

Solution: Calculation of the enthalpy H of the feed is as follows:

Component No.	x_{Fi}	y_{Fi}	h_{Fi} @ $T_F = 100°F$	$h_{Fi} x_{Fi}$	H_{Fi} @ $T_F = 100°F$	$H_{Fi} y_{Fi}$
1	0.701	0.234	13,000	9,113	20,000	4,680
2	0.187	0.374	10,000	1,870	15,000	5,610
3	0.112	0.392	600	67	900	353
				$h_F = 11,050$		$H_F = 10,643$

Thus,

$$H = \frac{V_F H_F}{F} + \frac{L_F h_F}{F} = (0.787)(10,643) + (0.213)(11,050) = 10,740 \text{ Btu/lb mole}$$

*Taken from Holland, *Introduction to the Fundamentals of Distillation.* Courtesy Instrument Society of America.

ADIABATIC FLASH. The term *adiabatic flash* is used to describe the problem wherein the following specifications are made: P, $Q = 0$ (no heat is added at the flash drum), H, $\{X_i\}$, and F. In this case there are $2c + 3$ unknowns $[T_F, V_F, L_F, \{y_{Fi}\}, \{x_{Fi}\}]$. The independent equations are also $2c + 3$ in number, the $2c + 2$ given by Equation (3-26) plus the enthalpy balance given by Equation (3-32), that is,

$$\text{Equilibrium relationships} \begin{cases} y_{Fi} = K_{Fi}x_{Fi} & (1 \leq i \leq c) \\[2mm] \displaystyle\sum_{i=1}^{c} y_{Fi} = 1 \\[4mm] \displaystyle\sum_{i=1}^{c} x_{Fi} = 1 \end{cases}$$

(3-34)

$$\text{Material balances } \{ FX_i = V_F y_{Fi} + L_F x_{Fi} \qquad (1 \leq i \leq c)$$

$$\text{Enthalpy balance } \{ FH = V_F H_F + L_F h_F$$

One relatively simple method for solving an adiabatic flash problem consists of the repeated use of the procedure described above whereby an H_n is computed for an assumed T_{Fn}. The problem then reduces to finding a T_{Fn} such that the resulting H_n is equal to the specified value H, that is, it is desired to find the T_{Fn} such that $\delta(T_{Fn}) = 0$, where

$$\delta(T_{Fn}) = \delta_n = H_n - H \tag{3-35}$$

One numerical method for solving such a problem is called interpolation *regula falsi* (4, 8). This method consists of the linear interpolation between the most recent pairs of points, (T_{Fn}, δ_n) and $(T_{F,n+1} \delta_{n+1})$ by use of the following formula (see Appendix A),

$$T_{F,n+2} = \frac{T_{F,n+1}\delta_n - T_{Fn}\delta_{n+1}}{\delta_n - \delta_{n+1}} \tag{3-36}$$

To initiate this interpolation procedure, it is necessary to evaluate δ for each of two assumed temperatures T_{F1} and T_{F2}. Then Equation (3-36) is applied to obtain T_{F3}. After δ_3 has been obtained, the new temperature T_{F4} is found by interpolation between the points (T_{F2}, δ_2) and (T_{F3}, δ_3). When $|\delta|$ has been reduced to a value less than some arbitrary, preassigned positive number, the desired solution is said to have been obtained.

It should be pointed out that the equations required to describe the adiabatic flash are of precisely the same form as those required to describe the separation process that occurs on the plate of a distillation column in the process of separating a multicomponent mixture.

The procedures described above as well as others for solving bubble point, dew point, and flash problems have been described in greater detail elsewhere (8). It is, however, informative to demonstrate briefly the use of the Newton-Raphson method (see Appendix A) for solving an adiabatic flash problem because this method has also been applied in various ways in the solution of problems involving distillation columns.

Solution of the adiabatic flash problem by use of the Newton-Raphson method

To solve a problem by using the Newton-Raphson method, it is, of course, necessary that the number of independent functions be equal to the number of independent variables. To illustrate one application of this procedure to the adiabatic flash problem, consider again the set of equations given by Equation (3-34). The first step in the application of the Newton-Raphson method is the restatement of the independent equations in functional form as follows:

$$
\text{Equilibrium relationships}
\begin{cases}
f_1 = K_{F1}x_{F1} - y_{F1} \\
\quad \cdot \quad \cdot \quad \cdot \quad \cdot \\
\quad \cdot \quad \cdot \quad \cdot \quad \cdot \\
\quad \cdot \quad \cdot \quad \cdot \quad \cdot \\
f_c = K_c x_{Fc} - y_{Fc} \\
f_{c+1} = \sum_{i=1}^{c} y_{Fi} - 1 \\
f_{c+2} = \sum_{i=1}^{c} x_{Fi} - 1
\end{cases}
\tag{3-37}
$$

$$
\text{Material balances}
\begin{cases}
f_{c+3} = V_F y_{F1} + L_F x_{F1} - FX_1 \\
\quad \cdot \quad \cdot \quad \cdot \quad \cdot \\
\quad \cdot \quad \cdot \quad \cdot \quad \cdot \\
f_{2c+2} = V_F y_{Fc} + L_F x_{Fc} - FX_c
\end{cases}
$$

$$
\text{Enthalpy balance} \quad \{ f_{2c+3} = V_F H_F + L_F h_F - FH
$$

As demonstrated in Chapter 2, the application of the Newton-Raphson method to this set of equations may be represented by the following matrix equation.

$$
J \, \Delta X = -f \tag{3-38}
$$

The Jacobian matrix J and the column vectors ΔX and f are defined as follows:

$$
J =
\begin{bmatrix}
\dfrac{\partial f_1}{\partial y_{F1}} & \cdots & \dfrac{\partial f_1}{\partial y_{Fc}} & \dfrac{\partial f_1}{\partial x_{F1}} & \cdots & \dfrac{\partial f_1}{\partial x_{Fc}} & \dfrac{\partial f_1}{\partial V_F} & \dfrac{\partial f_1}{\partial L_F} & \dfrac{\partial f_1}{\partial T_F} \\
\cdot & & \cdot & \cdot & & \cdot & \cdot & \cdot & \cdot \\
\cdot & & \cdot & \cdot & & \cdot & \cdot & \cdot & \cdot \\
\cdot & & \cdot & \cdot & & \cdot & \cdot & \cdot & \cdot \\
\dfrac{\partial f_{2c+3}}{\partial y_{F1}} & \cdots & \dfrac{\partial f_{2c+3}}{\partial y_{Fc}} & \dfrac{\partial f_{2c+3}}{\partial x_{F1}} & \cdots & \dfrac{\partial f_{2c+3}}{\partial x_{Fc}} & \dfrac{\partial f_{2c+3}}{\partial V_F} & \dfrac{\partial f_{2c+3}}{\partial L_F} & \dfrac{\partial f_{2c+3}}{\partial T_F}
\end{bmatrix}
$$

$$
\Delta X = [\Delta y_{F1} \ldots \Delta y_{Fc} \, \Delta x_{F1} \ldots \Delta x_{Fc} \, \Delta V_F \, \Delta L_F \, \Delta T_F]^T \tag{3-39}
$$

$$
f = [f_1 \ldots f_c f_{c+1} \ldots f_{2c} f_{2c+1} f_{2c+c} f_{2c+3}]^T
$$

where each element of ΔX is equal to the new predicted value of the variable minus the assumed value; for example, $\Delta y_{F1} = y_{F1,n+1} - y_{F1,n}$. To initiate the calculational procedure, a complete set of values for the variables must be assumed; say,

$$(y_{F1,n}, \ldots, y_{Fc,n}, x_{F1,n}, \ldots, x_{Fc,n}, V_{F,n}, L_{F,n}, T_{F,n})$$

The functions and all of their partial derivatives are evaluated on the basis of this set of assumed values for the variables. Then Equation (3-38) is solved for the elements of ΔX from which the values of the variables to be used for the next trial are computed as described above. [This method for solving the adiabatic flash problem was presented because it is analogous to the method first proposed by Greenstadt et al. (7) for solving problems involving distillation columns.]

   In many instances it is possible to reduce the number of equations by the simultaneous elimination of some of the variables before applying the Newton-Raphson method. For example, the expressions given by Equation (3-34) may be reduced to two equations in two unknowns as demonstrated previously (8, 10). As one might expect, the reduction of the number of equations and the number of variables generally results in a set of equations which are of a more complex form than the original set. But, as the number of equations and the number of variables are reduced, it is generally easier to pick an initial set of values of the variables for which the Newton-Raphson method will converge to the desired solution. The convergence of the Newton-Raphson equations is considered in Appendix A.

Part 3. Separation of Binary Mixtures by Use of Multiple Stages

Many of the concepts of distillation may be illustrated by use of the graphical method of design proposed by McCabe and Thiele (13). In the description of this process, the following symbols are used in addition to those explained above (see Figure 3-6). The mole fraction of the most volatile component in the feed is represented by X, in the distillate by X_D, and in the bottoms by x_B. The subscript j is used as the counting integer for the number of the plates. Since the distillate is withdrawn from the accumulator ($j = 0$) and the bottoms is withdrawn from the reboiler ($j = N + 1$), the mole fractions in the distillate and bottoms have double representation, that is, $X_{Di} = x_{0i}$ (for a column having a total condenser) and $x_{Bi} = x_{N+1,i}$. When the column has a partial condenser (D is withdrawn as a vapor), $X_{Di} = y_{0i}$.

   The *rectifying section* consists of the partial or total condenser and all plates down to the feed plate. The *stripping section* consists of the feed plate and all plates below it, including the reboiler. When the total flow rates do not vary from plate to plate within each section of the column, they are denoted by V_r (vapor) and L_r (liquid) in the rectifying section and by V_s and L_s in the stripping section. The feed rate (F), distillate rate (D), bottoms rate (B), and reflux rate (L_0) are all expressed in moles per unit time.

The design method of McCabe and Thiele (13) is best described by solving the following numerical example.

ILLUSTRATIVE EXAMPLE 3-6*

Suppose it is desired to find the minimum number of perfect plates required to effect the separation ($X_D = 0.95$ and $x_B = 0.05$) of component A from the feed mixture $X_A = X_B = 0.5$ at the following set of operating conditions: (1) The column pressure is 1 atm, and a total condenser is to be used (D is a liquid); (2) the thermal condition of the feed is such that the liquid rate L_s leaving the feed plate is given by $L_s = L_r + 0.583F$; and (3) a reflux ratio $L_0/D = 0.52$ is to be employed. The equilibrium sets (x_A, y_A) are given by the equilibrium curve in Figure 3-10.

This set of specifications fixes the system, that is, the number of independent equations that describe the system is equal to the number of unknowns. Before solving this problem, the equations needed are developed. First, the equilibrium pairs (x, y) satisfying the equilibrium relationship $y = Kx$ may be read from a boiling point diagram and plotted in the form of y versus x to give the equilibrium curve (see Figure 3-10). Observe that the equilibrium pairs (x, y) are those mole fractions connected by the tie lines of the boiling point diagram (see Figure 3-8).

A component-material balance enclosing the top of the column and plate j (see Figure 3-6) is given by

$$y_{j+1} = \left(\frac{L_r}{V_r}\right)x_j + \frac{DX_D}{V_r} \tag{3-40}$$

Similarly, for the stripping section, the component-material balance (see Figure 3-6) is given by

$$y_{j+1} = \left(\frac{L_s}{V_s}\right)x_j - \frac{Bx_B}{V_s} \tag{3-41}$$

The component-material balance enclosing the entire column is given by

$$FX = DX_D + Bx_B \tag{3-42}$$

The total flow rates within each section of the column are related by the following defining equation for q, namely,

$$L_s = L_r + qF \tag{3-43}$$

By means of a total-material balance enclosing plates $f - 1$ and f, it is readily shown through the use of Equation (3-43) that

$$V_r - V_s = (1 - q)F \tag{3-44}$$

By means of energy balances, it can be shown that q is approximately equal to the

*Taken from C. D. Holland and J. D. Lindsay, *Encyclopedia Chemical Technology*, Vol. 7, 2nd ed. (New York: John Wiley & Sons, Inc., 1965), pp. 204–48.

heat required to vaporize one mole of feed divided by the latent heat of vaporization of the feed.

Since Equations (3-40) and (3-41) are straight lines, they intersect at some point (x_I, y_I), provided, of course, they are not parallel. When the point of intersection is substituted into Equations (3-40) and (3-41), and L_r, V_r, L_s, V_s, x_B, and X_D are eliminated by use of Equations (3-42) through (3-44), the following equation for the *q-line* is obtained.

$$y_I = -\left(\frac{q}{1-q}\right)x_I + \left(\frac{1}{1-q}\right)X \qquad (3\text{-}45)$$

Solution: With the aid of the above equations, the number of plates required to effect the specified separation may be determined. To plot the operating line (Equation 3-40) for the rectifying section, the y intercept (DX_D/V_r) is computed in the following manner: Since $V_r = L_r + D$ and $L_0 = L_r$, it follows that

$$\frac{DX_D}{V_r} = \frac{X_D}{\dfrac{L_r}{D} + 1} = \frac{0.95}{1.52} = 0.63$$

Since $y_1 = X_D$ (for a total condenser), the point (y_1, X_D) lies on the 45° diagonal. The y intercept and the point (y_1, X_D) locate the operating line for the rectifying section as shown in Figure 3-10.

When $x_I = X$ is substituted in Equation (3-45), the result $y_I = X$ is obtained, and hence the q-line passes through the point (X, X) which in this case is the point $(0.5, 0.5)$. Since $q = 0.583$, the y intercept of the q-line [Equation (3-45)] is computed as follows:

$$\frac{X}{1-q} = \frac{0.5}{(1 - 0.583)} = 1.2$$

Since the operating line for the stripping section [Equation (3-41)] passes through the point $(x_B, x_B) = (0.05, 0.05)$ and the intersection of the q-line with the operating line for the rectifying section, it may be constructed simply by connecting these two points as shown in Figure 3-10.

The number of perfect plates required to effect the specified separation may be determined graphically as indicated in Figure 3-10. It is readily confirmed that the construction shown in Figure 3-10 gives the desired solution. Since $y_1 = X_D = x_0$ (for a total condenser) and since y_1 is in equilibrium with x_1, the desired value of x_1 is determined by the point of intersection of line 1 and the equilibrium curve as shown in Figure 3-10. Line 1 also represents plate 1. When x_1 is substituted into Equation (3-40), the value of y_2 is obtained. Since (x_1, y_2) lies on the operating line for the rectifying section, this point is located by passing a vertical line through (x_1, y_1). The ordinate y_2 obtained is displayed graphically in Figure 3-10. When the first opportunity to change operating lines is taken, the minimum number of total plates needed to effect the specified separation at the specified operating conditions is obtained. As seen in Figure 3-10, a total of five equilibrium stages is required, four plates plus the reboiler, when the feed enters on the third plate from the top of the column.

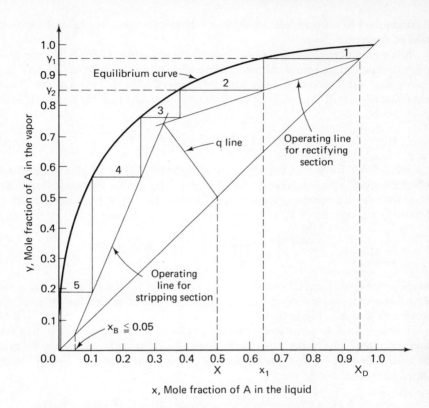

x, Mole fraction of A in the liquid

Figure 3-10. Graphical solution of Illustrative Example 3-6 by the McCabe-Thiele method. (Taken from Holland and Lindsay, *Encyclopedia Chemical Technology*, Vol. 7, 2nd ed., p. 217.)

It should be noted that if the operating line for the rectifying section is used indefinitely instead of changing to the operating line for the stripping section, the specified value of $x_B = 0.05$ can never be attained even though infinitely many plates are employed.

If the specified value of the reflux ratio (L_0/D) is decreased, the intersection of the two operating lines would be closer to the equilibrium line and the minimum number of plates required to effect the specified separation $(x_B = 0.05, X_D = 0.95)$ increases. But, as L_0/D is decreased, the condenser and reboiler duties decrease. The *minimum reflux ratio* is the smallest one that can be used to effect the specified separation. This reflux ratio requires infinitely many plates in each section as demonstrated in Figure 3-11. It should be noted that here the plates at and adjacent to the feed plate have the same composition.

At total reflux, the operating lines approach indefinitely closer to the

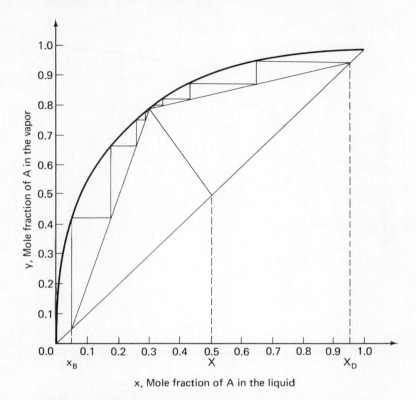

Figure 3-11. At the minimum reflux ratio (L_0/D), it is seen that infinitely many plates are required to effect the specified separation (X_D, x_B). (Taken from Holland and Lindsay, *Encyclopedia Chemical Technology*, p. 219.)

45° line. This gives the smallest number of plates needed to effect the separation. As pointed out by Robinson and Gilliland (17), two physical interpretations of total reflux are possible. From a laboratory or operational point of view, total reflux is attained by introducing an appropriate quantity of feed to the column and then operating so that $F = D = B = 0$. From the standpoint of design, total reflux can be thought of as a column of infinite diameter operating at infinite vapor and liquid rates, and with a feed that enters at a finite rate F and with distillate and bottoms that leave at the rates D and B, where $F = D + B$. At total reflux, two plates and the reboiler are required to effect the specified separation, as shown in Figure 3-12.

In order to demonstrate that the set of equations required to describe a distillation column in the process of separating a binary mixture is merely an extension of the sets stated previously for the boiling point diagram [Equation (3-3)], bubble point and dew point temperatures [Equation (3-12)], and the flash process [Equation (3-26)], the complete set of equations solved

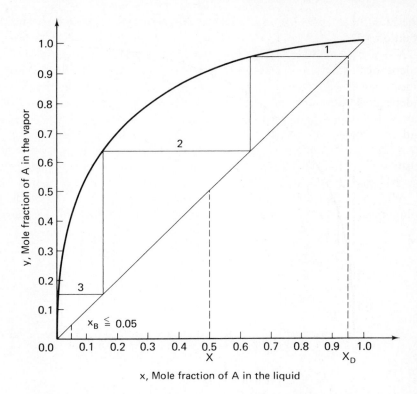

Figure 3-12. Determination of the total number of plates required to effect the specified separation at total reflux. At this reflux the minimum number of plates is required. (Taken from Holland and Lindsay, *Encyclopedia Chemical Technology*, p. 220.)

above by the McCabe-Thiele method are summarized as follows for purposes of comparison.

$$
\text{Equilibrium relationships}
\begin{cases}
y_{ji} = K_{ji}x_{ji} & \begin{pmatrix} i = 1, 2 \\ 0 \leq j \leq N + 1 \end{pmatrix} \\[2mm]
\displaystyle\sum_{i=1}^{2} y_{ji} = 1 & (0 \leq j \leq N + 1) \\[2mm]
\displaystyle\sum_{i=1}^{2} x_{ji} = 1 & (0 \leq j \leq N + 1)
\end{cases}
$$

$$
\text{Material balances}
\begin{cases}
V_r y_{j+1,i} = L_r x_{ji} + DX_{Di} & \begin{pmatrix} i = 1, 2 \\ 0 \leq j \leq f - 1 \end{pmatrix} \\[2mm]
V_s y_{j+1,i} = L_s x_{ji} - Bx_{Bi} & \begin{pmatrix} i = 1, 2 \\ f \leq j \leq N \end{pmatrix} \\[2mm]
FX_i = DX_{Di} + Bx_{Bi} & (i = 1, 2)
\end{cases}
$$

(3-46)

The counting integer j for stage number takes on only integral values, and the notation $(0 \leq j \leq N + 1)$ is used here to mean $j = 0, 1, 2, \ldots, N - 1$, $N, N + 1$. Examination of Equation (3-46) shows that it consists of $6(N + 2)$ independent equations. This result could have been predicted as follows: Since a single equilibrium stage [Equation (3-26)] is represented by $2c + 2$ independent equations and since the column represented by Equation (3-46) has $N + 2$ equilibrium stages [the condenser $j = 0$, plates $j = 1, 2, \ldots, N$, and the reboiler $j = N + 1$], then one would expect to obtain $(2c + 2)$ $(N + 2)$ independent equations that reduce to $6(N + 2)$ for a binary mixture. Also, in the McCabe-Thiele method as presented above, it is assumed that the behavior on the feed plate may be represented by Model 1 of Figure 3-13.

When the total flow rates V_j and L_j vary throughout each section of the column, these flow rates may be determined by solving the enthalpy balances

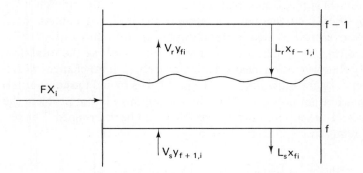

Model 1.   Assumed in the McCabe-Thiele method

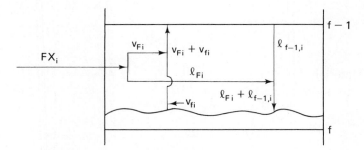

Model 2.   Behavior assumed on the feed plate.

Figure 3-13. Models for the behavior of the feed plate. (Taken from Holland, *Introduction to the Fundamentals of Distillation*, Proceedings of the Fourth Annual Education Symposium of the Instrument Society of America, April 5–7, 1972, Wilmington, Delaware.)

simultaneously with the above set of equations. For binary mixtures, the desired solution may be found by using either graphical methods (15, 18) or the numerical methods proposed below for multicomponent systems.

Part 4. Separation of Multicomponent Mixtures by Use of Conventional Distillation Columns with Multiple Stages

A conventional distillation column is defined as one that has one feed and two product streams, the distillate D and the bottoms B. This column has the same configuration as the one shown in Figure 3-14. First consider the case in which the following specifications are made for a column at steady-state operation: (1) number of plates in each section of the column, (2) quantity, composition, and thermal condition of the feed, (3) column pressure, (4) type of overhead condenser (total or partial), (5) reflux ratio, L_0/D, or V_1 or L_0, and (6) one specification on the distillate such as the total flow rate D. Steady-state operation means that no process variable changes with time. For this set of operating conditions, the problem is to find the compositions of the top and bottom products. Thus, by solving this kind of problem the characteristics of the top and bottom products can be determined. The set of equations required to represent such a system is as follows:

$$
\text{Equilibrium relationships}
\begin{cases}
y_{ji} = K_{ji}x_{ji} & \left(\begin{array}{l}1 \leq i \leq c \\ 0 \leq j \leq N+1\end{array}\right) \\[2mm]
\displaystyle\sum_{i=1}^{c} y_{ji} = 1 & (0 \leq j \leq N+1) \\[2mm]
\displaystyle\sum_{i=1}^{c} x_{ji} = 1 & (0 \leq j \leq N+1)
\end{cases}
$$

$$
\text{Material balances}
\begin{cases}
V_{j+1}y_{j+1,i} = L_j x_{ji} + DX_{Di} & \left(\begin{array}{l}1 \leq j \leq c \\ 0 \leq j \leq f-2\end{array}\right) \\[2mm]
V_f y_{fi} + V_F y_{Fi} = L_{f-1}x_{f-1,i} + DX_{Di} & (1 \leq i \leq c) \\[2mm]
V_{j+1}y_{j+1,i} = L_j x_{ji} - Bx_{Bi} & \left(\begin{array}{l}1 \leq i \leq c \\ f \leq j \leq N\end{array}\right) \\[2mm]
FX_i = DX_{Di} + Bx_{Bi} & (1 \leq i \leq c)
\end{cases}
$$

$$
\text{Enthalpy balances}
\begin{cases}
V_{j+1}H_{j+1} = L_j h_j + DH_D + Q_c & (0 \leq j \leq f-2) \\[2mm]
V_f H_f + V_F H_F = L_{f-1}h_{f-1} + DH_D + Q_c & \\[2mm]
V_{j+1}H_{j+1} = L_j h_j - Bh_B + Q_R & (f \leq j \leq N) \\[2mm]
FH = Bh_B + DH_D + Q_c - Q_R &
\end{cases}
$$

$$(3\text{-}47)$$

Inspection of this set of equations shows that the equations are a logical

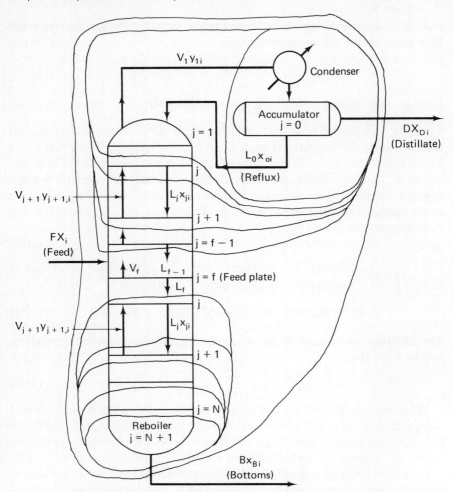

Figure 3-14. Representation of the component-material balances given by Equation (3-47). (Taken from Holland, *Introduction to the Fundamentals of Distillation*.)

extension of those stated above for the binary system. A schematic representation of the component-material balances is shown in Figure 3-14. The behavior assumed on the feed plate is demonstrated by Model 2 of Figure 3-13.

The above enthalpy balances may be represented by the same enclosures shown in Figure 3-14. As in the case of the material balances for any one component, the number of independent energy balances is equal to the number of stages ($j = 0, 1, 2, \ldots, N, N + 1$). In this case the total number of independent equations is equal to $(2c + 3)(N + 2)$, as might be expected from the fact that an adiabatic flash is represented by $2c + 3$ equations. When it is supposed that the vapor and liquid streams form ideal solutions, the enthalpy per mole of vapor and the enthalpy per mole of liquid leaving plate j are given by the following expressions:

$$H_j = \sum_{i=1}^{c} H_{ji} y_{ji} \qquad \text{(vapor)} \tag{3-48}$$

$$h_j = \sum_{i=1}^{c} h_{ji} x_{ji} \qquad \text{(liquid)} \tag{3-49}$$

where the enthalpy of each pure component i in the vapor and liquid streams leaving plate j are represented by H_{ji} and h_{ji}, respectively. These enthalpies are, of course, evaluated at the temperature and pressure of plate j. The meaning of H_D depends on the condenser used. For a total condenser (D is withdrawn from the accumulator as a liquid at its bubble point temperature T_0 at the column pressure, and $y_{1i} = x_{0i} = X_{Di}$),

$$H_D = \sum_{i=1}^{c} h_{0i} X_{Di} = \sum_{i=1}^{c} h_{0i} x_{0i} = h_0 \tag{3-50}$$

For a partial condenser (D is withdrawn from the accumulator as a vapor at its dew point temperature T_0 at the column pressure, and $y_{0i} = X_{Di}$),

$$H_D = \sum_{i=1}^{c} H_{0i} X_{Di} = \sum_{i=1}^{c} H_{0i} y_{0i} = H_0 \tag{3-51}$$

The enthalpy per mole of bottoms has double but equivalent representation, h_B and h_{N+1}, that is,

$$h_B = \sum_{i=1}^{c} h_{Bi} x_{Bi} = \sum_{i=1}^{c} h_{N+1,i} x_{N+1,i} = h_{N+1} \tag{3-52}$$

The symbols Q_c and Q_R are used to denote the condenser and reboiler duties, respectively. The condenser duty Q_c is equal to the net amount of heat removed per unit time by the condenser and the reboiler duty Q_R is equal to the net amount of heat introduced to the reboiler per unit time.

A wide variety of numerical methods have been proposed for solving the set of equations represented by Equation (3-47). The calculational procedure described below consists of an iterative technique which employs the θ method of convergence (12), the tridiagonal formulation of the component-material balances and equilibrium relationships, the K_b method for the determination of temperatures, and the constant-composition method for the determination of the total flow rates. Following Thiele and Geddes (21), the temperatures are taken to be the independent variables. This choice of independent variables has come to be known as the Thiele and Geddes method.

The Thiele and Geddes method plus the θ method, K_b method, and constant-composition method

Merely the statement that the Thiele and Geddes choice of independent variables (or the Thiele and Geddes method) has been employed to solve a problem does not sufficiently describe the calculational procedure. In the solution of a set of nonlinear equations by iterative techniques, the convergence or divergence of a given calculational procedure depends not only on the initial choice of the independent variables but also on the precise ordering and arrangement of each equation of the set. Over a period of years the author has investigated a variety of arrangements and combinations of Equation (3-47). Of these the most successful combination consisted of the following procedure that contains certain improvements over the original θ method as stated by Lyster *et al.* (12). These improvements were summarized by Nartker *et al.* (14). The development of the proposed calculational procedure is initiated by the restatement of the component-material balances and equilibrium relationships in tridiagonal matrix form.

STATEMENT OF THE COMPONENT-MATERIAL BALANCES AND EQUILIBRIUM RELATIONSHIPS AS A TRIDIAGONAL MATRIX. Although the equations used in this procedure differ in form from those presented by Equation (3-47), they are an equivalent independent set.

In the case of the component-material balances, a new set of variables, the component flow rates in the vapor and liquid phases, are introduced, namely,

$$v_{ji} = V_j y_{ji} \quad \text{and} \quad l_{ji} = L_j x_{ji} \tag{3-53}$$

Also, the flow rates of component i in the distillate and bottoms are represented by

$$d_i = D X_{Di} \quad \text{and} \quad b_i = B x_{Bi} \tag{3-54}$$

and the flow rates of component i in the vapor and liquid parts of the feed are represented by

$$v_{Fi} = V_F y_{Fi} \quad \text{and} \quad l_{Fi} = L_F x_{Fi} \tag{3-55}$$

The equilibrium relationship $y_{ji} = K_{ji} x_{ji}$ may be restated in an equivalent form in terms of the component flow rates v_{ji} and l_{ji} as follows. First, observe that the expression $y_{ji} = K_{ji} x_{ji}$ may be restated in the form:

$$V_j y_{ji} = \left(\frac{V_j K_{ji}}{L_j} \right) L_j x_{ji} \tag{3-56}$$

and from Equation (3-53), it follows that

$$v_{ji} = S_{ji} l_{ji} \quad \text{and} \quad l_{ji} = A_{ji} v_{ji} \tag{3-57}$$

where

$$A_{ji} = \frac{1}{S_{ji}} = \frac{L_j}{K_{ji} V_j}$$

Instead of enclosing the end of the column and the respective plates in each section of the column as demonstrated by Equation (3-47) and Figure 3-14, an equivalent set of component-material balances is obtained by enclosing each plate ($j = 0, 1, 2, \ldots, N, N + 1$) by a component-material balance as demonstrated in Figure 3-15. The corresponding set of material

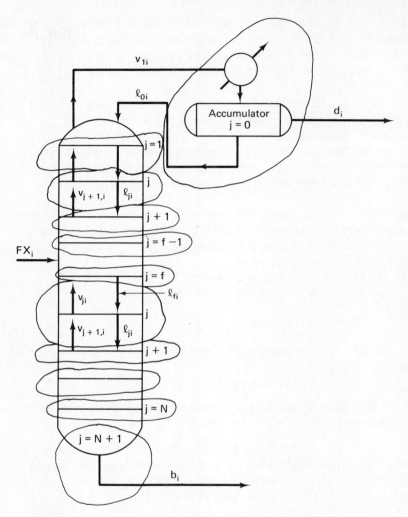

Figure 3-15. Representation of the component-material balances given by Equation (3-58). (Taken from Holland, *An Introduction to the Fundamentals of Distillation.*)

balances for each component i are as follows:

$$\text{Material balances}\quad \begin{cases} -l_{0i} - d_i + v_{1i} = 0 \\ l_{j-1,i} - v_{ji} - l_{ji} + v_{j+1,i} = 0 \qquad (1 \leqq j \leqq f-2) \\ l_{f-2,i} - v_{f-1,i} - l_{f-1,i} + v_{fi} = -v_{Fi} \\ l_{f-1,i} - v_{fi} - l_{fi} + v_{f+1,i} = -l_{Fi} \\ l_{j-1,i} - v_{ji} - l_{ji} + v_{j+1,i} = 0 \qquad (f+1 \leqq j \leqq N) \\ l_{Ni} - v_{N+1,i} - b_i = 0 \end{cases} \qquad (3\text{-}58)$$

Except for the first expression of Equation (3-58), the l_{ji}'s may be eliminated from Equation (3-58) by use of the equilibrium relationship, Equation (3-57). For the case of a total condenser, l_{0i} and d_i have the same composition, and thus,

$$l_{0i} = \left(\frac{L_0}{D}\right)d_i \qquad (3\text{-}59)$$

For a partial condenser, $y_{0i} = X_{Di}$, and hence

$$DX_{Di} = \left(\frac{DK_{0i}}{L_0}\right)L_0 x_{0i} \qquad (3\text{-}60)$$

Thus,

$$l_{0i} = A_{0i}d_i \qquad (3\text{-}61)$$

where

$$A_{0i} = \frac{L_0}{K_{0i}D}$$

The expression given by Equation (3-61) may be used to represent both a partial condenser and a total condenser, provided A_{0i} is set equal to L_0/D for a total condenser.* Also, the form of $A_{N+1,i}$ differs slightly from that for A_{ji} because of the double representation of the reboiler by the subscripts "$N + 1$" and "B," that is,

$$V_{N+1}y_{N+1,i} = \left(\frac{V_{N+1}K_{N+1,i}}{B}\right)Bx_{Bi} \quad \text{or} \quad b_i = A_{N+1,i}v_{N+1,i} \qquad (3\text{-}62)$$

where

$$A_{N+1,i} = \frac{B}{K_{N+1,i}V_{N+1}}$$

When the l_{ji}'s and b_i are eliminated from Equation (3-58) by use of Equations (3-57), (3-59), (3-61), and (3-62), the following result is obtained.

$$
\begin{array}{l}
\text{Material} \\
\text{balances} \\
\text{and} \\
\text{equilibrium} \\
\text{relationships}
\end{array}
\left\{
\begin{array}{ll}
-(A_{0i} + 1)d_i + v_{1i} = 0 & \\
A_{j-1,i}v_{j-1,i} - (A_{ji} + 1)v_{ji} + v_{j+1,i} = 0 & (1 \leq j \leq f - 2) \\
A_{f-2,i}v_{f-2,i} - (A_{f-1,i} + 1)v_{f-1,i} + v_{fi} = -v_{Fi} & \\
A_{f-1,i}v_{f-1,i} - (A_{fi} + 1)v_{fi} + v_{f+1,i} = -l_{Fi} & \\
A_{j-1,i}v_{j-1,i} - (A_{ji} + 1)v_{ji} + v_{j+1,i} = 0 & (f + 1 \leq j \leq N) \\
A_{Ni}v_{Ni} - (A_{N+1,i} + 1)v_{N+1,i} = 0 &
\end{array}
\right.
$$

$$(3\text{-}63)$$

*This symbolic operation of convenience should not be taken to mean that $K_{0i} = 1$ for a total condenser, for the boiling point temperature T_0 of the distillate leaving a total condenser is computed by use of the equation:

$$\sum_{i=1}^{c} y_{0i} = 1 = \sum_{i=1}^{c} K_{0i}X_{Di}$$

This set of equations may be stated in the matrix form

$$A_i v_i = -\boldsymbol{f}_i \tag{3-64}$$

where

$$
A_i =
\begin{bmatrix}
-\rho_{0i} & 1 & 0 & 0 & \cdots\cdots\cdots\cdots\cdots\cdots & 0 \\
A_{0i} & -\rho_{1i} & 1 & 0 & \cdots\cdots\cdots\cdots\cdots & 0 \\
\multicolumn{6}{c}{\cdots\cdots\cdots\cdots\cdots\cdots\cdots\cdots} \\
0 \cdots 0 & A_{f-2,i} & -\rho_{f-1,i} & 1 & 0 & 0 \cdots 0 \\
0 \cdots 0 & 0 & A_{f-1,i} & -\rho_{fi} & 1 & 0 \cdots 0 \\
\multicolumn{6}{c}{\cdots\cdots\cdots\cdots\cdots\cdots\cdots\cdots} \\
0 & \cdots\cdots\cdots\cdots\cdots & 0 & A_{N-1,i} & -\rho_{Ni} & 1 \\
0 & \cdots\cdots\cdots\cdots\cdots & 0 & 0 & A_{Ni} & -\rho_{N+1,i}
\end{bmatrix}
$$

$$v_i = [d_i v_{1i} v_{2i} \ldots v_{f-1,i} v_{fi} \ldots v_{Ni} v_{N+1,i}]^T$$

$$\boldsymbol{f}_i = [0\ 0\ 0 \ldots v_{Fi} l_{Fi} \ldots 0\ 0]^T$$

$$\rho_{ji} = A_{ji} + 1$$

The remainder of the development of the calculational procedure is ordered in the same sequence in which the calculations are carried out. The calculational procedure is initiated by the assumption of a set of temperatures $\{T_j\}$ and a set of vapor rates $\{V_j\}$ from which the corresponding set of liquid rates $\{L_j\}$ is found by using the total-material balances enumerated below. This particular choice of independent variables was first proposed by Thiele and Geddes (21). On the basis of the assumed temperatures and total flow rates, the absorption factors $\{A_{ji}\}$ appearing in Equation (3-64) may be evaluated for component i on each plate j. Since matrix A_i in Equation (3-64) is of tridiagonal form, this matrix equation may be solved for the calculated values of the vapor rates for component i [denoted by $(v_{ji})_{ca}$] by use of the Thomas algorithm (4, 8) that follows. Consider the following set of linear equations in the variables $x_0, x_1, \ldots, x_N, x_{N+1}$ whose coefficients form a tridiagonal matrix.

$$B_0 x_0 + C_0 x_1 = D_0$$

$$A_1 x_0 + B_1 x_1 + C_1 x_2 = D_1$$

$$A_2 x_1 + B_2 x_2 + C_2 x_3 = D_2$$

$$\cdots$$

$$A_N x_{N-1} + B_N x_N + C_N x_{N+1} = D_N$$

$$A_{N+1} x_N + B_{N+1} x_{N+1} = D_{N+1}$$

The following recurrence formulas are applied in the order stated:

$$f_0 = \frac{C_0}{B_0}, \quad g_0 = \frac{D_0}{B_0}$$

$$f_k = \frac{C_k}{B_k - A_k f_{k-1}} \qquad (k = 1, 2, \ldots, N)$$

$$g_k = \frac{D_k - A_k g_{k-1}}{B_k - A_k f_{k-1}} \qquad (k = 1, 2, \ldots, N+1)$$

After the f's and g's have been computed, the values of $x_{N+1}, x_N, \ldots, x_1, x_0$ are computed as follows:

$$v_{N+1} = g_{N+1}$$

$$x_k = g_k - f_k x_{k+1} \qquad (k = N, N-1, \ldots, 2, 1, 0)$$

The development of the recurrence formulas is outlined in Problem 3-14. As pointed out by Boston and Sullivan (2), the above recurrence formulas are subject to round-off error for columns which have both a large number of plates and components whose respective absorption factors are less than unity in one section of the column and greater than unity in another section. The modified version of the above recurrence formulas suggested by Boston and Sullivan (2) was shown to reduce the round-off errors to insignificant levels.

After these recurrence formulas have been applied for each component i and the complete set of vapor rates $\{(v_{ji})_{ca}\}$ has been found, the corresponding set of liquid rates $\{(l_{ji})_{ca}\}$ is then found by use of Equation (3-57). These sets of calculated flow rates are used in conjunction with the θ method of convergence and the K_b method in the determination of an improved set of temperatures.

FORMULATION OF THE θ METHOD OF CONVERGENCE. In this application of the θ method of convergence, it is used to weight the mole fractions that are employed in the K_b method for computing a new temperature profile. The corrected set of product rates is used as weight factors in the calculation of improved sets of mole fractions. The corrected terminal rates are selected so that they are both in overall component-material balance and in agreement with the specified value of D, that is,

$$FX_i = (d_i)_{co} + (b_i)_{co} \tag{3-65}$$

and

$$\sum_{i=1}^{c} (d_i)_{co} = D \tag{3-66}$$

These two conditions may be satisfied simultaneously by suitably choosing the multiplier θ, which is defined by

$$\left(\frac{b_i}{d_i}\right)_{co} = \theta \left(\frac{b_i}{d_i}\right)_{ca} \tag{3-67}$$

(The subscripts "co" and "ca" are used throughout this discussion to distin-

guish between the corrected and calculated values of a variable, respectively.)
Elimination of $(b_i/d_i)_{co}$ from Equation (3-65) and (3-67) yields the formula
for $(d_i)_{co}$, namely,

$$(d_i)_{co} = \frac{FX_i}{1 + \theta \left(\dfrac{b_i}{d_i} \right)_{ca}} \tag{3-68}$$

Since the specified values of $(d_i)_{co}$ are to have a sum equal to the specified
value of D, the desired value of θ is that $\theta > 0$ that makes $g(\theta) = 0$, where

$$g(\theta) = \sum_{i=1}^{c} (d_i)_{co} - D \tag{3-69}$$

A graph of this function is shown in Figure 3-16.

In the determination of θ by Newton's method, the following formula
for the first derivative, $g'(\theta)$, is needed.

$$g'(\theta) = \sum_{i=1}^{c} \frac{d(d_i)_{co}}{d\theta} = -\sum_{i=1}^{c} \frac{\left(\dfrac{b_i}{d_i} \right)_{ca} FX_i}{\left[1 + \theta \left(\dfrac{b_i}{d_i} \right)_{ca} \right]^2} \tag{3-70}$$

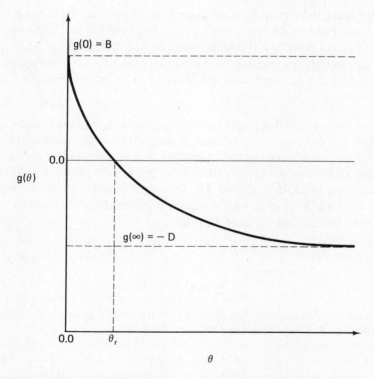

Figure 3-16. Geometrical representation of the function $g(\theta)$ in the
neighborhood of the positive root θ_r. [Taken from Holland, *Multicomponent Distillation* (Englewood Cliffs, N.J.: Prentice-Hall, Inc., 1963), p. 84.]

After the desired value of θ has been obtained, $(b_i)_{co}$ may be computed by using Equation (3-67). [Note that Newton's method converges to the positive root of $g(\theta)$, provided that $\theta = 0$ is taken to be the first trial value (see Problem 3-15)].

The corrected mole fractions for the liquid and vapor phases are computed as follows:

$$x_{ji} = \frac{\left(\dfrac{l_{ji}}{d_i}\right)_{ca} (d_i)_{co}}{\sum\limits_{i=1}^{c} \left(\dfrac{l_{ji}}{d_i}\right)_{ca} (d_i)_{co}} \tag{3-71}$$

$$y_{ji} = \frac{\left(\dfrac{v_{ji}}{d_i}\right)_{ca} (d_i)_{co}}{\sum\limits_{i=1}^{c} \left(\dfrac{v_{ji}}{d_i}\right)_{ca} (d_i)_{co}} \tag{3-72}$$

These expressions are consistent with the definition of θ given by Equation (3-67); see Problem 3-21.

DETERMINATION OF AN IMPROVED SET OF TEMPERATURES BY USE OF THE K_b METHOD. On the basis of the mole fractions given by Equations (3-71) and (3-72) and the last temperature profile (the one assumed to make the nth trial), the new temperature profile is found using the K_b method (8, 14) in the following manner: For any plate j, Equations (3-21) or (3-22) may be applied as follows:

$$K_{jb}|_{T_{j,n+1}} = \frac{1}{\sum\limits_{i=1}^{c} \alpha_{ji}|_{T_{jn}} x_{ji}} \quad \text{or} \quad K_{jb}|_{T_{j,n+1}} = \sum\limits_{i=1}^{c} \frac{y_{ji}}{\alpha_{ji}|_{T_{jn}}} \tag{3-73}$$

where

$\alpha_{ji} = K_{ji}/K_{jb}$, the relative volatility of component i at the temperature of plate j. The rate of convergence of the entire calculational procedure is dependent upon the precise choice of K_{jb}.

It can be shown that the x_{ji}'s and y_{ji}'s defined by Equations (3-71) and (3-72), respectively, form a consistent set in that they give the same value of K_{jb} (8) (see also Problem 3-16). Component b represents a hypothetical base component whose K value is given by

$$\log_e K_{jb} = \frac{a}{T_j} + b \tag{3-74}$$

where the constants a and b are evaluated on the basis of the values of K at the upper and lower limits of the curve fits of the mid-boiling component of the mixture or one just lighter. Thus, after K_{jb} has been computed by use of Equation (3-73), the temperature $T_{j,n+1}$ to be assumed for the next trial is calculated directly by use of Equation (3-74) or an improved variation of it (3).

The corrected compositions and the new temperatures are used in the enthalpy balances to determine the total flow rates to be used for the next trial through the column.

DETERMINATION OF AN IMPROVED SET OF TOTAL FLOW RATES BY USE OF THE CONSTANT-COMPOSITION METHOD. In the constant-composition method (8), one total flow rate (V_j or L_j) is eliminated from each of the enthalpy balances given by Equation (3-47) by use of one set of the component-material balances given by Equation (3-47). To illustrate the development of these equations, consider the enthalpy balance enclosing any plate j of the rectifying section, namely,

$$V_{j+1}H_{j+1} = L_j h_j + DH_D + Q_c \qquad (1 \leq j \leq f - 2) \qquad (3\text{-}75)$$

The total flow rate V_{j+1} is eliminated from Equation (3-75) by use of component-material balance enclosing plate j

$$v_{j+1,i} = l_{ji} + d_i \qquad (1 \leq j \leq f - 2)$$

as follows:

$$V_{j+1}H_{j+1} = \sum_{i=1}^{c} H_{j+1,i} v_{j+1,i} = \sum_{i=1}^{c} H_{j+1,i}(l_{ji} + d_i)$$

$$= L_j \sum_{i=1}^{c} H_{j+1,i} x_{ji} + D \sum_{i=1}^{c} H_{j+1,i} X_{Di}$$

$$= L_j H(x_j)_{j+1} + DH(X_D)_{j+1} \qquad (3\text{-}76)$$

where

$$H(x_j)_{j+1} = \sum_{i=1}^{c} H_{j+1,i} x_{ji}; \quad H(X_D)_{j+1} = \sum_{i=1}^{c} H_{j+1,i} X_{Di}$$

Elimination of $V_{j+1}H_{j+1}$ from Equations (3-75) and (3-76) yields

$$L_j = \frac{D[H_D - H(X_D)_{j+1}] + Q_c}{[H(x_j)_{j+1} - h_j]} \qquad (1 \leq j \leq f - 2) \qquad (3\text{-}77)$$

Similarly,

$$L_{f-1} = \frac{D[H_D - H(X_D)_f] + V_F[H(y_F)_f - H_F] + Q_c}{[H(x_{f-1})_f - h_{f-1}]} \qquad (3\text{-}78)$$

and

$$Q_c = L_0[H(x_0)_1 - h_0] + D[H(X_D)_1 - H_D] \qquad (3\text{-}79)$$

where the enthalpy expressions appearing in these equations are defined in a manner analogous to those stated below Equation (3-76).

The flow rates in the stripping section may be determined by using enthalpy balances that enclose either the top or the bottom of the column. When the reboiler is enclosed, the following formula for the computation of the vapor rates is developed in a manner analogous to that demonstrated above:

$$V_{j+1} = \frac{B[h(x_B)_j - h_B] + Q_R}{[H_{j+1} - h(y_{j+1})_j]} \qquad (f \leq j \leq N) \qquad (3\text{-}80)$$

where

$$h(x_B)_j = \sum_{i=1}^{c} h_{ji} x_{Bi}; \quad h(y_{j+1})_j = \sum_{i=1}^{c} h_{ji} y_{j+1,i}$$

The reboiler duty Q_R is found by use of the overall enthalpy balance [the last expression given by Equation (3-47)]. When the vapor and liquid phases do not form ideal solutions, the pure component enthalpies h_{ji} and H_{ji} in the above expressions should be replaced by the corresponding partial molar enthalpies. Formulas for the calculation of suitable values for the partial molar enthalpies are developed in Appendix D.

The total flow rates of the vapor and liquid streams are related by the following total-material balances, which are readily obtained by use of the enclosures shown in Figure 3-14.

$$
\begin{aligned}
V_{j+1} &= L_j + D \qquad (0 \leq j \leq f - 2) \\
V_f + V_F &= L_{f-1} + D \\
L_j &= V_{j+1} + B \qquad (f \leq j \leq N) \\
F &= D + B
\end{aligned}
\qquad (3\text{-}81)
$$

After the L_j's for the rectifying section and the V_j's for the stripping section have been determined by use of the enthalpy balances, the remaining total flow rates are found by use of Equation (3-81). These most recent sets of values of the variables $\{T_{j,n+1}\}$, $\{V_{j,n+1}\}$, and $\{L_{j,n+1}\}$ are used to make the next trial through the column. The procedure described is repeated until values of the desired accuracy have been obtained. A summary of the steps of the proposed calculational procedure follow:

1. Assume a set of temperatures $\{T_j\}$ and a set of vapor rates $\{V_j\}$. [The set of liquid rates corresponding to the set of assumed vapor rates are found by use of the total-material balances as in Equation (3-81).]
2. On the basis of the temperatures and flow rates assumed in Step 1, compute the component-flow rates by use of Equation (3-64) for each component i.
3. Find the $\theta > 0$ that makes $g(\theta) = 0$ [see Equations (3-68) through (3-70)]. [Newton's method (8) always converges to the desired θ, provided that the first assumed value of θ is taken to be equal to zero.]
4. Use Equation (3-71) to compute the set of corrected x_{ji}'s for each component i and plate j.
5. Use the results of Step 4 to compute the K_{jb} for each j by use of the first expression of Equation (3-73). Use the K_{jb}'s so obtained to compute a new set of temperatures $\{T_{j,n+1}\}$ by use of Equation (3-74).
6. Use the results of Steps 4 and 5 to compute new sets of total flow rates, $\{V_{j,n+1}\}$ and $\{L_{j,n+1}\}$, by use of Equations (3-77) through (3-81).
7. If θ, the T_j's and V_j's are within the prescribed tolerances, convergence

has been achieved; otherwise, repeat Steps 2 through 6 on the basis of the most recent set of T_j's and V_j's.

The solution of the component-material balances and equilibrium relationships by using the above recurrence formulas is demonstrated by the following numerical example.

ILLUSTRATIVE EXAMPLE 3-7

On the basis of the initial set of temperatures ($T_0 = T_1 = T_2 = T_3 = T_4 = 100°F$) and the total flow rates displayed in Figure 3-17, solve Equation (3-64) for the component flow rates by use of the above recurrence formulas for tridiagonal matrices.

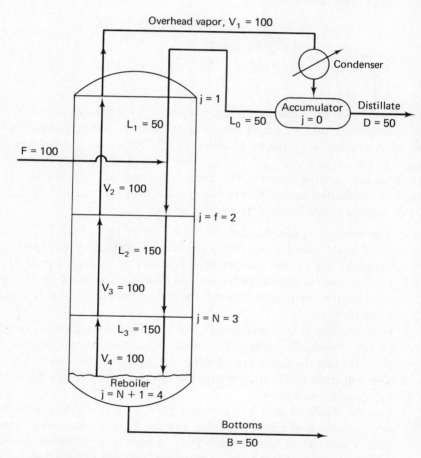

Figure 3-17. Flow diagram for Illustrative Example 3-7. (Taken from Holland, *Multicomponent Distillation*, p. 55.)

Given:

Component No.	FX_i	K_i	SPECIFICATIONS
1	33.3	$0.01T/P*$	Total condenser, $P = 1$ atm, boiling point
2	33.3	$0.02T/P$	liquid feed ($l_{Fi} = FX_i$), $N = 3, f = 2,$
3	33.4	$0.03T/P$	$F = 100$ moles/hr, $D = L_0 = L_1 = 50$
			moles/hr, $V_1 = V_2 = V_3 = V_4 = 100$
			moles/hr, $L_2 = L_3 = 150$ moles/hr.

*T is in °F and P is in atm.

Solution: Correspondence of the symbols in the recurrence formulas and the elements of A_i and $\mathscr{f}_i$ for the above example is as follows:

$$B_0 = -(A_{0i} + 1),\ C_0 = 1,\ D_0 = 0,\ A_1 = A_{0i}$$

$$B_1 = -(A_{1i} + 1),\ C_1 = 1,\ D_1 = 0,\ A_2 = A_{1i}$$

$$B_2 = -(A_{2i} + 1),\ C_2 = 1,\ D_2 = -FX_i,\ A_3 = A_{2i}$$

$$B_3 = -(A_{3i} + 1),\ C_3 = 1,\ D_3 = 0,\ A_4 = A_{3i}$$

$$B_4 = -(A_{4i} + 1),\ D_4 = 0$$

Calculation of the A_{ji}'s is as follows:

Component No.	$A_{0i} = \dfrac{L_0}{D}$	K_{1i}, K_{2i}, K_{3i} K_{4i} @ 100°F and 1 atm	$A_{1i} = \dfrac{L_1}{K_{1i}V_1} = \dfrac{1}{2K_{1i}}$	$A_{2i} = \dfrac{L_2}{K_{2i}V_2} = \dfrac{3}{2K_{2i}}$
1	1	1	$\dfrac{1}{2}$	$\dfrac{3}{2}$
2	1	2	$\dfrac{1}{4}$	$\dfrac{3}{4}$
3	1	3	$\dfrac{1}{6}$	$\dfrac{1}{2}$

Component No.	$A_{3i} = \dfrac{L_3}{K_{3i}V_3} = \dfrac{3}{2K_{3i}}$	$A_{4i} = \dfrac{B}{K_{4i}V_4} = \dfrac{1}{2K_{4i}}$
1	$\dfrac{3}{2}$	$\dfrac{1}{2}$
2	$\dfrac{3}{4}$	$\dfrac{1}{4}$
3	$\dfrac{1}{2}$	$\dfrac{1}{6}$

Application of the recurrence formulas for tridiagonal matrix equations is as follows:

Comp. No.	B_0	C_0	D_0	A_1	B_1	C_1	D_1	A_2	B_2	C_2	D_2
1	-2	1	0	1	-1.50000	1	0	0.50000	-2.5	1	33.3
2	-2	1	0	1	-1.25000	1	0	0.25000	-1.75	1	33.3
3	-2	1	0	1	-1.16667	1	0	0.16667	-1.5	1	33.4

Component No.	A_3	B_3	C_3	D_3	A_4	B_4	D_4
1	1.50	−2.50	1	0	1.50	−1.50000	0
2	0.75	−1.75	1	0	0.75	−1.25000	0
3	0.50	−1.50	1	0	0.50	−1.16667	0

Comp. No.	$f_0 = \dfrac{C_0}{B_0}$	g_0	$A_1 f_0$
1	−0.5	0	−0.5
2	−0.5	0	−0.5
3	−0.5	0	−0.5

Comp. No.	$B_1 - A_1 f_0$	$f_1 = \dfrac{C_1}{B_1 - A_1 f_0}$	$D_1 - A_1 g_0$	$g_1 = \dfrac{D_1 - A_1 g_0}{B_1 - A_1 f_0}$	$A_2 f_1$	$B_2 - A_2 f_1$	$A_2 g_1$
1	−1.00000	−1.00000	0	0	−0.50000	−2.00000	0
2	−0.75000	−1.33333	0	0	−0.33333	−1.41667	0
3	−0.66667	−1.50000	0	0	−0.25000	−1.25000	0

Comp. No.	$D_2 - A_2 g_1$	$f_2 = \dfrac{C_2}{B_2 - A_2 f_1}$	$g_2 = \dfrac{D_2 - A_2 g_1}{B_2 - A_2 f_1}$	$A_3 f_2$	$B_3 - A_3 f_2$
1	−33.3	−0.50000	16.65000	−0.75000	−1.75000
2	−33.3	−0.70588	23.50583	−0.52941	−1.22059
3	−33.4	−0.80000	26.72000	−0.40000	−1.10000

Comp. No.	$A_3 g_2$	$D_3 - A_3 g_2$	$f_3 = \dfrac{C_3}{B_3 - A_3 f_2}$	$g_3 = \dfrac{D_3 - A_3 g_2}{B_3 - A_3 f_2}$
1	24.97500	−24.97500	−0.57143	14.27142
2	17.62937	−17.62937	−0.81928	14.44331
3	13.36000	−13.36000	−0.90909	12.14547

Comp. No.	$A_4 f_3$	$B_4 - A_4 f_3$	$A_4 g_3$	$v_{4i} = x_4 = g_4 = \dfrac{D_4 - A_4 g_3}{B_4 - A_4 f_3}$
1	−0.85714	−0.64286	21.40714	33.29985
2	−0.61446	−0.63554	10.83248	17.04453
3	−0.45454	−0.71212	6.07274	8.52768

Comp. No.	$f_3 x_4$	$v_{3i} = x_3 = g_3 - f_3 x_4$	$f_2 x_3$	$v_{2i} = x_2 = g_2 - f_2 x_3$
1	−19.02853	33.29995	−16.64998	33.29998
2	−13.96424	28.40759	−20.05247	43.55830
3	−7.75244	19.89790	−15.91833	42.63833

Comp. No.	$v_{1i} = x_1 = g_1 - f_1 x_2$	$d_i = x_0 = g_0 - f_0 x_1$	$b_i = A_{4i} v_{4i}$
1	33.29998	16.64999	16.64993
2	58.07759	29.03880	4.26113
3	63.95750	31.97875	1.42128

The student may confirm his ability to apply the proposed calculational procedure in its entirety (the recurrence formulas, the θ method, the K_b method, and the constant-composition method) by solving Problems 3-11 through 3-13.

The tridiagonal formulation of the component-material balances and equilibrium relationships is generally preferred in computer applications because the method is readily applied to other kinds of columns such as absorbers and strippers as demonstrated in Chapter 4. For making calculations for conventional distillation columns by hand, however, the use of nesting equations as originally suggested by Thiele and Geddes (21) is perhaps the more convenient of the two methods.

SOLUTION OF THE COMPONENT-MATERIAL BALANCES AND EQUILIBRIUM RELATIONSHIPS BY USE OF NESTING EQUATIONS. Nesting equations are obtained by first restating those given by Equation (3-47) in terms of the component flow rates as follows:

$$
\begin{aligned}
v_{j+1,i} &= l_{ji} + d_i \qquad (0 \leq j \leq f - 2) \\
v_{fi} + v_{Fi} &= l_{f-1,i} + d_i \\
v_{j+1,i} &= l_{ji} - b_i \qquad (f \leq j \leq N) \\
FX_i &= d_i + b_i
\end{aligned}
\qquad (3\text{-}82)
$$

Elimination of l_{ji} from the first expression of Equation (3-82) by means of the equilibrium relationship $l_{ji} = A_{ji}v_{ji}$ [Equation (3-57)] yields the following expression upon rearrangement:

$$
\frac{v_{j+1,i}}{d_i} = A_{ji}\left(\frac{v_{ji}}{d_i}\right) + 1 \qquad (3\text{-}83)
$$

for $j = 1, 2, \ldots, f - 1$. For $j = 0$ (the condenser-accumulator section) and for a total condenser, the first expression of Equation (3-82) becomes

$$
\frac{v_{1i}}{d_i} = \frac{l_{0i}}{d_i} + 1 = \frac{L_0 x_{0i}}{DX_{Di}} + 1 = \frac{L_0}{D} + 1 \qquad (3\text{-}84)
$$

since $x_{0i} = X_{Di}$.

For a partial condenser, $y_{0i} = X_{Di}, y_{0i} = K_{0i}x_{0i}$ or $l_{0i} = A_{0i}d_i$, and the first expression of Equation (3-82) reduces to

$$
\frac{v_{1i}}{d_i} = A_{0i} + 1 \qquad (3\text{-}85)
$$

where $A_{0i} = L_0/K_{0i}D$. By use of Equation (3-84) or (3-85) and Equation (3-83), the nesting calculations are initiated at the top of the column and continued down toward the feed plate. For the case of boiling point liquid and subcooled feed, the nesting calculations are discontinued as soon as v_{fi}/d_i has been obtained. For the case of dew point vapor and superheated feeds, the nesting calculations are ceased as soon as $v_{f-1,i}/d_i$ has been determined, as discussed below.

The nesting equations for the stripping section are initiated at the reboiler. Since $y_{N+1,i} = K_{N+1,i}x_{N+1,i} = K_{N+1,i}x_{Bi}$ or $v_{N+1,i} = S_{N+1i,i}b_i$, the component-material balance enclosing the reboiler [given by Equation (3-82) for $j = N$] reduces to

$$\frac{l_{Ni}}{b_i} = \left(\frac{S_{N+1,i}}{b_i}\right)b_i + 1 = S_{N+1,i} + 1 \tag{3-86}$$

where $S_{N+1,i} = B/(V_{N+1}K_{N+1,i})$. After a number value has been obtained for l_{Ni}/b_i, it is used to compute $l_{N-1,i}/b_i$ by use of the following equation which is obtained by eliminating $v_{j+1,i}/b_i$ from the expressions ($f \leqq j \leqq N$) of Equation (3-84) by use of the equilibrium relationship $v_{ji} = S_{ji}l_{ji}$, that is,

$$\frac{l_{ji}}{b_i} = S_{j+1,i}\left(\frac{l_{j+1,i}}{b_i}\right) + 1 \tag{3-87}$$

which holds for $j = f, f+1, \ldots, N-1$. After l_{fi}/b_i has been computed, the nesting calculations are ceased and the quantity v_{fi}/b_i is computed by use of the equilibrium relationship, namely,

$$\frac{v_{fi}}{b_i} = S_{fi}\left(\frac{l_{fi}}{b_i}\right)$$

For the case of a boiling point liquid or subcooled feed, $v_{Fi} = 0$, $l_{Fi} = FX_i$, and hence the moles of vapor entering plate $f-1$ is equal to the moles of vapor leaving plate j. Thus, b_i/d_i may be computed from the number values found for v_{fi}/d_i and v_{fi}/b_i by the nesting calculations as follows:

$$\frac{b_i}{d_i} = \frac{\dfrac{v_{fi}}{d_i}}{\dfrac{v_{fi}}{b_i}} \tag{3-88}$$

Next, the overall component-material balance of Equation (3-82) may be solved for d_i in terms of b_i/d_i in the following manner:

$$FX_i = d_i\left(1 + \frac{b_i}{d_i}\right) \quad \text{and} \quad d_i = \frac{FX_i}{1 + \dfrac{b_i}{d_i}} \tag{3-89}$$

After d_i has been obtained, the complete set of component flow rates $\{b_i, v_{ji}, l_{ji}\}$ may be obtained from previously calculated results in an obvious manner.

For the general case of a partially vaporized feed, the expression for computing b_i/d_i is obtained by commencing with the second expression of Equation (3-82) and rearranging it to give

$$\left(\frac{v_{fi}}{b_i}\right)\left(\frac{b_i}{d_i}\right) + \left(\frac{v_{Fi}}{FX_i}\right)\left(\frac{FX_i}{d_i}\right) = \frac{l_{f-1,i}}{d_i} + 1 \tag{3-90}$$

Since

$$\frac{v_{Fi}}{FX_i} = 1 - \frac{l_{Fi}}{FX_i} \quad \text{and} \quad \frac{FX_i}{d_i} = 1 + \frac{b_i}{d_i}$$

Equation (3-90) may be solved for b_i/d_i to give

$$\frac{b_i}{d_i} = \frac{\dfrac{l_{f-1,i}}{d_i} + \dfrac{l_{Fi}}{FX_i}}{\dfrac{v_{fi}}{b_i} + \dfrac{v_{Fi}}{FX_i}} \tag{3-91}$$

When the appropriate values for l_{Fi} and v_{Fi} are employed, Equation (3-91) may be used to calculate b_i/d_i for a feed of any thermal condition. For bubble point liquid and subcooled feeds, $l_{Fi} = FX_i$ and $v_{Fi} = 0$. For feeds that enter the column as dew point and superheated vapors, $v_{Fi} = FX_i$ and $l_{Fi} = 0$.

The use of the nesting equations for solving the component-material balances and equilibrium relationships is demonstrated by the following example.

ILLUSTRATIVE EXAMPLE 3-8

Use the nesting equations developed above to solve Illustrative Example 3-7.

Solution: Below are calculations for the rectifying section.

Comp. No.	$\dfrac{L_0}{D}$	$\dfrac{v_{1i}}{d_i} = \dfrac{L_0}{D} + 1$	K_{1i} @ 100°F and 1 atm	A_{1i} $= \dfrac{L_1}{K_{1i}V_1}$	$\dfrac{l_{1i}}{d_i}$ $= A_{1i}\left(\dfrac{v_{1i}}{d_i}\right)$	$\dfrac{v_{2i}}{d_i}$ $= \dfrac{l_{1i}}{d_i} + 1$
1	1.0	2.0	1.0	0.5	1.0	2.0
2	1.0	2.0	2.0	0.25	0.5	1.5
3	1.0	2.0	3.0	0.1666667	0.33333334	1.33333333

Below are calculations for the stripping section and for the d_i's.

Comp. No.	K_{4i} @ 100°F and 1 atm	S_{4i} $= \dfrac{K_{4i}V_4}{B}$	$\dfrac{l_{3i}}{b_i}$ $= S_{4i} + 1$	K_{3i} @ 100°F and 1 atm	S_{3i} $= \dfrac{K_{3i}V_3}{L_3}$	$\dfrac{v_{3i}}{b_i}$ $= S_{3i}\left(\dfrac{l_{3i}}{b_i}\right)$
1	1.0	2.0	3.0	1.0	0.66666666	1.9999998
2	2.0	4.0	5.0	2.0	1.3333333	6.6666660
3	3.0	6.0	7.0	3.0	1.9999999	13.999999

Product distribution

Comp. No.	$\dfrac{l_{2i}}{b_i}$ $= \dfrac{v_{3i}}{b_i} + 1$	K_{2i} @ 100°F and 1 atm	S_{2i} $= \dfrac{K_{2i}V_2}{L_2}$	$\dfrac{v_{2i}}{b_i}$ $= S_{2i}\dfrac{l_{2i}}{b_i}$	$\dfrac{b_i}{d_i} = \dfrac{\dfrac{v_{2i}}{d_i}}{\dfrac{v_{2i}}{b_i}}$	$d_i = \dfrac{FX_i}{1 + \dfrac{b_i}{d_i}}$
1	2.9999999	1	0.66666666	1.9999999	1.0000000	16.649999
2	7.6666666	2	1.3333333	10.222222	0.14673913	29.038862
3	14.999999	3	1.9999999	29.999995	0.04444444	31.978724

$$D_{ca} = \overline{77.667585.}$$

Convergence characteristics of the θ method

The θ method and the associated calculational procedure described above is one of the fastest known methods for solving problems involving distillation columns (3), and it converges for almost all problems of this type.

Some of the convergence characteristics of the θ method may be demonstrated by comparing it with the method of direct iteration. The method of direct iteration differs from the θ method only by the procedure used to compute the compositions. Instead of Equations (3-71) and (3-72), the following expressions are used in the method of direct iteration:

$$x_{ji} = \frac{\left(\dfrac{l_{ji}}{d_i}\right)_{ca} (d_i)_{ca}}{\sum\limits_{i=1}^{c} \left(\dfrac{l_{ji}}{d_i}\right)_{ca} (d_i)_{ca}} \tag{3-92}$$

$$y_{ji} = \frac{\left(\dfrac{v_{ji}}{d_i}\right)_{ca} (d_i)_{ca}}{\sum\limits_{i=1}^{c} \left(\dfrac{l_{ji}}{d_i}\right)_{ca} (d_i)_{ca}} \tag{3-93}$$

By comparison of these expressions with Equations (3-71), (3-72), and (3-68), it is evident that the method of direct iteration amounts to setting $\theta = 1$ in Equation (3-68) for all trials. The results obtained for Illustrative Example 3-7 by the θ method are presented in Table 3-1. When the method of direct iteration was used, a calculated value of 52.14 was obtained for D at the end of the third trial, and eleven trials were required to obtain temperatures that did not change in the eighth digit (8). The outstanding convergence characteristics of the θ method result in part, perhaps, from the fact that the θ method constitutes an exact solution to certain problems at total reflux. In order to demonstrate this important result, it is necessary to develop the well-known Fenske equation (6) for a conventional distillation column at total reflux.

SEPARATION OF MULTICOMPONENT MIXTURES IN CONVENTIONAL DISTILLATION COLUMNS AT TOTAL REFLUX. In the following developments, the concept of total reflux from the design point of view is utilized, that is, the total flow rates $[L_j\ (0 \leqq j \leqq N),\ V_j\ (1 \leqq j \leqq N + 1)]$ are unbounded while the feed and product rates are finite. More precisely,

$$\lim_{V_{j+1} \to \infty} \left(\frac{L_j}{V_{j+1}}\right) = 1 - \lim_{V_{j+1} \to \infty} \left(\frac{D}{V_{j+1}}\right) = 1$$

and

$$F = D + B$$

where F, D, and B are all nonzero, finite, and positive. The corresponding component-material balances are given by

$$y_{j+1,i} = x_{ji}\left[\lim_{V_{j+1} \to \infty} \left(\frac{L_j}{V_{j+1}}\right)\right] + X_{Di}\left[\lim_{V_{j+1} \to \infty} \left(\frac{D}{V_{j+1}}\right)\right] = x_{ji} \tag{3-94}$$

TABLE 3-1

SOLUTION OF ILLUSTRATIVE EXAMPLE 3-7 BY USE OF THE THIELE AND GEDDES METHOD, THE θ METHOD, AND THE K_b METHOD*

Calculated Values of the Temperatures, °F

Plate No.	Trial No.						
	1	2	3	4	5	6	7
0	41.086575	41.362833	41.372024	41.372407	41.372426	41.372427	41.372427
1	46.121292	46.719933	46.73821	46.739097	46.739133	46.739136	46.739136
2	48.755212	51.157089	51.154290	51.154505	51.154520	51.154523	51.154522
3	55.069408	55.398386	55.401216	55.401135	55.401126	55.401126	55.401126
4	63.770834	63.11654	63.095157	63.094266	63.094222	63.094220	63.094220
D	77.667585	49.006364	49.999530	49.999902	49.999993	49.999998	49.999999
(Calculated) θ	5.6215101	0.94749173	0.99997452	0.99999469	0.99999961	0.99999987	0.99999998

Final Flow Rates

Comp. No.	b	d
1	27.764577	5.535423
2	15.224283	18.075717
3	7.011140	26.388860

*Taken from C.D. Holland, *Multicomponent Distillation* (Englewood Cliffs, N.J.: Prentice Hall, Inc., 1963), p. 87.

The final equality follows from the limits stated above. By the alternate application of Equation (3-94) and the equilibrium relationship $y_{ji} = K_{ji}x_{ji}$, the Fenske equation (6),

$$\frac{b_i}{d_i} = \frac{\dfrac{B}{D}}{\displaystyle\prod_{j=1}^{N+1} K_{ji}} \tag{3-95}$$

is obtained (for a column having a total condenser). An abbreviated development of this equation follows. The component-material balance enclosing the condenser-accumulator section is given by $y_{1i} = x_{0i}$, and for a total condenser $y_{1i} = x_{0i} = X_{Di}$. Elimination of y_{1i} from the second expression by use of the equilibrium relationship for plate 1 ($y_{1i} = K_{1i}x_{1i}$) gives

$$x_{1i} = \frac{X_{Di}}{K_{1i}} \tag{3-96}$$

For plate 2, the component-material balance and equilibrium relationship for component i are as follows:

$$y_{2i} = x_{1i} \quad \text{and} \quad y_{2i} = K_{2i}x_{2i}$$

Elimination of x_{1i} and y_{2i} from these expressions and Equation (3-96) yields

$$x_{2i} = \frac{X_{Di}}{K_{1i}K_{2i}} \tag{3-97}$$

Continuation of this procedure for plates $j = 3$ through $j = N + 1$ (the reboiler) yields

$$x_{N+1,i} = \frac{X_{Di}}{K_{1i}K_{2i}\ldots K_{Ni}K_{N+1,i}} \tag{3-98}$$

Since $x_{N+1,i} = x_{Bi}$, it is evident that Equation (3-95) is obtained by multiplying both sides of Equation (3-98) by B/D. [It should be noted that the alternate use of material balances and equilibrium relationships in the above derivation is the same procedure used to obtain the graphical solution for a binary mixture (see Figure 3-12).]

An alternate form of Equation (3-95) which reduces to an exact solution when the relative volalities are constant is obtained as follows. First, state Equation (3-95) for the base component b, and then divide the members of Equation (3-95) by the corresponding members for component b and rearrange the result so obtained to give

$$\frac{b_i}{d_i} = \frac{\dfrac{b_b}{d_b}}{\alpha_{1i}\alpha_{2i}\ldots\alpha_{Ni}\alpha_{N+1,i}} \tag{3-99}$$

If the α_{ij}'s are independent of temperature, Equation (3-99) reduces to

$$\frac{b_i}{d_i} = \frac{b_b}{d_b}\alpha_i^{-(N+1)} \tag{3-100}$$

For the case of a partial condenser ($y_{0i} = X_{Di}$), the appropriate expressions for b_i/d_i are obtained by replacing the exponent $(N + 1)$ in the above expressions by the exponent $(N + 2)$, that is, the partial condenser counts as an additional equilibrium stage.

At a fixed number of plates N, the set of b_i/d_i's relative to b_b/d_b may be computed for a given system by use of Equation (3-100). Then for any specified value of b_b/d_b, the corresponding set d_i's and D may be computed by use of Equation (3-89) and the following formula for D, which is obtained by summing each member of Equation (3-89) over all components

$$D = \sum_{i=1}^{c} \frac{FX_i}{1 + \dfrac{b_i}{d_i}} \tag{3-101}$$

In summary, Equations (3-89), (3-100), and (3-101) may be used to compute the best possible separation (the lightest possible distillate and heaviest bottoms) which may be achieved with a fixed number of plates at the limiting condition of total reflux, provided, of course, that the α_i's are constant throughout the column. At this limiting condition of total reflux, the column diameter as well as the reboiler and condenser duties become infinite. Problem 3-19 requires the use of Equation (3-100) for the computation of the b_i/d_i's for different specified values of b_b/d_b.

A plot of Equation (3-100) for an example such as the one described in Problem 3-19 where two different distillate rates are employed is plotted in Figure 3-18. The distance between these two lines has been denoted by $|\log_e \theta|$. The equation of the upper line is

$$\log_e \left(\frac{b_i}{d_i}\right)_2 = \log_e \left(\frac{b_b}{d_b}\right)_2 - (N + 1) \log_e \alpha_i \tag{3-102}$$

and for the lower line

$$\log_e \left(\frac{b_i}{d_i}\right)_1 = \log_e \left(\frac{b_b}{d_b}\right)_1 - (N + 1) \log_e \alpha_i \tag{3-103}$$

where the subscripts "1" and "2" refer to problems analogous to those posed by Parts (a) and (b), respectively, of Problem 3-19. When the members of Equation (3-103) are subtracted from the corresponding members of Equation (3-102), one obtains the following result upon rearrangement

$$\log_e \left(\frac{b_i}{d_i}\right)_2 = \log_e \left(\frac{b_i}{d_i}\right)_1 + \log_e \theta \tag{3-104}$$

where

$$\log_e \theta = \log_e \frac{\left(\dfrac{b_b}{d_b}\right)_2}{\left(\dfrac{b_b}{d_b}\right)_1}$$

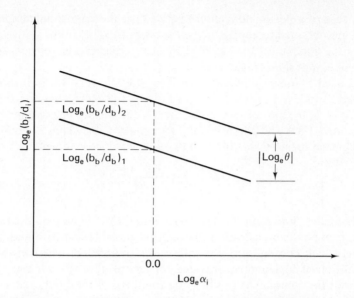

Figure 3-18. A graphical representation of θ is obtained by considering two arbitrarily specified values for a base component, b, in a column at total reflux.

Restatement of Equation (3-104) in product form yields

$$\left(\frac{b_i}{d_i}\right)_2 = \theta \left(\frac{b_i}{d_i}\right)_1 \tag{3-105}$$

Thus, if the b_i/d_i's are known at one condition (a given distillate rate), they may be determined at any other specified distillate rate by use of the θ method. Once θ has been determined, the b_i/d_i's at the new distillate rate will have been determined; thus, the θ method constitutes an exact solution. Since the component flow rates at condition 2 must be in material balance and in agreement with the specified value of D, the g function is obtained by replacing the subscripts "ca" and "co" in Equations (3-65) through (3-69) by "1" and "2," respectively. That the θ method constitutes an exact solution to problems of the type mentioned is illustrated by solving Problem 3-20.

Since many of the methods proposed for solving problems involving conventional distillation columns are analogous to those presented in the next chapter for absorbers and strippers, further discussion of these methods is delayed until after absorbers and strippers have been considered.

PROBLEMS

3-1(a). Complete the construction of the boiling point diagram initiated in Illustrative Example 3-1.

3-1(b). From the plot obtained in Problem 3-1 (a), construct the equilibrium curve.

3-2. Find the minimum reflux ratio (L_0/D) required to effect the following separation of a mixture of 50 mole % A and 50 mole % B. The feed is 55% vaporized at the column pressure of 1 atm.

$$X_{D,A} = 0.95$$

$$x_{B,A} = 0.05$$

Given the following equilibrium data for component A.

x	y
0.1	0.57
0.2	0.7
0.3	0.8
0.5	0.905
0.9	0.995

Ans. $\left(\dfrac{L_0}{D}\right)_{min} = 0.435$

3-3. Find the minimum total number of plates required to effect the separation of Problem 3-2 for the following conditions:
   1. Column pressure = 1 atm
   2. Total condenser
   3. Use an operating reflux ratio L_0/D equal to two times the minimum reflux ratio found in Problem 3-2.

Ans. Three plates plus the reboiler. The feed plate is the second plate from the top of the column.

3-4. For the case in which the distillate is withdrawn as a vapor, the partial condenser represents an additional separation stage. In this case

$$X_D = y_0 = K_0 x_0$$

where y_0 is the vapor in equilibrium with the liquid reflux x_0 in the accumulator. The material balance enclosing the condenser-accumulator section is represented by

$$y_1 = \frac{L_r}{V_r} x_0 + \frac{D X_D}{V_r}$$

Repeat Problem 3-3 for the case in which a partial condenser instead of a total condenser is used and the distillate is withdrawn as a vapor with the vapor composition $X_D = 0.95$ rather than as a liquid with this same composition.

Ans. Two plates plus the reboiler plus the partial condenser. The feed plate is the top plate in the column.

3-5(a). Repeat Illustrative Example 3-2 for the case in which the first assumed value of T is taken to be equal to 40°F.
   Ans. 50°F.

3-5(b). Repeat Illustrative Example 3-3 for the case in which the assumed value of T is taken to be equal to 60°F.
   Ans. 50°F.

3-5(c). After the bubble point temperature has been determined by each of the above methods, compute the corresponding values of y_i's which are in equilibrium with the x_i's.

Ans. $y_1 = \frac{1}{6}$, $y_2 = \frac{1}{3}$, $y_3 = \frac{1}{2}$.

3-6. Repeat Problem 3-5 where the following vapor compositions are known instead of the liquid compositions. In this case determine the dew point temperature at a specified total pressure of $P = 1$ atm. (If Newton's method should fail for $T_1 = 100°F$, try taking $T_1 = 80°F$.) After the dew point temperature has been determined by each of the methods, compute the corresponding x_i's that are in equilibrium with the y_i's.

Component No.	K_i	y_i
1	$K_1 = \dfrac{0.01T^*}{P^*}$	$\dfrac{1}{6}$
2	$K_2 = \dfrac{0.02T}{P}$	$\dfrac{1}{3}$
3	$K_3 = \dfrac{0.03T}{P}$	$\dfrac{1}{2}$

*T is in °F and P is in atm.

Ans. $T_{D.P.} = 50°F$, $x_1 = \frac{1}{3}$, $x_2 = \frac{1}{3}$, $x_3 = \frac{1}{3}$.

3-7. Repeat Problem 3-5 for the case in which the K_i's vary with temperature in the following manner:

$$K_i = C_i e^{-E_i/T}$$

where T is in °R.

Component No.	C_i	E_i
1	$2.35 \times \dfrac{10^3}{P^*}$	4.6×10^3
2	$4.7 \times \dfrac{10^3}{P}$	4.6×10^3
3	$9.4 \times \dfrac{10^3}{P}$	4.6×10^3

*P is in atm.

Ans. 74.3°F.

3-8. If the α_i's are not independent of temperature, then the temperature found at the end of the first trial by use of the K_b method depends on the component selected as the base component as well as on the temperature assumed to evaluate the α_i's. These facts are illustrated by solving the following problems:

Component No.	K_i at $P = 1$ atm	x_i
1	$0.01T^*$	$\dfrac{1}{3}$
2	$0.0002T^2$	$\dfrac{1}{3}$
3	$0.03T$	$\dfrac{1}{3}$

*T is in °F.

(a). Find the correct bubble point temperature at a specified total pressure of 1 atm.
Ans. 58.1°F.

(b). Make one trial by use of the K_b method. Evaluate the α_i's at an assumed temperature of 100°F. Take component 1 as the base component.
Ans. 50°F.

(c). Repeat Part (b) for the case in which component 2 is selected as the base component.
Ans. 70.7°F.

(d). Repeat Part (b) for the case in which the α_i's are evaluated at an assumed temperature of 50°F rather than 100°F.
Ans. 60°F.

3-9(a). On the basis of the solutions obtained for Illustrative Example 3-4 and 3-5, determine the temperature that the feed must have in order for it to possess the enthalpy H found in Illustrative Example 3-5. Assume the feed is at a pressure such that it is all liquid.
Ans. 269°F.

3-9(b). Find the smallest pressure the feed can be under at the temperature found in Problem 3-9 (a) and be in the liquid state.
Ans. 5.23 atm.

3-10. Use the results found in Problem 3-9 to restate this problem as an adiabatic flash problem. Initiate the calculational procedure outlined in the text by making two complete trials on the basis of

$$T_{F1} = T_{\text{B.P.}} \text{ of the feed}$$
$$T_{F2} = T_{\text{D.P.}} \text{ of the feed}$$

Use the corresponding values $\delta(T_{F1})$ and $\delta(T_{F2})$ to predict the improved value of T_F, namely, T_{F3}.
Ans. $T_{F3} = 103°F$.

3-11. This problem and Problems 3-12 and 3-13 are formulated in such a way that their solutions require that one complete trial be made by use of the calculational procedure proposed in the text for distillation columns.

On the basis of the set of assumed temperatures,

$$T_0 = T_1 = T_2 = T_3 = 100°F$$

and the set of assumed flow rates

$$V_1 = V_2 = V_3 = 100 \text{ lb-moles/hr}$$
$$L_1 = 50, \text{ and } L_2 = 150 \text{ lb-moles/hr}$$

solve the component-material balances and equilibrium relationships by use of the recurrence formulas given below Equation (3-64), and show that

Comp. No.	$\left(\dfrac{b_i}{d_i}\right)_{\text{ca}}$
1	1.000000
2	0.225000
3	0.095238

The feed, K values, and column specifications are as follows:

Comp. No.	X_i	K_i	Specifications
1	$\frac{1}{3}$	$\frac{T^*}{P^*}$	The feed $\{X_i\}$ enters the column as a liquid at its bubble point at the column pressure of $P = 100$ psia. $F = 100$
2	$\frac{1}{3}$	$\frac{2T}{P}$	lb-moles/hr, $D = L_0 = 50$ lb-moles/hr, $N = 2$, $f = 2$, and a total condenser is used.
3	$\frac{1}{3}$	$\frac{3T}{P}$	

$^*P$ is the total pressure in psia and T is the temperature in °F.

3-12(a). Find the θ that makes $g(\theta) = 0$ for the set of calculated b_i/d_i's found in Problem 3-11.
Ans. $\theta = 3.69$.

3-12(b). On the basis of the θ that makes $g(\theta) = 0$, calculate the corrected sets of the x_{ji}'s and y_{ji}'s [see Equations (3-71) and (3-72)]. Use these x_{ji}'s or y_{ji}'s to compute K_{jb} for plate $j = 0$, 1, 2, and 3. Base the α_{ji}'s on the K-value for component No. 1, that is, $K_{jb} = K_{j1}$.
Ans. $K_{0b} = 0.425$, $K_{1b} = 0.489$, $K_{2b} = 0.522$, $K_{3b} = 0.606$.

3-12(c). From the results of Problem 3-12 (b), find the new set of temperatures, T_0, T_1, T_2, and T_3.
Ans. $T_0 = 42.5$°F, $T_1 = 48.9$°F, $T_2 = 52.2$°F, and $T_3 = 60.6$°F.

3-13. On the basis of the new sets of compositions and temperatures found in Problem 3-12, find the new set of total flow rates to be used for the next trial through the column. Use the relationships given by Equations (3-77) through (3-81) and the enthalpy data given in Illustrative Example 3-5.
Ans. $V_2 = 87$, $V_3 = 70$, $L_1 = 37$, $L_2 = 120$ mole/hr.

3-14. The recurrence formulas given below Equation (3-64) for solving equations that are tridiagonal in form may be developed as outlined below by use of the Gaussian elimination. Consider the system of linear equations represented by the following matrix equation.

$$
\begin{bmatrix}
B_0 & C_0 & 0 & 0 & 0 & \cdots & 0 \\
A_1 & B_1 & C_1 & 0 & 0 & \cdots & 0 \\
0 & A_2 & B_2 & C_2 & 0 & \cdots & 0 \\
\multicolumn{7}{c}{\cdots\cdots\cdots\cdots\cdots\cdots\cdots} \\
0 & \cdots & 0 & A_N & B_N & C_N \\
0 & \cdots & 0 & 0 & A_{N+1} & B_{N+1}
\end{bmatrix}
\begin{bmatrix}
x_0 \\ x_1 \\ x_2 \\ \cdots \\ x_N \\ x_{N+1}
\end{bmatrix}
=
\begin{bmatrix}
D_0 \\ D_1 \\ D_2 \\ \cdots \\ D_N \\ D_{N+1}
\end{bmatrix}
\quad \text{(A)}
$$

(a). By using the following definitions of f_0, g_0, f_k, and g_k given in the text, show that Equation (A) may be transformed to the following form.

$$
\begin{bmatrix}
1 & f_0 & 0 & 0 & 0 & \cdots & 0 \\
0 & 1 & f_1 & 0 & 0 & \cdots & 0 \\
0 & 0 & 1 & f_2 & 0 & \cdots & 0 \\
\multicolumn{7}{c}{\cdots\cdots\cdots\cdots\cdots\cdots} \\
0 & \cdots & 0 & 1 & f_N \\
0 & \cdots & 0 & 0 & 1
\end{bmatrix}
\begin{bmatrix}
x_0 \\ x_1 \\ x_2 \\ \cdots \\ x_N \\ x_{N+1}
\end{bmatrix}
=
\begin{bmatrix}
g_0 \\ g_1 \\ g_2 \\ \cdots \\ g_N \\ g_{N+1}
\end{bmatrix}
\quad \text{(B)}
$$

(b). Commencing with the bottom row of Equation (B), show that the matrix multiplication rule may be applied to give

$$x_{N+1} = g_{N+1}, \qquad x_k = g_k - f_k x_{k+1},$$
$$(k = N, N - 1, N - 2, \ldots, 0) \qquad \text{(C)}$$

3-15(a). Show that the branch of $g(\theta)$ that contains the positive root is as depicted in Figure 3-16.

3-15(b). Show that if $\theta = 0$ is selected as the first assumed value of θ, then Newton's method always converges to the positive root of $g(\theta)$.

3-16. To prove that the same value for K_{jb} is obtained by use of either the x_{ji}'s or y_{ji}'s, begin with one of the expressions of Equation (3-73) for K_{jb} and Equations (3-71), (3-72), (3-57) and produce the other expression given for K_{jb} in Equation (3-73).

3-17. A variety of forms for the flash function have been proposed. Alternate but equivalent forms such as the one presented in the next chapter are obtained by different choices of independent variables and function formulations. For example, if in the development of $P(\Psi)$, the x_{Fi}'s are eliminated from the component-material balances instead of the y_{Fi}'s by use of $y_{F1} = K_{Fi} x_{Fi}$, show that the following form of the flash function is obtained.

$$p(\psi) = \sum_{i=1}^{c} \frac{X_i}{1 - \psi \left(1 - \frac{1}{K_{Fi}}\right)} - 1$$

where

$$y_{Fi} = \frac{X_i}{1 - \psi \left(1 - \frac{1}{K_{Fi}}\right)}; \qquad \psi = \frac{L_F}{F}$$

3-18. If 100 moles/hr of the following mixture are to be flashed at 150°F and 50 psia, find the moles of vapor and the moles of liquid formed per hour.

Component	X_i	K_i* @ 150°F and 50 psia
C_2H_6	0.0079	16.2
C_3H_8	0.1321	5.2
$i\text{-}C_4H_{10}$	0.0849	2.6
$n\text{-}C_4H_{10}$	0.2690	1.98
$i\text{-}C_5H_{12}$	0.0589	0.91
$n\text{-}CH_{12}$	0.1321	0.72
$n\text{-}C_6H_{k4}$	0.3151	0.28

*The values of the K's were taken from the fifth edition of the *Technical Manual* (1946) prepared by the Natural Gasoline Supply Men's Association, 422 Kennedy Bldg., Tulsa 3, Oklahoma.

Ans. $L_F = 45.11$ and $V_F = 55.89$ moles/hr.

3-19. Given a column which has two plates, a reboiler, a total condenser, a reboiler, and a feed F of 100 moles per hour. The composition of the feed and the relative volatilities (which are independent of temperature) are as follows:

Component	X_i	α_i
1	$\frac{1}{3}$	1
2	$\frac{1}{3}$	2
3	$\frac{1}{3}$	3

Find the distillate rates D that must be employed for the above column in order to achieve the following separations of the base component (component 1)

(a). $\dfrac{b_b}{d_b} = 8$

(b). $\dfrac{b_b}{d_b} = 16$

at total reflux.

Ans. (a) $D = 56.2634$; (b) $D = 46.4632$.

3-20. Given a column that has two plates, a reboiler, a total condenser, and a feed F of 100 moles per hour. The composition of the feed and the relative volatilities are the same as those stated in Problem 3-19. Initially, the column is operating at total reflux at a distillate rate $D_1 = 56.2634$ and with the following set of b_i/d_i's

Component No.	$\left(\dfrac{b_i}{d_i}\right)_1$
1	8.0000
2	0.5000
3	0.09875

If the distillate rate is changed to $D_2 = 46.4632$ while the column remains at total reflux operation, find the corresponding steady state values of b_i/d_i at D_2 by use of the θ method of convergence.

3-21. Let the corrected component flow rates be defined in terms of the undetermined multipliers η_j and σ_j.

$$(l_{ji})_{co} = \eta_j \left(\frac{l_{ji}}{d_i}\right)_{ca} (d_i)_{co}$$

$$(v_{ji})_{co} = \sigma_j \left(\frac{v_{ji}}{d_i}\right)_{ca} (d_i)_{co}$$

(a). If the multipliers η_j and σ_j are picked such that

$$\sum_{i=1}^{c} (l_{ji})_{co} = (L_j)_{co}$$

$$\sum_{i=1}^{c} (v_{ji})_{co} = (V_j)_{co}$$

show that the expressions given by Equations (3-71) and (3-72) follow from the above expressions and the definition of a mole fraction.

(b). Show that if the undetermined multiplier η_{N+1} for the reboiler is called θ, then the defining equation for θ [Equation (3-67)] follows from the definition of $(l_{ji})_{co}$ for $j = N + 1$ (the reboiler).

NOTATION

A_{ji} = absorption factor; defined by Equation (3-57).

b_i = flow rate of component i in the bottoms lb-mole/hr.

B = total flow rate of bottoms, lb-mole/hr.

c = total number of components.

d_i = flow rate of component i in the distillate, lb-mole/hr.

D = total flow rate of the distillate, lb-mole/hr.

$\bar{f}_i^L, \bar{f}_i^V$ = fugacities of components i in the liquid and vapor phases (composed of any number of components), respectively; evaluated at the total pressure and temperature of the two-phase system, atm.

f_i^L, f_i^V = fugacities of pure component i in the liquid and vapor phases, respectively; evaluated at the total pressure and temperature of the two-phase system, atm.

$f(T)$ = bubble point function; defined by Equation (3-14).

$\mathcal{f}_i$ = feed vector.

$F(T)$ = dew point function; defined by Equation (3-17).

F = total flow rate of the feed, lb-mole/hr.

$g(\theta)$ = a function of θ; defined by Equation (3-69).

h_{Fi}, H_{Fi} = enthalpies of pure component i; evaluated at the temperature T_F and pressure P of the flash, Btu/lb-mole.

$h_j = \sum\limits_{i=1}^{c} h_{ji}x_{ji}$, for an ideal solution; evaluated at the temperature T_j, pressure and composition of the liquid leaving the jth plate, Btu/lb-mole.

$H_j = \sum\limits_{i=1}^{c} H_{ji}y_{ji}$, for an ideal solution; evaluated at the temperature T_j, pressure, and composition of the vapor leaving the jth plate, Btu/lb-mole.

H = enthalpy per mole of feed, regardless of state, Btu/lb-mole.

$H(x_j)_k = \sum\limits_{i=1}^{c} H_{ki}x_{ji}$, for an ideal solution; evaluated at the temperature and pressure of the vapor leaving the kth stage and at the composition of the liquid leaving the jth stage.

$h(y_j)_k = \sum\limits_{i=1}^{c} h_{ki}y_{ji}$, for an ideal solution; evaluated at the temperature and pressure of the liquid leaving the kth stage and at the composition of the vapor leaving the jth stage.

K_{ji} = equilibrium vaporization constant; evaluated at the temperature and pressure of the liquid leaving the jth stage.

l_{ji} = flow rate at which component i in the liquid phase leaves the jth mass transfer section, lb-mole/hr.

l_{0i} = flow rate of component i in the liquid reflux, lb-mole/hr.

l_{Fi}, v_{Fi} = flow rates of component i in the liquid and vapor parts, respectively, of a partially vaporized feed, lb-mole/hr. For bubble point liquid and subcooled feeds, $l_{Fi} = FX_i$ and $v_{Fi} = 0$. For dewpoint vapor and superheated feeds; $v_{Fi} = FX_i$ and $l_{Fi} = 0$.

L_j = total flow rate at which liquid leaves the jth stage, lb-mole/hr.

N = total number of plates.

P_i = vapor pressure of component i, atm.

P = total pressure, atm.

$P(\Psi)$ = flash function; defined by Equation (3-30).

q = a factor related to the thermal condition of the feed; defined by Equation (3-43).

Q_c = condenser duty, Btu/hr.

Q_R = reboiler duty, Btu/hr.

$S_{ji} = K_{ji}V_j/L_j$, stripping factor for component i; evaluated at the conditions of the liquid leaving the jth stage.

T = temperature. $T_{B.P.}$ = bubble point temperature; and $T_{D.P.}$ = dewpoint temperature.

T_F = flash temperature.

v_{ji} = flow rate at which component i in the vapor phase leaves the jth stage, lb-mole/hr.

V_j = total flow rate of vapor leaving the jth stage, lb-mole/hr.

x_{Fi} = mole fraction of component i in the liquid leaving a flash process.

x_{ji} = mole fraction of component i in the liquid leaving the jth stage.

x_{Bi} = mole fraction of component i in the bottoms.

x_I = abscissa of the point of intersection of the operating lines for a binary mixture.

X_i = total mole fraction of component i in the feed (regardless of state).

X_{Di} = total mole fraction of component i in the distillate (regardless of state).

y_{ji} = mole fraction of component i in the vapor leaving plate j.

y_I = ordinate of the point of intersection of the operating lines for a binary mixture.

GREEK LETTERS

α_{ji} = relative volatility, $\alpha_{ji} = K_{ji}/K_{jb}$.

δ = function of T_F; defined by Equation (3-35).

γ_i^L, γ_i^V = activity coefficients for component i in the liquid and vapor, respectively.

Ψ = fraction of the feed converted to vapor by a flash process, $\Psi = V_F/F$.

θ = a multiplier defined by Equation (3-67).

SUBSCRIPTS

    ca = calculated value.

    co = corrected value.

    f = feed plate.

    F = variables associated with a partially vaporized feed.

    i = component number, $i = 1, 2, \ldots, c$ or $(1 \leq i \leq c)$.

    j = plate number; $j = 0$ for the accumulator; for the top plate $j = 1$, for the feed plate $j = f$, for the bottom plate $j = N$, and for the reboiler $j = N + 1$, that is, $j = 0, 1, 2, \ldots, f, \ldots, N, N + 1$, or $(0 \leq j \leq N + 1)$.

    n = trial number.

    N = total number of plates.

    r = rectifying section.

    s = stripping section.

SUPERSCRIPTS

    L = liquid phase.

    V = vapor phase.

MATHEMATICAL SYMBOLS

$\sum_{i=1}^{c} x_i$ = sum over all values x_i, $i = 1, 2, \ldots, c$, or $(1 \leq i \leq c)$.

$\{x_j\}$ = set of all values x_j belonging to the particular set under consideration.

$\prod_{j=1}^{c} x_j = x_1 x_2 \ldots x_{c-1} x_c$, product of the x_j's from $j = 1$ through $j = c$.

REFERENCES

1. Anderson, E. V., R. Brown, and C. E. Bolton, "Styrene-crude Oil to Polymer," *Ind. Eng. Chem.*, **52**, (1960), 550.

2. Boston, J. F. and S. L. Sullivan, Jr., "An Improved Algorithm for Solving Mass Balance Equations in Multistage Separation Processes," *Can. J. Chem. Eng.* **50**, (1972), 663.

3. Billingsley, D. S., "On the Equations of Holland in the Solution of Problems in Multicomponent Distillation," *IBM J. Res. Develop.*, **14**, (1970), 33. See also, Billingsley, D. S., *A. I. Ch. E. Journal,* **16**, (1970), 441.

4. Carnahan, Brice, H. A. Luther, and J. O. Wilkes, *Applied Numerical Methods.* New York: John Wiley & Sons, Inc., 1964.

5. Denbigh, Kenneth, *The Principles of Chemical Equilibrium.* New York: Cambridge University Press, 1955.

6. Fenske, M. R., "Fractionation of Straight-Run Pennsylvania Gasoline," *Ind. Eng. Chem.*, **24**, (1932), 482.

7. Greenstadt, John, Yonathan Bard, and Burt Morse, "Multicomponent Distillation Calculation on the IBM 704," *Ind. Eng. Chem.*, **50**, (1958), 1944.

8. Holland, C. D., *Multicomponent Distillation*. Englewood Cliffs, N.J.: Prentice-Hall, Inc., 1963. See also Holland, C. D., *Unsteady State Processes with Applications in Multicomponent Distillation*. Englewood Cliffs, N.J.: Prentice-Hall, Inc., 1966.

9. ————, *Introduction to the Fundamentals of Distillation*, Proceedings of the Fourth Annual Education Symposium of the ISA, April 5–7, 1972, Wilmington, Delaware.

10. Holland, C. D., and J. D. Lindsay, *Encyclopedia Chemical Technology*, Vol. 7, 2nd ed. New York: John Wiley & Sons, Inc., 1965, pp. 204–48.

11. Lewis, W. J., "Theory of Fractional Distillation," *J. Ind. Chem.*, **1**, (1909), 522.

12. Lyster, W. N., S. L. Sullivan, Jr., D. S. Billingsley, and C. D. Holland, "Figure Distillation This New Way: Part 1—New Convergence Method Will Handle Many Cases," *Petroleum Refiner*, **38**, 6 (1959), 221.

13. McCabe, W. L., and E. W. Thiele, "Graphical Design of Fractionating Columns," *Ind. Eng. Chem.*, **17**, (1925), 605.

14. Nartker, T. A., J. M. Srygley, and C. D. Holland, "Solution of Problems Involving Systems of Distillation Columns," *Can. J. Chem. Engr.*, **44**, (1966), 217.

15. Ponchon, Marcel, "Graphical Study of Fractional Distillation," *Tech. moderne*, **13**, (1921), 20.

16. Rayleigh, Lord (J. Strutt), "On the Distillation of Binary Mixtures," *Phil. Mag.*, **4**, (1902), 527.

17. Robinson, C. S., and E. R. Gilliland, *Elements of Fractional Distillation*, 4th ed. New York: McGraw-Hill Book Co., Inc., 1950.

18. Savarit, R., *Arts et metiers*, "Definition of Distillation, Simple Discontinuous Distillation, Theory and Operation of Distillation Column," and "Exhausting and Concentrating Columns for Liquid and Gaseous Mixtures and Graphical Methods for Their Determination" (1922), pp. 65, 142, 178, 241, 266, and 307.

19. Sorel, E., *La Rectification de l'Alcool*. Paris: Gauthier-Villars et fils, 1893.

20. Sujata, A. D., "Absorber Stripper Calculations Made Easier," *Hydrocarbon Processing* and *Petroleum Refiner*, **40**, (1961), 137.

21. Thiele, E. W., and R. L. Geddes, "Computation of Distillation Apparatus for Hydrocarbon Mixtures," *Ind. Eng. Chem.*, **25**, (1933), 289.

Absorbers and Strippers 4

From a light gas stream such as natural gas that contains primarily methane plus small quantities of say ethane through n-pentane, the desired quantities of the components heavier than methane may be removed by contacting the natural gas stream with a heavy oil stream (say n-octane or heavier) in a countercurrent, multiple stage column such as the one shown in Figure 4-1. Since absorption is a heat liberating process, the lean oil is customarily introduced at a temperature below the average temperature at which the column is expected to operate. The flow rate of the lean oil is denoted by L_0, and the lean oil enters at the top of the column as implied by Figure 4-1. The *rich gas* (which is sometimes called the *wet gas*) enters at the bottom of the column at a temperature equal to or above its dew point temperature at the column pressure but generally below the average operating temperature of the column. The total flow rate of the rich gas is denoted by V_{N+1}. The absorber oil plus the material that it has absorbed leave at the bottom of the column; this stream is called the *rich oil*. The treated gas leaving the top of the column is called the *lean gas* (or the *stripped gas*).

Strippers are used to remove relatively light gases from a heavy oil stream by contacting it with a relatively light gas stream such as steam. The sketch of a typical stripper is shown in Figure 4-1.

In Part 1 of this chapter, a relatively simple model is presented for approximating the behavior of absorbers and strippers. In Part 2, calculational procedures are developed for more exact models for absorbers and strippers. A brief description of the application of certain of the methods developed

107

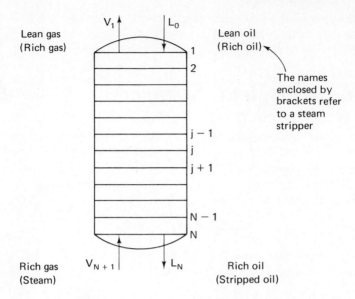

Lean gas
(Rich gas)

V_1 L_0

Lean oil
(Rich oil)

The names
enclosed by
brackets refer
to a steam
stripper

Rich gas V_{N+1} L_N Rich oil
(Steam)   (Stripped oil)

Figure 4-1. Sketch of an absorber (or steam stripper).

in Part 2 to other types of columns is also presented. In addition, other methods which have been proposed for solving various types of distillation problems are described briefly.

Part 1. Solution of Absorber and Stripper Problems by Use of the Absorption Factor Method

This well-known classical model called the *absorption factor method* is based on the following assumptions (4, 13).

1. Each plate of the column is perfect plate, that is, the vapor leaving each plate is in equilibrium with the liquid leaving.
2. The absorption factors for each component i do not vary with plate number, that is, $A_{1i} = A_{2i} = \cdots = A_{Ni} = A_i$, where,

$$A_{ji} = \frac{L_j}{K_{ji}V_j}$$

The component-material balances enclosing each plate of a typical absorber or stripper may be reduced to a set of equations containing either the set of unknown liquid rates (l_{ji}) or the set of unknown vapor rates (v_{ji}) by use of the equilibrium relationship $y_{ji} = K_{ji}x_{ji}$, which may be restated in the following form:

$$l_{ji} = A_{ji}v_{ji}; \qquad v_{ji} = S_{ji}l_{ji} \tag{4-1}$$

When the first of the two relationships given by Equation (4-1) is used to eliminate the l_{ji}'s from the material balances for plates $j = 1$ through $j = N$, the following set of equations is obtained for each component i,

$$v_{2i} - (1 + A_{1i})v_{1i} = -l_{0i}$$

$$v_{j+1,i} + A_{j-1,i}v_{j-1,i} - (1 + A_{ji})v_{ji} = 0 \qquad (j = 2, 3, \ldots N) \qquad (4\text{-}2)$$

$$A_{N-1,i}v_{N-1,i} - (1 + A_{Ni})v_{Ni} = -v_{N+1,i}$$

In the remainder of this development, the subscript i is dropped in the interest of simplicity and the A_{ji}'s are taken to be equal for each stage j in accordance with the second assumption listed above. When the first expression of Equation (4-2) is solved for v_1, the second for v_2, the third for $v_3, \ldots$, and the Nth for v_N, the following set of equations is obtained:

$$v_1 = \frac{l_0 + v_2}{1 + A}, \qquad v_2 = \frac{Av_1 + v_3}{1 + A}, \qquad \ldots, \qquad v_N = \frac{Av_{N-1} + v_{N+1}}{1 + A} \qquad (4\text{-}3)$$

When v_1 is eliminated from the first and second expressions and the result so obtained is solved for v_2, it is found that

$$v_2 = \frac{Al_0 + (1 + A)v_3}{1 + A + A^2} \qquad (4\text{-}4)$$

By continuation of this elimination process, it is found that for the Nth plate

$$v_N = \frac{A^{N-1}l_0 + (1 + A + A^2 + \cdots + A^{N-1})v_{N+1}}{1 + A + A^2 + \cdots + A^N} \qquad (4\text{-}5)$$

When the geometric series in Equation (4-5) are replaced by their respective sums, one obtains

$$v_N = \frac{A^{N-1}l_0 + \left(\dfrac{1 - A^N}{1 - A}\right)v_{N+1}}{\dfrac{1 - A^{N+1}}{1 - A}} \qquad (4\text{-}6)$$

The component flow rate v_N may be expressed in terms of v_1, l_0, and v_{N+1} by use of the component-material balance that encloses the entire column, namely,

$$v_{N+1} + l_0 - v_1 - l_N = 0 \qquad (4\text{-}7)$$

Elimination of l_N from Equation (4-7) by use of the equilibrium relationship $l_N = A_N v_N = Av_N$ gives

$$v_N = \frac{v_{N+1} - v_1 + l_0}{A} \qquad (4\text{-}8)$$

Elimination of v_N from Equations (4-6) and (4-8) followed by rearrangement gives

$$\frac{v_{N+1,i} - v_{1,i}}{v_{N+1,i} - \dfrac{l_{0i}}{A_i}} = \frac{A_i^{N+1} - A_i}{A_i^{N+1} - 1} \qquad (4\text{-}9)$$

The following alternate but equivalent form of Equation (4-9) is in common usage,

$$\frac{Y_{N+1,i} - Y_{1i}}{Y_{N+1,i} - Y_{0i}} = \frac{A_i^{N+1} - A_i}{A_i^{N+1} - 1} \tag{4-10}$$

where

$$Y_{N+1,i} = \frac{v_{N+1,i}}{V_{N+1}};$$

$$Y_{1i} = \frac{v_{1i}}{V_{N+1}};$$

$$Y_{0i} = \frac{l_{0i}}{A_i V_{N+1}}.$$

A formula that involves the stripping factors may be developed by solving the component-material balances simultaneously in terms of the l_{ji}'s in a manner analogous to that demonstrated above. Alternately, one may begin with Equation (4-9) and replace A_i by its equivalent $1/S_i$ and then eliminate v_{1i} from the left-hand side by use of Equation (4-7) to obtain the following result upon rearrangement:

$$\frac{l_{0i} - l_{Ni}}{l_{0i} - \dfrac{v_{N+1,i}}{S_i}} = \frac{S_i^{N+1} - S_i}{S_i^{N+1} - 1} \tag{4-11}$$

Again, the left-hand side of this equation may be stated in the following alternate form:

$$\frac{X_{0i} - X_{Ni}}{X_{0i} - X_{N+1,i}} = \frac{S_i^{N+1} - S_i}{S_i^{N+1} - 1} \tag{4-12}$$

where

$$X_{0i} = \frac{l_{0i}}{L_0};$$

$$X_{Ni} = \frac{l_{Ni}}{L_0};$$

$$X_{N+1,i} = \frac{v_{N+1,i}}{S_i L_0}.$$

The approximate forms that these expressions take for components for which $A_i \gg 1$, $A_i \ll 1$, or $A_i \cong 1$ are helpful in the application of these equations (see Problem 4-2).

ILLUSTRATIVE EXAMPLE 4-1

An absorber with three perfect plates is to be used to treat a gas stream (V_{N+1}) whose composition is listed below. The temperatures of the inlet gas and lean oil are to be selected such that the average operating temperature of the column is 90°F. The column is to operate at a pressure of 300 psia. A rich gas rate V_{N+1} of 100 moles per hour and a lean oil rate L_0 of 20 moles per hour are to be used. Use the absorption factor method to determine the flow rate of each component in the treated gas stream V_1.

Given:

Component	$v_{N+1,i}$	l_{0i}	K_i^* @ 90°F and 300 psia
CH_4	70	0	12.99100
C_2H_6	15	0	2.18080
C_3H_8	10	0	0.63598
$n\text{-}C_4H_{10}$	4	0	0.18562
$n\text{-}C_5H_{12}$	1	0	0.05369
$n\text{-}C_8H_{18}$	0	20	0.00136

*Taken from Appendix C-3.

Solution: Assume an average V of 95 moles/hr and an average L_0 of 25 moles/hr to give an L/V of 0.263

Component	$A_i = \dfrac{L}{K_i V}$	A_i^{N+1}	$\dfrac{A_i^{N+1} - A_i}{A_i^{N+1} - 1}$	v_{1i} (at end of first trial)
CH_4	0.02026	0.00000	0.02026	68.852
C_2H_6	0.12066	0.00022	0.12048	13.192
C_3H_8	0.41377	0.02932	0.39607	6.039
$n\text{-}C_4H_{10}$	1.41760	4.03900	0.86255	0.550
$n\text{-}C_5H_{12}$	4.90090	5.7693×10^2	0.99323	0.007
$n\text{-}C_8H_{18}$	193.5200	1.4025×10^8	1.00000	0.103
				$V_1 = 88.473$

The calculated value of V_1 at the end of the first trial was 88.473 and the corresponding average value of V was $(100 + 88.473)/2 = 94.236$. Since these values differ slightly from the corresponding assumed values, the calculational procedure described above was repeated on the basis of the most recently calculated values [the method of direct-iteration (see Appendix A)] until the solution set of values had been found. At the end of the fourth trial, the following set of v_{1i}'s (the solution) was obtained:

Component	Solution Set of v_{1i}'s
CH_4	68.513
C_2H_6	13.105
C_3H_8	5.869
$n\text{-}C_4H_{10}$	0.501
$n\text{-}C_5H_{12}$	0.006
$n\text{-}C_8H_{18}$	0.099
	$V_1 = 88.093$ and $V_{av} = 94.046$.

It is evident that since the initial V_{av} of 95 was close to the solution value of 94.046, the values of v_{1i}'s calculated at the end of the first trial were close to the values of the solution set.

Although the absorption factor method is based on a rather restrictive set of assumptions, this method is useful for purposes of orientation where approximate solutions are satisfactory. More exact solutions to problems of this type may be obtained by use of any one of the calculational procedures presented in Part 2.

Part 2. Calculational Procedures for Solving Absorber and Stripper Problems

The procedures developed and demonstrated below constitute different methods for solving the component-material balances, the equilibrium relationships, and the enthalpy balances which are required to describe absorbers and strippers.

The methods are developed for existing absorbers that is, the number of plates is fixed and it is desired to determine the separation that can be effected at a specified set of operating conditions. The set of specifications considered are as follows:

Specifications: $\underline{P, V_{N+1}, \{y_{N+1,i}\}, L_0, \{x_{0i}\}, T_0, T_{N+1}, \text{ and } N}$

The problem to be solved consists of finding the solution sets of total flow rates, compositions, and temperatures corresponding to a given set of specifications. Four procedures are presented for solving problems of this type. In the presentation of these procedures, both perfect plates and perfect heat transfer are assumed. Perfect heat transfer means that the vapor and liquid streams leaving each plate possess the same temperature. The temperatures may vary, however, from plate to plate.

Procedures 2 and 4 make direct use of the Newton-Raphson method (see Appendix A) while Procedures 1 and 3 consist of an approximation of this method. The following relative simple algebraic example illustrates the two exact applications of the Newton-Raphson method. In the first application, which is analogous to Method II of Procedure 2 and Procedure 4, both x and y are regarded as independent variables, and in the second application, which is analogous to Method I of Procedure 2, y is regarded as the independent variable and x as the dependent variable.

ILLUSTRATIVE EXAMPLE 4-2

It is desired to find the pair of positive roots that make $f_1(x, y) = 0$ and $f_2(x, y) = 0$, simultaneously,

$$f_1(x, y) = x^2 - xy^2 - 2$$
$$f_2(x, y) = 2x^2 - 3xy^2 + 3$$

Make one trial by use of the Newton-Raphson method for each of the following cases:

(a) Both x and y are to be regarded as independent variables and the initial values are $x_n = 1$ and $y_n = 1$.

(b) Take y to be the independent variable and x the dependent variable, and take the initial value of y to be $y_n = 1$.

Solution.

(a) In this case the Newton-Raphson equations are given by

$$\frac{\partial f_1}{\partial x} \Delta x + \frac{\partial f_1}{\partial y} \Delta y = -f_1$$

$$\frac{\partial f_2}{\partial x} \Delta x + \frac{\partial f_2}{\partial y} \Delta y = -f_2$$

or in matrix form

$$\begin{bmatrix} \dfrac{\partial f_1}{\partial x} & \dfrac{\partial f_1}{\partial y} \\ \dfrac{\partial f_2}{\partial x} & \dfrac{\partial f_2}{\partial y} \end{bmatrix} \begin{bmatrix} \Delta x \\ \Delta y \end{bmatrix} = - \begin{bmatrix} f_1 \\ f_2 \end{bmatrix}$$

where

$$\frac{\partial f_1}{\partial x} = 2x - y^2; \qquad \frac{\partial f_1}{\partial y} = -2xy;$$

$$\frac{\partial f_2}{\partial x} = 4x - 3y^2; \qquad \frac{\partial f_2}{\partial y} = -6xy.$$

When the functions $f_1(x, y)$ and $f_2(x, y)$ and their partial derivatives are evaluated at $x_n = 1$ and $y_n = 1$, the Newton-Raphson equations may be solved to give

$$\Delta x = 2; \qquad x_{n+1} = x_n + \Delta x = 5$$
$$\Delta y = 1; \qquad y_{n+1} = y_n + \Delta y = 2$$

(b) In this case, one of the functions, say f_1, is set equal to zero,

$$x^2 - xy^2 - 2 = 0$$

to give the constraining condition that is used to compute the value of x corresponding to each choice of the independent variable y. Then for $y_n = 1$, it is found that $x_n = 2$. In the event that it is impossible to solve the constraining equation explicitly for the dependent variable, it may be solved by successive applications of the Newton-Raphson method (which is also referred to herein as Newton's method for the case of a single variable)

$$\frac{\partial f_1(x, y_n)}{\partial x} \Delta x = -f_1(x, y_n)$$

On the basis of $x_n = 2$ and $y_n = 1$, an improved value of y is found by applying the Newton-Raphson method one time to the function $f_2(x, y)$ as follows:

$$\frac{df_2}{dy} \Delta y = -f_2$$

where

$$\frac{df_2}{dy} = \frac{\partial f_2}{\partial x} \frac{dx}{dy} + \frac{\partial f_2}{\partial y}$$

The expressions for $\partial f_2/\partial x$ and $\partial f_2/\partial y$ are given in Part (a). The termwise differentiation of the constraining equation with respect to the independent variable y may be represented as follows:

$$0 = \frac{\partial f_1}{\partial x}\frac{dx}{dy} + \frac{\partial f_1}{\partial y}$$

After $\partial f_1/\partial x$ and $\partial f_1/\partial y$ have been eliminated by use of the expressions given in Part (a), the resulting equation may be rearranged to give

$$\frac{dx}{dy} = \frac{2xy}{2x - y^2}$$

Then for $x_n = 2$ and $y_n = 1$, one obtains

$$\Delta y = \frac{15}{16}, \qquad y_{n+1} = y_n + \Delta y = \frac{31}{16}$$

A comparison of these two methods for applying the Newton-Raphson method in Illustrative Example 4-2 shows that the Jacobian matrices of Part (a) are four times as large as the corresponding ones in Part (b). If convergence is to be assured for the first method, both of the initial values (x_n and y_n) must be picked within the region of convergence of the complete set of equations. However, in the second method only y_n must be picked within the region of convergence of the complete set of equations, and for x_n the region of convergence within which it must be picked is the one for constraining equations. In the second method it may be necessary to apply the Newton-Raphson method successively in order to find the x_n that satisfies the constraining equation for the given choice of y_n. Furthermore, values of y_n for which solution values of x_n exist must be selected in the second method, but not in the first method.

Calculational Procedures 1 and 3 may be described in terms of two approximations of the second method shown in Part (b) of Illustrative Example 4-2 for applying the Newton-Raphson method. First, the constraining equations are solved by the method of direct iteration in Procedure 3 and a variation of it in Procedure 1. In Appendix A, it is demonstrated that the method of direct iteration fails for some problems for which the Newton-Raphson method converges. The second approximation contained in Procedures 1 and 3 consists of neglecting the dependency of the dependent variables on the independent variables in the calculation of a new set of the latter. In Illustrative Example 4-2, this approximation amounts to setting $dx/dy = 0$ in the expression for df_2/dy.

Procedures 1 and 2 are based on the θ method of convergence which has been successfully used to solve problems in which the independent product rates such as the distillate rate of a conventional column are specified. For absorbers and strippers, however, the set of specifications listed above, which contains neither product rate (V_1 or L_N), is generally made. Procedures 1 and 2 consist of two applications of the θ method of convergence for solving problems of this type. Both of these applications are shown to be plausible extensions of the θ method of convergence for an absorber with one plate (the adiabatic flash problem). The first application (Procedure 1) of the θ method makes use of a single θ and is called the *single-θ method* and the second application (Procedure 2) of the θ method makes use of a θ for each plate j and is called the *multi-θ method.*

The development of these two applications of the θ method of convergence is initiated by showing that they may be deduced as logical extensions of the θ method for an isothermal flash.

The special case of an absorber with one plate

The special case of an absorber with one plate where the specifications are as enumerated above reduces to the adiabatic flash problem (see Chapter 3). One of the methods for solving the adiabatic flash problem consists of regarding the flash temperature as the independent variable and the remaining variables as the dependent variables. On the basis of an assumed flash temperature, the component-material balances and equilibrium relationships are solved simultaneously for the corresponding values of the dependent variables. The values of the total flow rates and compositions so obtained are then used to compute an improved temperature by use of the enthalpy balance as demonstrated in Chapter 3.

First consider the problem in which the temperature of plate 1 (the one and only plate) is regarded as fixed, and a solution to the corresponding component-material balances and equilibrium relationships is to be found. [This problem is commonly called the isothermal flash problem (see Chapter 3).] The component-material balance for each component is given by

$$v_{N+1,i} + l_{0i} = v_{1i} + l_{1i} \tag{4-13}$$

By use of the equilibrium relationship $y_{1i} = K_{1i}x_{1i}$ and an assumed value for L_1/V_1 (denoted by the subscript "a"), the following formula for calculating the corresponding values of l_{1i}/v_{1i} is readily obtained:

$$\left(\frac{l_{1i}}{v_{1i}}\right)_{ca} = \frac{1}{K_{1i}}\left(\frac{L_1}{V_1}\right)_a \tag{4-14}$$

Corresponding to this set of calculated values of l_{1i}/v_{1i}, the corresponding calculated values for v_{1i} may be found by rearranging Equation (4-13) to the following form:

$$(v_{1i})_{ca} = \frac{v_{N+1,i} + l_{0i}}{1 + \left(\dfrac{l_{1i}}{v_{1i}}\right)_{ca}}$$

The θ method of convergence is applied to this problem by first defining the multiplier θ as follows:

$$\frac{l_{1i}}{v_{1i}} = \theta\left(\frac{l_{1i}}{v_{1i}}\right)_{ca} \tag{4-15}$$

The customary subscript "co" used in Chapter 3 to identify the corrected values has been omitted in the interest of simplicity. By requiring that the corrected flow rates v_{1i} and l_{1i} be in component-material balance, the following formula for v_{1i} is obtained:

$$v_{1i} = \frac{v_{N+1,i} + l_{0i}}{1 + \theta \left(\dfrac{l_{1i}}{v_{1i}}\right)_{ca}} \tag{4-16}$$

Also, the expression for the correct value of the total flow rate V_1 may be obtained from the above definition of θ as follows. By use of Equation (4-14), it is possible to restate the defining equation for θ [Equation (4-15)] in the following form:

$$\left(\frac{L_1}{V_1}\right)\frac{1}{K_{1i}} = \theta\left(\frac{L_1}{V_1}\right)_a \frac{1}{K_{1i}} \tag{4-17}$$

and since T_1 is regarded as fixed (at the assumed value), it is evident that

$$\frac{L_1}{V_1} = \theta\left(\frac{L_1}{V_1}\right)_a \tag{4-18}$$

Since the correct values of the total flow rates must satisfy the material balance,

$$V_{N+1} + L_0 = V_1 + L_1 \tag{4-19}$$

it follows from this expression and Equation (4-18) that

$$V_1 = \frac{V_{N+1} + L_0}{1 + \theta\left(\dfrac{L_1}{V_1}\right)_a} \tag{4-20}$$

Obviously, the desired value of $\theta > 0$ is the one that makes the sum of the v_{1i} given by Equation (4-16) equal to the value of V_1 given by Equation (4-20). Also, the dew point (or bubble point) expression must be satisfied by the corrected flow rates. All of these conditions are contained in the following form of the dew point function:

$$F_1(\theta) = \frac{1}{V_1} \sum_{i=1}^{c} \left(\frac{1}{K_{1i}} - 1\right) v_{1i} \tag{4-21}$$

where v_{1i} is given by Equation (4-16) and V_1 is given by Equation (4-20). Now observe that when the correct value of θ has been found, Equation (4-21) reduces to the familiar form of the dew point function, namely,

$$F_1 = \sum_{i=1}^{c} \frac{y_{1i}}{K_{1i}} - 1 = 0$$

A solution exists [a $\theta > 0$ that makes $F_1(\theta) = 0$] provided that the assumed value of T_1 lies between the bubble point and dew point temperatures of the combined feed $(V_{N+1} + L_0)$. The value of θ so obtained represents the solution to the component-material balances and equilibrium relationships at the fixed value of T_1. Observe that the particular value of $(L_1/V_1)_a$ selected to initiate the trial procedure is immaterial because the value found for θ compensates for the choice of $(L_1/V_1)_a$. The truth of this statement is evident upon examination of Equation (4-18). Thus, the θ method of convergence constitutes an exact solution to the isothermal flash problem in that once θ has been determined, the solution of the equations for the isothermal flash has been obtained.

The remainder of the calculational procedure for solving adiabatic flash problems consists of finding a new temperature by use of the enthalpy balance. This temperature may be found by any one of several methods such as interpolation *regula falsi* or Newton's method. The use of the method of interpolation *regula falsi* was described in Chapter 3 and Newton's method follows as a special case of the multi-θ method presented below for a column with any number of plates.

For $N > 1$, the definition given by Equation (4-15) may be extended by supposing that the ratio pertains to the terminal streams leaving the column to give

$$\frac{l_{Ni}}{v_{1i}} = \theta \left(\frac{l_{Ni}}{v_{1i}} \right)_{\text{ca}} \tag{4-22}$$

where θ is to be selected so that an overall component-material balance is satisfied and so that the bubble point function for each plate j is satisfied as discussed below. Alternately, it may be supposed that the ratio given by Equation (4-15) pertains to the streams leaving each plate j to obtain,

$$\frac{l_{ji}}{v_{ji}} = \theta_j \left(\frac{l_{ji}}{v_{ji}} \right)_{\text{ca}} \qquad (1 \leqq j \leqq N) \tag{4-23}$$

where $(l_{ji}/v_{ji})_{\text{ca}}$ is equal to $(L_j/V_j)_{\text{a}}/K_{ji}$ and where the set of θ_j's is to be picked so that all of the bubble point or dew point functions are satisfied simultaneously. The definition given by Equation (4-22) forms the basis for the *single-θ method* of convergence, and the definition given by Equation (4-23) forms the basis for the *multi-θ method*.

The single-θ method and Method I of the multi-θ method of convergence are analogous to the procedure described above for the adiabatic flash in that the temperatures are regarded as the independent variables and the remaining variables are regarded as the dependent variables. For each set of assumed temperatures $\{T_j\}$, a solution to the constraining equations (the component-material balances and equilibrium relationships) is found. The flow rates so obtained are used in the enthalpy balance functions in the prediction of an improved set of temperatures by one application of the Newton-Raphson method.

Material-balance equations

Each proposed calculational procedure is initiated by the assumption of a set of vapor rates $\{V_j\}$ and a set of temperatures $\{T_j\}$. After the liquid rates $\{L_j\}$ corresponding to the set of assumed vapor rates have been computed by use of the total-material balances,

$$V_{j+1} + L_{j-1} - V_j - L_j = 0 \qquad (1 \leq j \leq N) \qquad (4\text{-}24)$$

the component-material balances may be solved for the component flow rates. After the l_{ji}'s have been eliminated from the component-material balances by use of the equilibrium relationship [Equation (4-1)], the resulting equations so obtained [Equation (4-2)] may be stated in matrix form as follows:

$$\mathbf{A}_i \mathbf{v}_i = -\boldsymbol{\mathscr{f}}_i \qquad (4\text{-}25)$$

where $\mathbf{A}_i$ is a tridiagonal matrix, $\mathbf{v}_i$ is a column vector that contains the v_{ji}'s, and $\boldsymbol{\mathscr{f}}_i$ is a feed vector, that is,

$$\mathbf{A}_i = \begin{bmatrix} -(A_{1i}+1) & 1 & 0 & 0 & \dots & 0 \\ A_{1i} & -(A_{2i}+1) & 1 & 0 & \dots & 0 \\ \multicolumn{6}{c}{\dotfill} \\ 0 \cdots\cdots\cdots\cdots\cdots\cdots & 0 & A_{N-1,i} & -(A_{Ni}+1) \end{bmatrix}$$

$$\mathbf{v}_i = [v_{1i} v_{2i} \cdots v_{Ni}]^T \qquad (4\text{-}26)$$

$$\boldsymbol{\mathscr{f}}_i = [l_{0i} 0 \cdots 0 v_{N+1,i}]^T$$

Although Equation (4-25) represents a combination of the component-material balances and equilibrium relationships, it is sometimes referred to hereafter as simply the component-material balances. The solution set of component-flow rates $\{v_{ji}\}$ of Equation (4-25) was obtained by using the recurrence formulas (6, 10,), sometimes called the Thomas algorithm [given below Equation (3-64)]. For each component i, the corresponding liquid flow rates are computed by use of Equation (4-1). Let the flow rates so obtained be identified by the subscript "ca," that is, $\{(v_{ji})_{\text{ca}}\}$ and $\{(l_{ji})_{\text{ca}}\}$.

Up to this point, the two θ methods of convergence are identical. In the single-θ method, the calculated values of the component-flow rates are used to obtain an improved set of flow rates in the following manner.

Procedure 1 : The single-θ method of convergence

In this case, the multiplier θ is defined by Equation (4-22) with the qualification that the corrected rates must satisfy the component-material balances enclosing the entire column, namely,

$$v_{N+1,i} + l_{0i} = l_{Ni} + v_{1i} \qquad (4\text{-}27)$$

Equations (4-22) and (4-27) may be solved for v_{1i} to give

$$v_{1i} = \frac{v_{N+1,i} + l_{0i}}{1 + \theta\left(\dfrac{l_{Ni}}{v_{1i}}\right)_{\text{ca}}} \qquad (4\text{-}28)$$

For any choice of θ, the corresponding value of the total flow rate V_1 is found by summing the component flow rates given by Equation (4-28) over all components i. The corresponding flow rates l_{Ni} and L_N may be computed by use of Equations (4-22) and (4-28).

Next, formulas for the calculation of a new set of total flow rates which are consistent with the corrected flow rates V_1 and L_N are developed as follows. Let the corrected flow rates for all components i which appear in both phases be defined by

$$l_{ji} = \eta_j\left(\frac{l_{ji}}{v_{1i}}\right)_{\text{ca}} v_{1i} \quad \text{and} \quad v_{ji} = \sigma_j\left(\frac{v_{ji}}{v_{1i}}\right)_{\text{ca}} v_{1i} \tag{4-29}$$

where η_j and σ_j are at this point undetermined multipliers. The definitions given by Equation (4-29) do imply, however, that

$$L_j = \eta_j \mathcal{L}_j \quad \text{and} \quad V_j = \sigma_j \mathcal{V}_j \tag{4-30}$$

where

$$\mathcal{L}_j = \sum_{i=1}^{c} \left(\frac{l_{ji}}{v_{1i}}\right)_{\text{ca}} v_{1i};$$

$$\mathcal{V}_j = \sum_{i=1}^{c} \left(\frac{v_{ji}}{v_{1i}}\right)_{\text{ca}} v_{1i}.$$

From the definition of a mole fraction and the above relationships, it follows that

$$x_{ji} = \frac{\left(\frac{l_{ji}}{v_{1i}}\right) v_{1i}}{\mathcal{L}_j} \quad \text{and} \quad y_{ji} = \frac{\left(\frac{v_{ji}}{v_{1i}}\right)_{\text{ca}} v_{1i}}{\mathcal{V}_j} \tag{4-31}$$

The multipliers σ_{j+1} and η_j are related by a total material balance enclosing the top of the column and any plate j as follows:

$$\sigma_{j+1} \mathcal{V}_{j+1} - \eta_j \mathcal{L}_j - \mathcal{B} = 0 \tag{4-32}$$

where

$$\mathcal{B} = V_1 - L_0.$$

For any choice of θ, the values of the quantities V_1, $\mathcal{V}_{j+1}$, and $\mathcal{L}_j$ may be computed, and thus Equation (4-32) reduces to one equation in two unknowns, σ_{j+1} and η_j. Consequently, infinitely many choices of values of these variables exist which will satisfy Equation (4-32). Of the choices or relationships investigated (12), the following one

$$\sigma_{j+1} = \frac{1}{\eta_j} \quad (1 \leq j \leq N - 1) \tag{4-33}$$

gave the most satisfactory results for the absorber problems. This relationship is in agreement with the fact that at convergence $\sigma_{j+1} \times \eta_j = 1$ (also, at convergence $\sigma_{j+1} = \eta_j = 1$). Other characteristics possessed by this relationship are given in Reference (12). When Equations (4-32) and (4-33) are solved for η_j, the following formula is obtained:

$$\eta_j = \frac{-\mathcal{B} + \sqrt{\mathcal{B}^2 + 4\mathcal{L}_j \mathcal{V}_{j+1}}}{2\mathcal{L}_j} \tag{4-34}$$

Also, observe that Equations (4-22) and (4-29) require that $\eta_N = \theta$ and $\sigma_1 = 1$. Thus, for each choice of θ, corresponding sets of compositions and total flow rates may be computed by use of the above relationships.

Formulation of the g function

The formulation of the function $g(\theta)$ for the determination of θ is analogous to that demonstrated for the isothermal flash problem. Instead of only one bubble point function, however, there exist N bubble point functions to be satisfied by the choice of θ, namely,

$$f_j = \sum_{i=1}^{c} (K_{ji})_a x_{ji} - 1 \qquad (1 \leq j \leq N) \qquad (4\text{-}35)$$

(Again, as in the case of the flash problem, the dew point functions could have been used instead of the bubble point functions.) By use of the definitions of x_{ji}, y_{ji}, and the fact that $(l_{ji})_{ca} = (A_{ji})_a (v_{ji})_{ca}$, the assumed K values may be eliminated to give

$$f_j = \left(\frac{L_j}{V_j}\right)_a \left(\frac{\mathcal{V}_j}{\mathcal{L}_j}\right) - 1 \qquad (1 \leq j \leq N) \qquad (4\text{-}36)$$

The expressions given by Equation (4-36) are more rapidly evaluated than those given by Equation (4-35). The g function may be defined as the arithmetic average of the square of the Euclidean norm of the bubble point functions,

$$g(\theta) = \frac{1}{N} \sum_{j=1}^{N} f_j^2 \qquad (4\text{-}37)$$

Prior to convergence, it is generally impossible to find a single θ such that $f_j(\theta) = 0$, simultaneously, for all j. Thus, at the end of any given trial prior to convergence, a θ is picked such that $g(\theta)$ is minimized. .

The method used to find the θ closest to unity for which $g(\theta)$ is minimized should be regarded as peripheral to the θ method of convergence. Many suitable methods have been proposed for the minimization of a function of a single variable (20). The method used consisted of the successive approximation of the minimum through the use of a parabolic approximation.

In order to apply this method, it is first necessary to locate a finite interval that contains the minimum. The parabolic method is initiated by choosing three values of θ denoted by $\theta_{n-1} < \theta_n < \theta_{n+1}$, where $\theta_n = 1.0$. If the corresponding values of the g functions indicate that a minimum lies between θ_{n-1} and θ_{n+1}, the minimum is estimated by use of the expression obtained by curve fitting the equation of a parabola. If, however, the g function increases or decreases monotonically in part or all of the interval θ_{n-1} to θ_{n+1}, it is searched in the direction it decreases by varying θ by preassigned increments until either an interval containing the minimum has been located or a preassigned number of θ's has been tried. In the solution of the illustrative examples, a minimum of eight values of θ lying between 0.97 and 1.03 was tested. The rate of convergence of a given problem did not depend strongly on the size of the increment in which θ was employed, but it was necessary to make a sufficient number of trials to find θ to the desired accuracy as convergence of the problem was approached.

After the desired value of θ has been found, the corresponding sets of total flow rates $\{V_j\}$ and $\{L_j\}$ are computed as described above. On the basis of the total flow rates so obtained and the assumed temperature profile, the component-material balances [Equation (4-25)] are solved and the θ method applied again. This procedure is repeated until the convergence criteria for the material balance equations have been satisfied. The compositions and total flow rates so obtained are used in the enthalpy balances in the determination of an improved set of temperatures.

Determination of an improved set of temperatures

The energy balance for each plate j, $V_{j+1}H_{j+1} + L_{j-1}h_{j-1} - V_jH_j - L_jh_j = 0$, may be restated in either of the following functional forms:

$$G_j = \sum_{i=1}^{c} (v_{j+1,i}H_{j+1,i} + l_{j-1,i}h_{j-1,i} - v_{ji}H_{ji} - l_{ji}h_{ji}) \qquad (1 \leq j \leq N)$$

$$(4\text{-}38)$$

or

$$G_j = \sum_{i=1}^{c} [v_{j+1,i}(H_{j+1,i} - h_{ji}) + l_{j-1,i}(h_{j-1,i} - h_{ji}) - v_{ji}(H_{ji} - h_{ji})] \qquad (4\text{-}39)$$
$$(1 \leq j \leq N)$$

The second expression, a form of the constant-composition method, may be obtained from the first by eliminating l_{ji} by use of the component-material balance $l_{ji} = v_{ji} - v_{j+1,i} - l_{j-1,i}$.

In the prediction of a new set of temperatures by applying the Newton-Raphson method to the enthalpy balances, Sujata (18) regarded the total flow rates and compositions (or the component-flow rates) as the dependent variables, and he proposed that their dependencies on the temperatures be neglected. This approximation amounts to neglecting the dependency of x on y in Part (b) of Illustrative Example 4-2 to give $dx/dy = 0$ with the result that $df_2/dy = \partial f_2/\partial y$.

When the variations of the total flow rates and compositions (or component-flow rates) with the temperatures are neglected in the application of the Newton-Raphson method to the enthalpy balances, the Jacobian J_n of the Newton-Raphson equations,

$$J_n(G/T)\,\Delta T_n = -G_n \qquad (4\text{-}40)$$

reduces to a tridiagonal matrix as demonstrated by Sujata (18), (see Problem

4-4). The subscript "n" in Equation (4-40) denotes the nth iteration through the column, the Jacobian matrix $\boldsymbol{J}_n$ is a square matrix of order N, and $\Delta \boldsymbol{T}_n$ and $\boldsymbol{G}_n$ are conformable column vectors, that is,

$$
\boldsymbol{J}_n(G/T) = \begin{bmatrix} \dfrac{\partial G_1}{\partial T_1} & \cdots & \dfrac{\partial G_1}{\partial T_N} \\ \cdots\cdots\cdots\cdots \\ \dfrac{\partial G_N}{\partial T_1} & \cdots & \dfrac{\partial G_N}{\partial T_N} \end{bmatrix} \tag{4-41}
$$

$$
\boldsymbol{G}_n = [G_1 G_2 \cdots G_N]^T; \qquad \Delta \boldsymbol{T}_n = [\Delta T_1 \Delta T_2 \cdots \Delta T_N]^T; \qquad \Delta T_j = T_{jn} - T_{j,n-1}
$$

For all problems that could be solved by the single-θ method of covergence, the tridiagonal approximation of $\boldsymbol{J}_n$ gave satisfactory results provided the maximum change in $|\Delta T_j|$ was limited to 20°F. A summary of the steps of the proposed calculational procedures follows.

1. Assume a set of temperatures $\{T_j\}$ and a set of vapor rates $\{(V_j)_a\}$. Compute the corresponding set of liquid rates $\{(L_j)_a\}$ by use of the total material balances given by Equation (4-24). Solve Equation (4-25) for the component-flow rates for each component i and denote them by $\{(v_{ji})_{ca}\}$. Then compute the corresponding flow rates $\{(l_{ji})_{ca}\}$ for each component i in the liquid phase by use of relationship: $(l_{ji})_{ca} = (A_{ji})_a(v_{ji})_{ca}$.

2. On the basis of the sets of calculated flow rates $\{(v_{ji})_{ca}\}$ and $\{(l_{ji})_{ca}\}$ found in Step 1, the set of assumed vapor rates $\{(V_j)_a\}$ and corresponding liquid rates $\{(L_j)_a\}$ used in Step 1, find the θ closest to unity that minimizes $g(\theta)$. Then compute the corresponding corrected values of the total flow rates.

3. Repeat Steps 1 and 2 until the convergence criterion for the material balances has been satisfied. [In practice, a total of five iterations were made.]

4. Use the final sets of corrected compositions and total flow rates found in Steps 1 through 3 and the temperature profile assumed in Step 1 to evaluate the G_j functions and the partial derivatives appearing in the tridiagonal matrix approximation of $\boldsymbol{J}_n$.

5. Repeat the above procedure until convergence criteria for the temperatures and total flow rates have been satisfied.

Absorber and stripper problems may be characterized by the difference in boiling points of the two feeds (V_{N+1} and L_0). In typical absorber and stripper problems such as Examples 4-3 through 4-7 and 4-13 of Tables 4-1 and 4-2, the boiling point temperatures of the two feeds differ widely, and these systems are classified as *wide boiling mixtures*. When the boiling points of the two feeds approach each other more closely as demonstrated by Examples 4-8 through 4-12 of Tables 4-1 and 4-2, the systems are classified as *narrow boiling mixtures*.

For typical absorber and stripper problems (Examples 4-3 through 4-7

TABLE 4-1
STATEMENT OF EXAMPLES 4-3 AND 4-4

I. *Statement of Example 4-3*

Component	$v_{N+1,i}$ (lb-moles/hr)	l_{0i} (lb-moles/hr)
CO_2	14.08154	0.0
N_2	5.45767	0.0
CH_4	2655.8245	0.0
C_2H_6	199.85249	0.0
C_3H_8	83.19560	0.04345
$i\text{-}C_4H_{10}$	19.08945	0.01889
$n\text{-}C_4H_{10}$	10.94352	0.03778
$i\text{-}C_5H_{12}$	3.46664	0.20024
$n\text{-}C_5H_{12}$	1.51297	0.18324
C_6H_{14}	0.43565	4.47708
C_7H_{16}	0.24159	17.17535
C_8H_{18}	0.05973	54.53534
C_9H_{20}	0.00086	50.49841
$C_{10}H_{22}$	0.00042	61.73645
	2994.16259	188.9063

Initial temperature profile: $T_j = 25°F$ $(1 \leq j \leq N)$. Initial vapor rates are linear between $V_1 = 2,721$ and V_{N+1}. $T_0 = 2.9°F$, $T_{N+1} = 0°F$, $N = 8$, Column Pressure = 800 psia, and perfect plates. The equilibrium data are given in Table C-3 of Appendix C, and the enthalpy data are given in Table C-4 of Appendix C and corrected as indicated by Equations (10-33) and (10-34) and by Table 10-9.

II. *Statement of Example 4-4*
Same as Example 4-3 except that $N = 20$.

TABLE 4-2
STATEMENT OF EXAMPLES 4-5 THROUGH 4-13

Component (or Variable)	4-5		4-6		4-7		4-8		4-9	
	$v_{N+1,i}$	l_{0i}	$v_{N+1,i}$	l_{0i}	$v_{N+1,i}$	l_{0i}	$v_{N+1,i}$	l_{0i}	$v_{N+1,i}$	l_{0i}
Single-phase light	0	0	0	0	70	0	0	0	0	0
CH_4	70	0	70	0	0	0	1	0	0.1	0
C_2H_6	15	0	15	0	15	0	4	0	0.5	0
C_3H_8	10	0	10	0	10	0	10	0	2.4	0
$n\text{-}C_4H_{10}$	4	0	4	0	4	0	15	0	7.5	0
$n\text{-}C_5H_{12}$	1	0	1	0	1	0	20	0	44.5	0
C_6H_{14}	0	0	0	0	0	0	0	0	0	0
C_7H_{16}	0	0	0	0	0	0	0	0	0	0
C_8H_{18}	0	20	0	0	0	20	0	0	0	65
500	0	0	0	0	0	0	0	70	0	0
Single-phase heavy	0	0	0	20	0	0	0	0	0	0

TABLE 4-2 (*Cont.*)

Component (or Variable)	Example Number				
	4-5	4-6	4-7	4-8	4-9
T_0, °F	90	90	90	350	415
T_{N+1}, °F	-7.79	-6.34	-5.0	370	425
N	8	8	8	8	8
Initial$\{T_j\}$ °F	Linear between $T_1 = 100$ and $T_N = 80$	Linear between $T_1 = 100$ and $T_N = 80$	Linear between $T_1 = 100$ and $T_N = 80$	$T_j = 350$ ($1 \leqq j \leqq N$)	$T_j = 400$ ($1 \leqq j \leqq N$)
Initial$\{V_j\}$	Linear between $V_1 = 80$ and V_{N+1}	Linear between $V_1 = 80$ and V_{N+1}	Linear between $V_1 = 80$ and V_{N+1}	Linear between $V_1 = 80$ and V_{N+1}	Linear between $V_1 = 80$ and V_{N+1}

Component (or Variable)	Example Number			
	4-10, 4-11, 4-12		4-13	
	$v_{N+1,i}$	l_{0i}	$v_{N+1,i}$	l_{0i}
Single-phase light	0	0	13.47	0
CH_4	0.1	0	0	0.01
C_2H_6	0.5	0	0	0.17
C_3H_8	2.4	0	0	1.30
$n\text{-}C_4H_{10}$	7.5	0	0	2.83
$n\text{-}C_5H_{12}$	54.5	0	0	1.75
C_6H_{14}	0	0	0	2.35
C_7H_{16}	0	0	0	2.55
C_8H_{18}	0	55	0	0
500	0	0	0	82.24
T_0, °F	395		370	
T_{N+1}, °F	405		500	
N	$N = 8$ for Ex. 4-10 $N = 12$ for Ex. 4-11 $N = 20$ for Ex. 4-12		8	
Initial$\{T_j\}$	$T_j = 400$ ($1 \leqq j \leqq N$)		Linear between $T_1 = 340$, $T_8 = 375$	
Initial$\{V_j\}$	Linear between $V_1 = 80$ and V_{N+1}		Linear between $V_1 = 80$ and V_{N+1}	

Other Specifications

The column pressure for Examples 4-5 through 4-12 was 300 psia and 50 psia for Example 4-13. The equilibrium and enthalpy data were taken from Reference (10). The enthalpy of the single-phase light of Example 4-7 was taken to be equal to that of methane. The enthalpy of the single-phase heavy of Example 4-6 was taken to be equal to that of normal octane.

The feed entrance plates 1 and N were assumed to behave according to Model 1 in Chapter 3, that is, each feed (L_0 and V_{N+1}) was assumed to mix perfectly with the liquid on the respective entrance plates.

and 4-13), the single-θ method converged very rapidly. For problems involving relatively narrow boiling mixtures (Examples 4-8 through 4-12), it was necessary to modify the procedure described above in order to prevent over-corrections. In particular, if the corrected values of the total flow rates failed to give an improvement at the end of Step 1, this step was repeated on the basis of the calculated values of the total flow rates. An improvement was said to have been achieved if $g(\theta)$ at $\theta = 1$ at the end of Step 1 was less than it was for the previous trial.

Treatment of single-phase components

When the system contains single-phase components, only minor modifications of the above equations are required. A *single-phase light* component is defined as one that appears in the gas phase alone, and a *single-phase heavy* component is defined as one which is miscible in and appears in the liquid phase alone. It will be supposed, of course, that the single-phase lights enter in the stream V_{N+1} and that the single-phase heavies enter in the stream L_0. Let v_L denote the total flow rate of single-phase light components and l_H the total flow rate of single-phase heavy components. The component-material balances for the single-phase components are given by

$$v_{jL} = v_L; \qquad l_{jL} = 0$$
$$v_{jH} = 0; \qquad l_{jH} = l_H \tag{4-42}$$

while the bubble point function given by Equation (4-35) takes the following form:

$$f_j = \sum_{\substack{i=1 \\ i \neq H, L}}^{c} (K_{ji})_a x_{ji} - (1 - y_{jL})$$

where

$$x_{ji} = \frac{l_{ji}}{L_j} = \frac{\eta_j \left(\dfrac{l_{ji}}{v_{1i}}\right)_{ca} v_{1i}}{\eta_j \mathcal{L}_j + l_H};$$

$$y_{jL} = \frac{v_L}{V_j} = \frac{v_L}{\sigma_j \mathcal{V}_j + v_L}.$$

The definitions of $\mathcal{L}_j$ and $\mathcal{V}_j$ follow from those listed below Equation (4-30) by excluding the single-phase lights and heavies in the summation. The multipliers σ_j and η_j are computed in a manner analogous to that demonstrated above.

In order to demonstrate some of the characteristics of the calculational procedures described herein, a variety of examples were solved. Statements of Examples 4-3 through 4-13 are presented in Tables 4-1 and 4-2, and the numerical results are presented in Tables 4-3 through 4-5. This array of examples covers a wide range of feed mixtures (from wide to narrow boiling mix-

TABLE 4-3

SOLUTION SET OF TEMPERATURES AND VAPOR RATES FOR EXAMPLES 4-3 THROUGH 4-13

Plate No.	Example 4-3		Example 4-5		Example 4-6		Example 4-7		Example 4-8	
	$T_j(°F)$	V_j (lb-moles/hr)	$T_j(°F)$	V_j (lb-moles/hr)	$T_j(°F)$	V_j (lb-moles/hr)	$T_j(°F)$	V_j (lb-moles/hr)	$T_j(°F)$	V_j (lb-moles/hr)
1	27.929	2659.25	107.636	85.000	107.815	85.000	106.204	87.510	364.228	8.282
2	31.043	2813.22	111.436	89.813	111.976	89.851	110.114	90.691	367.777	27.234
3	30.900	2824.80	112.494	90.506	113.520	90.554	111.077	91.409	370.966	30.665
4	29.597	2833.72	112.131	90.785	113.761	90.883	110.548	91.683	374.494	33.420
5	27.665	2843.29	110.224	90.999	112.524	91.044	108.438	91.891	378.611	36.263
6	25.101	2855.30	105.816	91.284	108.693	91.329	103.847	92.163	382.796	39.673
7	21.468	2872.81	96.433	91.812	99.448	91.869	94.451	92.651	384.708	43.981
8	15.320	2904.61	74.365	93.156	76.559	93.246	72.995	93.893	377.632	49.387

Plate No.	Example 4-9		Example 4-10		Example 4-11		Example 4-13	
	$T_j(°F)$	V_j (lb-moles/hr)	$T_j(°F)$	V_j (lb-moles/hr)	$T_j(°F)$	V_j (lb-moles/hr)	$T_j(°F)$	V_j (lb-moles/hr)
1	458.479	9.122	456.300	17.544	457.116	17.446	368.007	23.268
2	465.935	48.604	461.257	65.791	463.131	65.540	367.392	19.120
3	470.484	56.258	462.474	71.850	465.907	71.786	366.909	17.762
4	472.582	61.963	461.265	74.862	467.054	75.090	366.628	17.083
5	472.094	66.063	457.751	76.230	467.204	76.856	366.361	16.629
6	468.204	68.670	451.218	76.621	466.562	77.767	365.029	16.235
7	458.837	69.798	439.823	76.399	465.084	78.166	365.403	15.791
8	438.797	69.454	419.217	76.019	462.507	78.216	359.832	15.102
9					458.285	77.990		
10					451.395	77.508		
11					439.838	76.807		
12					419.184	76.168		

TABLE 4-3. (CONT'D.)

Plate No.	Example 4-4 $T_f(°F)$	Example 4-4 V_j (lb-moles/hr)	Example 4-12 $T_f(°F)$	Example 4-12 V_j (lb-moles/hr)
1	29.084	2653.72	457.312	17.432
2	33.322	2802.99	463.545	65.517
3	34.344	2809.94	466.646	71.802
4	34.297	2813.55	468.276	75.168
5	33.881	2816.42	469.156	77.023
6	33.332	2819.01	469.626	78.065
7	32.742	2821.46	469.858	78.655
8	32.145	2823.83	469.936	78.988
9	31.554	2826.17	469.896	79.171
10	30.960	2828.51	469.744	79.263
11	30.358	2830.93	469.458	79.293
12	29.730	2833.48	468.990	79.276
13	29.058	2836.27	468.258	79.213
14	28.315	2839.43	467.123	79.097
15	27.455	2843.16	465.368	78.910
16	26.417	2847.82	462.635	78.626
17	25.072	2854.02	458.327	78.207
18	23.190	2863.02	451.395	77.614
19	20.246	2877.73	439.822	76.852
20	14.751	2906.94	419.167	76.181

TABLE 4-4
COMPARISON* OF CALCULATIONAL PROCEDURES 1, 2, AND 3

Example No.	Procedure 1: Single-θ method		Procedure 2: Multi-θ method		Procedure 3: Sujata Method (18)	
	Number of Trials	Time (sec)	Number of Trials	Time (sec.)	Number of Trials	Time (sec)
4-3	7	15	6	24	10	44
4-4	20	67	13	258	24	200
4-5	7	9	4	12	8	19
4-6	7	9	4	13	5	17
4-7	8	11	5	16	5	16
4-8	9	15	5	20	20	32
4-9	17	24	5	20	Diverged	—
4-10	11	16	5	20	Diverged	—
4-11	14	27	5	39	Diverged	—
4-12	20	48	5	104	Diverged	—
4-13	3	4	3	11	5	12

*Convergence Criteria: $\left|\frac{\Delta T}{T}\right| < 0.0001$ $\left|\frac{\Delta V}{V}\right| < 0.0001$. The times stated are for an IBM 360/65 computer. The initial conditions stated in Table 4-2 were not within the solution domain of the material balance functions for Examples 4-9 through 4-12, and the initial conditions based on an adiabatic flash of the combined feed ($V_{N+L} + L_0$) and one plate were used to obtain the results enclosed in brackets. In particular, for the first trial, it was assumed that $T_j = T_{1,\text{adiabatic}}$ ($1 \leq j \leq N$) and that V_j varied linearly between V_1 and V_{N+1}.

tures) as well as a significant variation in the number of plates. Example 4-13 was included in order to demonstrate the characteristics of the proposed procedures in the solution of separation problems involving the use of steam strippers (see Figure 4-1).

From Table 4-4 it is seen that the single-θ method is not only exceedingly fast, but that it also converges for all absorber problems which appear to be of commercial interest (Examples 4-3 through 4-7). Although convergence of Examples 4-8 through 4-12 could be achieved by using the forcing procedures described above, the single-θ method is not recommended for relatively narrow boiling mixtures such as Examples 4-8 through 4-12. For all problems of this type as well as those having lean oils lighter than their respective rich gas streams, the multi-θ method and other methods (1, 2, 14, 19, 20) based on exact applications of the Newton-Raphson method are recommended.

TABLE 4-5
INDEPENDENT COMPARISON OF THE METHODS
OF SUJATA (18) AND TOMICH (20)*

Example No.	Sujata[1,2] (Used curve fits stated in Tables 4-1 and 4-2)		Tomich[1,2] (Used curve fits stated in Tables 4-1 and 4-2)		Tomich[1,3,4] [Used data of Ref. (7)]	
	Number of Trials	Time (sec)	Number of Trials	Time (sec)	Number of Trials	Time (sec)
4-3	14	1.094	—	—	32	2.770
4-5	10	0.638	23	0.605	21	1.798
4-6	9	0.570	23	0.615	21	1.294
4-7	10	0.602	23	0.626	20	1.140
4-8	37	2.216	—	—	44	3.878
4-9	Diverged	—	—	—	24	1.699
4-10	Diverged	—	—	—	24	1.701
4-11	Diverged	—	—	—	36	3.950
4-12	6	0.431	—	—	—	—

Notes: 1. These solutions were obtained on a UNIVAC 708 Computer, and the following convergence criteria were used:

$$|\Delta T| \leq 0.001°F \quad \text{and} \quad \left|\frac{\Delta V}{V}\right| \leq 0.0001$$

between successive trials for both methods.
2. One iteration on the material balances and a maximum of five iterations were made on the enthalpy balances per trial.
3. K values which were independent of composition were used to get in the neighborhood of the solution, and then the composition dependent K values given in Reference (7) were used.
4. The program based on the Tomich method did not contain provisions for handling single-phase components, and to approximate these $n\text{-}C_{12}H_{26}$ was used in Example 4-5 and H_2 was used in Example 4-6.

*These results were provided through the courtesy of the ChemShare Corp., 730 Asp Street, Norman, Oklahoma. The program based on the Tomich method was written by Dr. A. D. Epperly.

Procedure 2:   The multi-θ method of convergence (17)

Although an absorber is used in the development of this method, the resulting method is not restricted to absorbers. This method is based on Equation (4-23), and it represents an exact application of the Newton-Raphson method as demonstrated in Parts (a) and (b) of Illustrative Example 4-2. First, the method analogous to Part (b) is presented and then the method analogous to Part (a) of Illustrative Example 4-2 is presented.

METHOD I: (The temperatures $\{T_j\}$ are selected as the independent variables)

This version of the multi-θ method of convergence differs from other proposed applications of the Newton-Raphson method (1, 2, 14, 19, 20) either by the number or choice of independent variables selected or the method used to calculate the partial derivatives.

For the assumed set of temperatures $\{T_j\}$, Equation (4-23) reduces to an equation of the same form as Equation (4-18), namely,

$$\frac{L_j}{V_j} = \theta_j \left(\frac{L_j}{V_j}\right)_a \qquad (1 \leq j \leq N) \tag{4-43}$$

The θ_j's are to be picked so that the equilibrium relationships and material balances are all satisfied simultaneously. The set of L_j/V_j's assumed to initiate the calculational procedure for the determination of the θ_j's is denoted by $\{(L_j/V_j)_a\}$. The new or corrected L_j/V_j's to be assumed for the next trial are found by use of the set $\{(L_j/V_j)_a\}$ and the most recent set of θ_j's. [It should be noted that since $(L_j/V_j)_a$ may be arbitrarily selected, the relationship given by Equation (4-43) amounts to replacing L_j/V_j by θ_j times an arbitrary constant.] The dependent variables are hereafter referred to as the θ_j's rather than the L_j/V_j's.) The total flow rates corresponding to a given set of θ_j's are found by solving the total material balances enclosing each plate j ($j = 1, 2, \ldots, N-1, N$) for the V_j's or L_j's. When solved for the V_j's, these equations may be stated as follows:

$$RV = -\mathfrak{F} \tag{4-44}$$

where R is a tridiagonal matrix, V is a column vector containing the V_j's, and $\mathfrak{F}$ is a feed vector containing L_0 and V_{N+1}, that is,

$$R = \begin{bmatrix} -(1+r_1) & 1 & 0 & 0 & \cdots & 0 \\ r_1 & -(1+r_2) & 1 & 0 & \cdots & 0 \\ \cdots\cdots\cdots\cdots\cdots\cdots\cdots\cdots\cdots\cdots\cdots\cdots\cdots \\ 0 & \cdots\cdots\cdots\cdots\cdots & 0 & r_{N-1} & -(1+r_N) \end{bmatrix} \tag{4-45}$$

$$r_j = \frac{L_j}{V_j} = \theta_j \left(\frac{L_j}{V_j}\right)_a$$

$$V = [V_1 V_2 \ldots V_{N-1} V_N]^T$$

$$\mathfrak{F} = [L_0 0 \ldots 0 V_{N+1}]^T$$

Then for a given trial, the V_j's corresponding to the assumed set of θ_j's are readily obtained by solving Equation (4-44) by use of the recurrence formulas given below Equation (3-64).

On the basis of an assumed set of θ_j's, the component-material balances, Equation (4-25) may be solved for the v_{ji}'s and then the corresponding set of l_{ji}'s found by using Equation (4-23), that is,

$$l_{ji} = A_{ji} v_{ji} = \frac{\theta_j}{K_{ji}} \left(\frac{L_j}{V_j}\right)_a v_{ji} \tag{4-46}$$

Obviously, when the sum of the v_{ji}'s and the sum of the l_{ji}'s computed by use of Equations (4-25) and (4-46) are equal to the respective total flow rates V_j and L_j [given by Equations (4-43) and (4-44)], the correct set of θ_j's has been assumed for the given temperature profile. Thus, it is desired to pick a set of θ_j's such that $F_j = 0$ for all j, where

$$F_j = \frac{\sum_{i=1}^{c} l_{ji}}{L_j} - \frac{\sum_{i=1}^{c} v_{ji}}{V_j} \qquad (1 \leq j \leq N) \tag{4-47}$$

Equation (4-46) may be used to reduce Equation (4-47) to either the bubble point or dew point function. The dew point function is given by

$$F_j = \frac{1}{V_j}\left[\sum_{i=1}^{c}\left(\frac{1}{K_{ji}} - 1\right)v_{ji}\right] \qquad (1 \leq j \leq N) \tag{4-48}$$

[Note that Equation (4-47) could have been stated in terms of the bubble point function by stating the v_{ji}'s in terms of the l_{ji}'s.]

The θ_j's for the nth trial are found by solving the Newton-Raphson equation,

$$J_n(F/\theta)\, \Delta\theta_n = -F_n \tag{4-49}$$

for the elements of $\Delta\theta_n$, where J_n is the Jacobian matrix,

$$J_n(F/\theta) = \begin{bmatrix} \dfrac{\partial F_1}{\partial \theta_1} & \cdots & \dfrac{\partial F_1}{\partial \theta_N} \\ \cdots\cdots\cdots\cdots \\ \dfrac{\partial F_N}{\partial \theta_1} & \cdots & \dfrac{\partial F_N}{\partial \theta_N} \end{bmatrix}$$

$$F_n = F(\theta_n) = [F_1(\theta_{1,n}, \ldots, \theta_{N,n}) \ldots F_N(\theta_{1,n}, \ldots, \theta_{N,n})]^T;$$

$$\Delta\theta_n = [\Delta\theta_{1,n} \ldots \Delta\theta_{N,n}]^T; \qquad \Delta\theta_{j,n} = \theta_{j,n+1} - \theta_{j,n}$$

The partial derivatives of F_j with respect to any $\theta_k (k = 1, 2, \ldots, N)$ which appear in J_n may be evaluated by use of the following equations:

$$\frac{\partial F_j}{\partial \theta_k} = \frac{1}{V_j}\left[\sum_{i=1}^{c}\left(\frac{1}{K_{ji}} - 1\right)\frac{\partial v_{ji}}{\partial \theta_k} - F_j\frac{\partial V_j}{\partial \theta_k}\right] \tag{4-50}$$

$$A_i\frac{\partial v_i}{\partial \theta_k} = C_{ki} \tag{4-51}$$

$$R\frac{\partial V}{\partial \theta_k} = E_k \tag{4-52}$$

where,

$$C_{ki} = -\frac{\partial A_i}{\partial \theta_k}v_i;$$

$$E_k = -\frac{\partial R}{\partial \theta_k}V$$

These equations are readily obtained by partial differentiation of Equations (4-48), (4-25), and (4-44), respectively. The matrices A_i, v_i, R, and V are defined by Equations (4-26) and (4-45), respectively, and C_{ki} and E_k are given by:

$$C_{1i} = [C_{1i}(-C_{1i})0 \ldots 0]^T; \qquad C_{2i} = [0C_{2i} = [0C_{2i}(-C_{2i})0 \ldots 0]^T;$$

$$C_{N-1,i} = [0 \ldots 0C_{N-1,i}(-C_{N-1,i})]^T; \qquad C_{Ni} = [0 \ldots 0C_{Ni}]^T;$$

$$C_{ki} = v_{ki}\frac{\partial A_{ki}}{\partial \theta_k} = \frac{v_{ki}}{K_{ki}}\left(\frac{L_k}{V_k}\right)_a$$

$$E_1 = [E_1(-E_1)0 \ldots 0]^T; \qquad E_1 = [0E_2(-E_2)0 \ldots 0]^T;$$

$$E_{N-1} = [0 \ldots 0E_{N-1}(-E_{N-1})]^T; \qquad E_N = [0 \ldots 0E_N]^T;$$

$$E_k = \left(\frac{L_k}{V_k}\right)_a V_k$$

After Equations (4-25) and (4-44) have been solved for the v_{ji}'s and V_j's, respectively, the partial derivatives appearing in Equation (4-50) may be found by use of Equations (4-51) and (4-52). Then the elements of J_n of Equation (4-49) may be evaluated by use of Equations (4-50) through (4-52). This procedure is repeated until the convergence criterion has been satisfied. In particular, it was required that

$$g(\theta_1, \theta_2, \ldots, \theta_N) = \frac{1}{N}\sum_{j=1}^{N} F_j^2(\theta_1, \theta_2, \ldots, \theta_N)$$

be less than δ, where δ is some small preassigned number, say 10^{-10}. After the solution set of θ_j's has been found, the corresponding flow rates so obtained are used in the enthalpy balances in the prediction of an improved set of temperatures.

In order to apply the Newton-Raphson exactly to the enthalpy balance functions, the dependency of the θ_j's on each T_k ($k = 1, 2, \ldots, N$) must be taken into account in the calculation of the partial derivatives of the G_j's with respect to each T_k just as the dependency of x on y was taken into account in the calculation of df_2/dy in Part (b) of Illustrative Example 4-2. To denote the fact that the dependency of the complete set of θ_j's on each T_k is to be taken into account in the partial differentiation of each G_j with respect to T_k (with the remaining temperatures held fixed), the notation $DG_j/\partial T_k$ is used. This notation serves to distinguish this partial derivative ($DG_j/\partial T_k$) from the conventional partial derivative $\partial G_j/\partial T_k$ in which all variables except T_k are held fixed in the partial differentiation of the function G_j.

When the dependency of the θ_j's on each T_k ($k = 1, 2, \ldots, N$) is taken into account in the partial differentiation of the G_j's, Equation (4-40) takes the form:

$$\mathcal{J}_n(G/T)\,\Delta T_n = -G_n \qquad (4\text{-}53)$$

$$\boldsymbol{\mathcal{J}}_n(\boldsymbol{G}/\boldsymbol{T}) = \begin{bmatrix} \dfrac{DG_1}{\partial T_1} & \cdots & \dfrac{DG_1}{\partial T_N} \\[1em] \cdot & & \cdot \\ \cdot & & \cdot \\ \cdot & & \cdot \\[1em] \dfrac{DG_N}{\partial T_1} & \cdots & \dfrac{DG_N}{\partial T_1} \end{bmatrix}$$

and where G_j is given by Equation (4-39),

$$\frac{DG_j}{\partial T_k} = \sum_{i=1}^{c} \left[v_{j+1,i} \frac{D(H_{j+1,i} - h_{ji})}{\partial T_k} + l_{j-1,i} \frac{D(h_{j-1,i} - h_{ji})}{\partial T_k} \right.$$
$$\left. - v_{ji} \frac{D(H_{ji} - h_{ji})}{\partial T_k} \right] + \sum_{i=1}^{c} \left[(H_{j+1,i} - h_{ji}) \frac{Dv_{j+1,i}}{\partial T_k} \right.$$
$$\left. + (h_{j-1,i} - h_{ji}) \frac{Dl_{j-1,i}}{\partial T_k} - (H_{ji} - h_{ji}) \frac{Dv_{ji}}{\partial T_k} \right]$$

Since H_{ji} and h_{ji} are functions of T_j alone, it follows that $DH_{ji}/\partial T_k = 0$ and $Dh_{ji}/\partial T_k = 0$, for $j \neq k$. The set of partial derivatives $(Dv_{ji}/\partial T_k)$ is readily obtained by partial differentiation of the component-material balances [Equation (4-25)] with respect to T_k.

$$A_i \frac{Dv_i}{\partial T_k} = C_{ki} \tag{4-54}$$

In this case, however, the vector C_{ki} $(k = 1, 2, \ldots, N)$ is given by

$$C_{ki} = -\frac{DA_i}{\partial T_k} v_i = [C_{1i} C_{2i} \cdots C_{Ni}]^T \tag{4-55}$$

where

$$C_{1i} = v_{1i} \frac{DA_{1i}}{\partial T_k};$$

$$C_{ji} = -v_{j-1,i} \frac{DA_{j-1,i}}{\partial T_k} + v_{ji} \frac{DA_{ji}}{\partial T_k} \qquad (j = 2, 3, \ldots, N);$$

$$\frac{DA_{ji}}{\partial T_k} = \frac{\partial A_{ji}}{\partial \theta_j} \frac{\partial \theta_j}{\partial T_k} + \frac{\partial A_{ji}}{\partial T_k}$$

or

$$\frac{DA_{ji}}{\partial T_k} = \frac{1}{K_{ji}} \left(\frac{L_j}{V_j}\right)_a \frac{\partial \theta_j}{\partial T_k} - \frac{\theta_j}{K_{ji}^2} \left(\frac{L_j}{V_j}\right)_a \frac{\partial K_{ji}}{\partial T_k}.$$

where $\partial K_{ji}/\partial T_k = 0$, for $j \neq k$. [The partial derivatives $\{\partial \theta_j/\partial T_k\}$ needed to compute the above C_{ji}'s are evaluated by use of Equation (4-57) below.]

After the partial derivatives $\{Dv_{ji}/\partial T_k\}$ have been evaluated by use of Equations (4-54) and (4-55), the corresponding partial derivatives $\{Dl_{ji}/\partial T_k\}$ for the liquid phase that appear in Equation (4-53) are evaluated as follows:

$$\frac{Dl_{ji}}{\partial T_k} = A_{ji} \frac{Dv_{ji}}{\partial T_k} + v_{ji} \frac{DA_{ji}}{\partial T_k} \tag{4-56}$$

where $DA_{ji}/\partial T_k$ is computed by use of the expression given beneath Equation (4-55).

The set of partial derivatives $\{\partial\theta_k/\partial T_k\}$ may be evaluated by use of the following equations which are developed on the basis of the fact that the θ_j's were selected as the dependent variables and the T_j's as the independent variables. Now suppose a particular temperature, say T_k, is changed while the remaining ones are held fixed. For each set of T_j's, the set of θ_j's is that particular solution set which makes $F_j = 0$ for all j [see Equation (4-48)]. Since for each set of T_j's each $F_j = 0$, it follows that for all j

$$\frac{DF_j}{\partial T_k} = 0 = \frac{\partial F_j}{\partial \theta_1}\frac{\partial \theta_1}{\partial T_k} + \frac{\partial F_j}{\partial \theta_2}\frac{\partial \theta_2}{\partial T_k} + \cdots + \frac{\partial F_j}{\partial \theta_N}\frac{\partial \theta_N}{\partial T_k} + \frac{\partial F_j}{\partial T_k}$$

The complete set of equations obtained by stating the above expression for $j = 1, 2, \ldots, N - 1, N$ may be restated in the following form:

$$J_n(F/\theta)\frac{\partial \theta}{\partial T_k} = -\frac{\partial F}{\partial T_k} \tag{4-57}$$

where J_n is the Jacobian matrix defined below Equation (4-49) and

$$\frac{\partial F}{\partial T_k} = \left[\frac{\partial F_1}{\partial T_k}\frac{\partial F_2}{\partial T_k}\cdots\frac{\partial F_N}{\partial T_k}\right]^T; \qquad \frac{\partial \theta}{\partial T_k} = \left[\frac{\partial \theta_1}{\partial T_k}\frac{\partial \theta_2}{\partial T_k}\cdots\frac{\partial \theta_N}{\partial T_k}\right]^T$$

The formula for $\partial F_j/\partial T_k$ is obtained by partial differentiation of Equation (4-48) with respect to T_k with the θ_j's held fixed to give

$$\frac{\partial F_j}{\partial T_k} = \frac{1}{V_j}\left[\sum_{i=1}^{c}\left(\frac{1}{K_{ji}} - 1\right)\frac{\partial v_{ji}}{\partial T_k} - \frac{v_{ji}}{K_{ji}^2}\frac{\partial K_{ji}}{\partial T_k}\right]$$

where $\partial K_{ji}/\partial T_k = 0, j \neq k$. Since the partial derivative $\partial F_j/\partial T_k$ is to be taken with the θ_j's held fixed, it follows that the partial derivatives $\partial v_{ji}/\partial T_k$ are also to be taken at this condition. Thus, the $\partial v_{ji}/\partial T_k$'s may be evaluated by use of the formulas given below Equation (4-52) for C_{ki} with the elements C_{ki} having the form:

$$C_{ki} = v_{ki}\frac{\partial A_{ki}}{\partial T_k} = -\left(\frac{v_{ki}\theta_k}{K_{ki}^2}\right)\left(\frac{L_k}{V_k}\right)_a\left(\frac{\partial K_{ki}}{\partial T_k}\right)$$

The equations presented above are also applicable for systems containing single-phase lights and heavies, provided that the flow rates of the single-phase lights, v_L, and the single-phase heavies, l_H, are contained in the definitions of the total flow rates V_j and L_j, respectively. For such systems, it is readily shown that the expressions for F_j and $\partial F_j/\partial \theta_k$ given by Equations (4-48) and (4-50) take the forms given by Equations (A) and (B), respectively, of Problem 4-5.

Obviously, Broyden's method (see Appendix A) may be applied to either or both the material-balance functions and the enthalpy balance functions. In this application, the initial values of the elements of the Jacobian matrices may be evaluated by use of the above expressions for the partial derivatives. When Broyden's method is applied, it must be remembered that for each set of assumed temperatures, the solution to the material balances must be found before a new set of temperatures is predicted by use of the enthalpy balances.

The results obtained for Method I of the multi-θ method of convergence (Procedure 2) are presented in Table 4-4. This method was very stable and it converged for all problems considered. In all applications of the multi-θ method, negative values of θ, if predicted, were discarded and replaced by 1/2 of their last positive value.

For problems in which the lean oil was lighter than the rich gas such as the test problem called Example 3 by Boyum (4), the single-θ method and Sujata's method (18) failed to converge; whereas, the multi-θ method converged. This problem was solved for a variety of starting temperature profiles, and it was necessary in some cases to restrict the change in temperatures predicted by use of Method I of the multi-θ method. In particular, if the set of predicted temperatures failed to give a solution to the material balances and equilibrium relationships [the convergence criterion on $g(\boldsymbol{\theta})$ stated below Equation (4-52) was not satisfied], a new set of predicted temperatures was computed on the basis of $\frac{1}{2}$ of the last predicted change. Although the program allowed this procedure to be repeated until a set of temperatures had been found which satisfied the convergence criterion, it was seldom if ever applied more than once.

The possibility of computing a new set of T_j's prior to obtaining a solution set of θ_j's led to the following question. Does there exist some method for correcting matrix Equation (4-53) for the fact that the last set of θ_j's used in the material balances and equilibrium relationships was not the solution set. The answer is "yes," and the resulting correction is equivalent to the formulas obtained by taking both the θ_j's and the T_j's as independent variables.

METHOD II. (*The temperatures $\{T_j\}$ and the $\{\theta_j\}$ are taken as the independent variable*)

In this method, the calculational procedure described above for Method I is used as stated with the exception that the material balances and equilibrium relationships are not necessarily converged for the assumed set of temperatures. Then the new set of temperatures is found by use of Equation (4-61), which differs from Equation (4-53) only by the correction term on the right-hand side. (Note that this correction term approaches $\mathbf{0}$ as $\boldsymbol{F}$ approaches $\mathbf{0}$.)

The development of Equation (4-61) is initiated by a reconsideration of the algebraic equations in Parts (a) and (b) of Illustrative Example 4-2 because the matrix results can be anticipated from the algebraic results. When the Newton-Raphson equations of Part (a) of this example are solved for Δy, the following result is obtained.

$$\Delta y = \left[-\frac{\partial f_2}{\partial x}\left(\frac{\partial f_1}{\partial x}\right)^{-1}\frac{\partial f_1}{\partial y} + \frac{\partial f_2}{\partial y} \right]^{-1}\left[-f_2 + \frac{\partial f_2}{\partial x}\left(\frac{\partial f_1}{\partial x}\right)^{-1}f_1 \right]$$

The set of equations used in Part (b) to compute Δy may be stated as a single equation as follows. When the expression,

$$0 = \frac{\partial f_1}{\partial x}\frac{dx}{dy} + \frac{\partial f_1}{\partial y}$$

is used to eliminate dx/dy from the expression for the total derivative df_2/dy and the result so obtained is substituted in the Newton-Raphson equation, the following expression is obtained upon solving for Δy, namely,

$$\Delta y = -\left[\frac{\partial f_2}{\partial x}\left(\frac{\partial f_1}{\partial x}\right)^{-1}\left(\frac{\partial f_1}{\partial y}\right) + \frac{\partial f_2}{\partial y}\right]^{-1}(-f_2)$$

By comparison of this equivalent expression for Δy as computed in Part (b) with the one stated above for Δy as computed in Part (a), it is seen that they differ in form only by term

$$\frac{\partial f_2}{\partial x}\left(\frac{\partial f_1}{\partial x}\right)^{-1}f_1$$

which compensates for the fact that the choice of x utilized does not necessarily make $f_1 = 0$.

There follows the development of the formula for the appropriate correction to Equation (4-53) for the case where $F \neq 0$ for the last set of θ_j's at the assumed set of T_j's. If, in a manner analogous to Part (a) of Illustrative Example 4-2, both the θ_j's and T_j's are regarded as independent variables, the Newton-Raphson equations for the F_j's and G_j's may be stated in the following matrix form:

$$0 = F + J(F/\theta)\,\Delta\theta + J(F/T)\,\Delta T$$
$$0 = G + J(G/\theta)\,\Delta\theta + J(G/T)\,\Delta T$$

When the first expression is solved for $\Delta\theta$, one obtains

$$\Delta\theta = -J^{-1}(F/\theta)[F + J(F/T)\,\Delta T] \tag{4-58}$$

After this expression for $\Delta\theta$ has been substituted into the second matrix equation, it is readily solved for ΔT to give

$$\Delta T = [-J(G/\theta)J^{-1}(F/\theta)J(F/T) + J(G/T)]^{-1}[-G + J(G/\theta)J^{-1}(F/\theta)F] \tag{4-59}$$

which is seen to be analogous in form to the first expression given above for Δy. Similarly, when F is set equal to 0, the following result is obtained.

$$\Delta T = [-J(G/\theta)J^{-1}(F/\theta)J(F/T) + J(G/T)]^{-1}(-G)$$

which is analogous to the second of the above expressions for Δy. A comparison of this expression with the following form of Equation (4-53)

$$\Delta T = \mathcal{J}^{-1}(G/T)[-G]$$

suggests that

$$\mathcal{J}^{-1}(G/T) = [-J(G/\theta)J^{-1}(F/\theta)J(F/T) + J(G/T)]^{-1} \tag{4-60}$$

An outline of the proof of this equality is given in Problem 4-9. Thus, it follows that if the assumed $\boldsymbol{\theta}$ vector does not make $\boldsymbol{F} = \boldsymbol{0}$, then the correction in the second set of brackets must be included in Equation (4-53); that is, Equation (4-53) becomes

$$\mathcal{J}_n(G/T)\,\Delta T = [-G + J_n(G/\boldsymbol{\theta})J_n^{-1}(F/\boldsymbol{\theta})F] \qquad (4\text{-}61)$$

The correction term in Equation (4-61) adds little difficulty to the calculational procedure outlined above since $J_n^{-1}(F/\boldsymbol{\theta})$ and $J_n(G/\boldsymbol{\theta})$ will have been evaluated in previous steps of the procedure. Whenever, Equation (4-61) is used to compute ΔT, the corresponding $\Delta\boldsymbol{\theta}$ for that trial is computed by use of Equations (4-58).

In the solution of the material balances (the component-material balances, the total-material balances, and the equilibrium relationships) at a fixed set of temperatures, the unknown θ_j's appear explicitly. However, in the solution of the enthalpy balances by the Newton-Raphson method, derivatives with respect to temperature are involved, and the temperatures appear implicitly in the K values and enthalpies. As these physical properties are generally expressed as polynomials in temperature, their evaluation is relatively time consuming. In addition, about twice as many equations are involved in the solution of the enthalpy balances as are involved in the solution of the material balances by the Newton-Raphson method. As an average of about five trials is required to converge the material balances for a given set of temperatures there is little difference in the amount of time required to solve problems by each of the two methods when say three to five trials are made on the material balances per trial on the enthalpy balances by Method II. Thus, only the results for Method I are presented in Table 4-4. When only one trial is made on the material balances per trial on the enthalpy balances by Method II, more total number of trials are generally required. When Example 4-12 was solved by Method II (with one material balance trial per enthalpy balance trial), a total of 8 complete trials were required; whereas, only 5 complete trials were required by Method I.

Since it is not necessary in Method II for a solution to the material balances and equilibrium relationships to exist for each set of assumed temperatures, poorer starting vectors $(\boldsymbol{\theta},\ \boldsymbol{T})$ may be used in Method II than in Method I. However, some peripheral method for adjusting the predicted changes $\Delta\theta_j$ and ΔT_j such that the Euclidean norms of the F_j's and G_j's are reduced from one trial to the next are generally required. For example, the cubic approximation [Equation (A-38)] proposed by Broyden (5) or some variation of it may be used.

Procedure 3:   The Sujata method (18)

This procedure, which is sometimes called the *sum rates method*, was first proposed by Sujata (18). For each set of assumed temperatures, the component-material balances [Equation (4-25)] are solved by the method of direct iteration to give $\{(v_{ji})_{ca}\}$. This step amounts to setting $\theta = 1$ in the single-θ method. The total flow rates are then equal to the sum of the respective component-flow rates for each plate j and the corresponding mole fractions are computed as follows:

$$(y_{ji})_{ca} = \frac{(v_{ji})_{ca}}{\sum\limits_{i=1}^{c} (v_{ji})_{ca}} \quad \text{and} \quad (x_{ji})_{ca} = \frac{(l_{ji})_{ca}}{\sum\limits_{i=1}^{c} (l_{ji})_{ca}}$$

The total flow rates and compositions so obtained are used in the enthalpy balances in the determination of a new set of temperatures by use of the tri-diagonal approximation of the Jacobian matrix J_n of Equation (4-41) (see Problem 4-4). The new set of temperatures is obtained by one application of the Newton-Raphson method. A summary of the steps of the calculational procedure follows:

1. Same as given for Procedure 1.
2. On the basis of the results of Step 1, compute the corresponding sets of total flow rates and compositions by the method of direct iteration.
3. On the basis of the results of Step 2, determine the temperatures for the next trial by one application of the Newton-Raphson method [Equation (4-41)] where the tridiagonal approximation of J_n is used.
4. If the T_j's found in Step 3 differ by no more than a preassigned amount from the values assumed in Step 1, convergence has been achieved; otherwise, Step 1 through 3 are repeated on the basis of the most recent sets of values for the L_j's, V_j's, and T_j's.

For problems involving the separation of wide boiling mixtures, Procedure 3 is both reliable and fast as shown in Table 4-3. This method, like the single-θ method, is not recommended for solving problems involving relatively narrow boiling mixtures. Also, as the number of plates in a given problem is increased, the tendency of this procedure to diverge is also increased as demonstrated by Friday *et al.* (8).

Procedure 4 : The Tomich method (20)

The following procedure suggested by Tomich (20) consists of an exact application of the Newton-Raphson method which is analogous to the procedure described in Part (a) of Illustrative Example 4-2. In this application, both the total flow rates $\{V_j\}$ and the temperatures $\{T_j\}$ are regarded as independent variables rather than the temperatures.

The first step of Procedure 4 is similar to that of the first two procedures presented in that a set of temperatures $\{T_j\}$ and a set of total flow rates of the vapor $\{V_j\}$ are assumed. The corresponding set of total flow rates of the liquid is found by use of the assumed V_j's and the total material balances [Equation (4-25)]. Instead of solving the component-material balances for either the component-vapor rates $\{v_{ji}\}$ or the component liquid rates $\{l_{ji}\}$, however, an equivalent set of variables was solved for, namely, the mole fractions of each component in the liquid phase. In this case, the equilibrium relationship in the form $y_{ji} = K_{ji}x_{ji}$ is used to eliminate each unknown y_{ji} from the component-material balances. The complete set of equations so obtained for any component i has the following tridiagonal form rather than that given by Equation (4-26).

$$
\begin{bmatrix}
-\rho_{1i} & V_2 K_{2i} & 0 & 0 & \cdots & 0 \\
L_1 & -\rho_{2i} & V_3 V_{3i} & 0 & \cdots & 0 \\
\multicolumn{6}{c}{\cdots\cdots\cdots\cdots\cdots\cdots\cdots} \\
0 & \cdots\cdots\cdots\cdots & 0 & L_{N-1} & & -\rho_{Ni}
\end{bmatrix}
\begin{bmatrix}
x_{1i} \\
x_{2i} \\
\cdots \\
x_{Ni}
\end{bmatrix}
= -
\begin{bmatrix}
P_{1i} \\
P_{2i} \\
\cdots \\
P_{Ni}
\end{bmatrix}
\tag{4-62}
$$

where

$$P_{1i} = L_0 x_{0i};$$
$$P_{ji} = 0 \qquad (2 \leq j \leq N - 1);$$
$$P_{Ni} = V_{N+1} y_{N+1,i};$$
$$\rho_{ji} = K_{ji} V_j - L_j.$$

After Equation (4-62) has been solved for the $\{x_{ji}\}$ for each component i, the corresponding y_{ji}'s are computed by use of the equilibrium relationship: $y_{ji} = K_{ji} x_{ji}$. The mole fractions so obtained are used in the enthalpy balances [see the $G_j (1 \leq j \leq N)$ functions given by Equation (4-39)]. The G_j functions are regarded as functions of the N unknown V_j's and the N unknown T_j's. Thus, in this case N additional independent functions must be introduced in order to obtain $2N$ equations in $2N$ unknowns. Tomich (20) elected to take these functions to be the difference between the sum of the x_{ji} and the sum of the y_{ji}'s for each plate j, that is,

$$F_j = \sum_{i=1}^{c} x_{ji} - \sum_{i=1}^{c} y_{ji} \qquad (1 \leq j \leq N)$$

The use of the Newton-Raphson method to solve the $2N$ functional expressions $(G_1, G_2, \ldots, G_N; F_1, F_2, \ldots, F_N)$ in the $2N$ unknowns $(T_1, T_2, \ldots, T_N; V_1, V_2, \ldots, V_N)$ may be represented by the following matrix equation:

$$
\begin{bmatrix}
\dfrac{\partial G_1}{\partial T_1} & \cdots & \dfrac{\partial G_1}{\partial T_N} & \dfrac{\partial G_1}{\partial V_1} & \cdots & \dfrac{\partial G_1}{\partial V_N} \\
\cdot & & \cdot & \cdot & & \cdot \\
\cdot & & \cdot & \cdot & & \cdot \\
\cdot & & \cdot & \cdot & & \cdot \\
\dfrac{\partial G_N}{\partial T_1} & \cdots & \dfrac{\partial G_N}{\partial T_N} & \dfrac{\partial G_N}{\partial V_1} & \cdots & \dfrac{\partial G_N}{\partial V_N} \\
\dfrac{\partial F_1}{\partial T_1} & \cdots & \dfrac{\partial F_N}{\partial T_N} & \dfrac{\partial F_1}{\partial V_1} & \cdots & \dfrac{\partial F_N}{\partial V_N} \\
\cdot & & \cdot & \cdot & & \cdot \\
\cdot & & \cdot & \cdot & & \cdot \\
\dfrac{\partial F_N}{\partial T_1} & \cdots & \dfrac{\partial F_N}{\partial T_N} & \dfrac{\partial F_N}{\partial V_N} & \cdots & \dfrac{\partial F_N}{\partial V_N}
\end{bmatrix}
\begin{bmatrix}
\Delta T_1 \\
\cdot \\
\cdot \\
\cdot \\
\Delta T_N \\
\Delta V_1 \\
\cdot \\
\cdot \\
\Delta V_N
\end{bmatrix}
= -
\begin{bmatrix}
G_1 \\
\cdot \\
\cdot \\
\cdot \\
G_N \\
F_1 \\
\cdot \\
\cdot \\
F_N
\end{bmatrix}
\tag{4-63}
$$

In the numerical application of this equation, Tomich (20) normalized the G_j's so that they ranged from zero to one. The functions G_j and F_j, as well as their partial derivatives appearing in Equation (4-63), were evaluated at the sets of variables, (T_{jn}) and (V_{jn}), assumed to make the nth trial. After the functions and their derivatives have been evaluated, Equation (4-63) may be solved by matrix inversion for the ΔT_{jn}'s and the ΔV_{jn}'s. Then the new values of the variables $\{T_{j,n+1}\}$ and $\{V_{j,n+1}\}$ for the $n+1$st trial are computed from the following equations:

$$T_{j,n+1} = T_{jn} + \mu_n \Delta T_{jn}$$
$$V_{j,n+1} = V_{jn} + \mu_n \Delta V_{jn}$$

(4-64)

The Broyden procedure (1) (described in Appendix A) searches for the value of μ_n which will meet the following condition:

$$\sum_{j=1}^{N} (G_{j,n+1}^2 + F_{j,n+1}^2) < \sum_{j=1}^{N} (G_{jn}^2 + F_{jn}^2)$$

(4-65)

Convergence is said to have been achieved when

$$\sum_{j=1}^{N} (G_{j,n+1}^2 + F_{j,n+1}^2) < \text{convergence tolerance}$$

(4-66)

If this condition is not satisfied, the entire procedure is repeated on the basis of the new set of T_j's and V_j's computed by use of Equation (4-64). As seen from the description in Appendix A, the Broyden procedure updates the inverse of the matrix of partial derivatives in Equation (4-63). Thus, there is no need to invert this matrix more than once per problem. To obtain the simultaneous solution of all of the equations, however new values of the L_j's x_{ji}'s, and y_{ji}'s must be calculated each time any one of the V_j's or T_j's is allowed to vary. A summary of the proposed calculational procedure follows:

1. On the basis of assumed values for V_j and T_j, solve Equation (4-62) for the x_{ji}'s. [Observe that the L_j's are again found by use of the assumed V_j's and the total material balances: Equation (4-24).] Use the x_{ji}'s so obtained to compute the corresponding y_{ji}'s by use of the equilibrium relationship: $y_{ji} = K_{ji} x_{ji}$.
2. On the basis of the results of Step 1, evaluate the G_j's and s_j's.
3. Apply Broyden's method (see Appendix A) for the determination of a new set of T_j's and V_j's which satisfy Equation (4-65).
4. If Equation (4-66) or some other suitable criterion is satisfied, convergence is said to have been achieved; otherwise, repeat Steps 1 through 3 until the convergence criterion has been satisfied.

The results obtained by use of the Tomich method (Procedure 4) are presented in Table 4-5. As noted there, a different type of computer was used to obtain these results than was used to obtain those presented in Table 4-4 for the first three procedures. However, since problems were solved by Procedure 3 by both computers, comparisons of the speeds of the various procedures may be made. The Tomich method like the single-θ method and multi-θ method converged for problems involving narrow boiling mixtures for which the Sujata method (Procedure 3) failed.

It should be pointed out that although the computer time and the total number of trials required to solve selected problems are presented in Tables 4-4 and 4-5, these variables are not necessarily valid criteria for choosing one method over another for the following reasons. First, the computer time required depends strongly on the efficiency with which the given method was programmed. Second, the total number of trials may be deceptive because one method, such as Method I of the multi-θ method, may contain an internal iterative loop, whereas, other methods such as Tomich's (20) and Method II of the multi-θ method do not. In the case of Tomich's method, however, the time consuming step of the first trial of the numerical evaluation of all of the partial derivatives appearing in the Jacobian matrix is not reflected by the total number of trials.

Calculational procedures for solving various types of distillation problems

Procedures 2 and 4 (the multi-θ method and the Tomich method, respectively) may be used to solve other types of distillation problems such as those involving reboiled absorbers and various types of distillation columns. However, Procedures 1 and 3 (the single-θ method and Subjata's method) are not recommended for general use because convergence cannot be assured. Procedures 1 and 3 are useful primarily for solving absorber and stripper problems involving the separation of wide boiling mixtures.

Procedures 2 and 4 stated above for absorbers and strippers are readily applied to other types of distillation columns. To illustrate the application of Method I of the multi-θ method to other types of columns, the matrix equations for a reboiled absorber (see Figure 4-2) are presented in Table 4-6. These methods may also be applied to conventional and complex distillation columns. Complex distillation columns are defined as those having any number of feeds and/or any number of sidestreams withdrawn in addition to the distillate and bottoms.

Also, the θ method presented in Chapter 3 for conventional distillation columns may be applied to complex columns as well as to systems of such columns. In spite of the fact that for problems of this type, the Newton-Raphson methods such as Procedures 2 and 4 have convergence characteristics which are superior to the θ method of Chapter 3, the latter is almost always faster, and it converges for virtually all such problems. The convergence characteristics of the θ method of Chapter 3 may be attributed in part

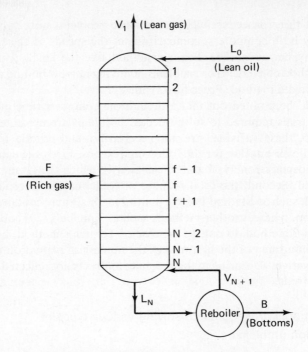

Figure 4-2. Sketch of a typical reboiled absorber.

TABLE 4-6
MATRIX EQUATIONS FOR A REBOILED ABSORBER

SPECIFICATIONS: P, F, $\{X_i\}$, thermal condition of F, L_0, $\{X_{0i}\}$, T_0, f, N, and Q_R

(The specifications are analogous to those stated in the text for the absorber except for the feeds and the addition of a reboiler. Since the reboiled absorber contains $N + 1$ stages rather than N, the matrices A_i, J_n, and R are of the same form as those stated in the text except that they are all of order $N + 1$ rather than N.)

1. $A_i v_i = -\mathscr{f}_i$

$v_i = [v_{1i} v_{2i} \cdots v_{N+1,i}]^T$

$\mathscr{f}_i = [l_{0i} 0 \cdots 0 v_{Fi} l_{Fi} 0 \cdots 0]^T$

$A_{ji} = \dfrac{\theta_j}{K_{ji}} \left(\dfrac{L_j}{V_j} \right)_a \quad (j = 1, 2, \cdots, N + 1)$

2. $RV = -\mathscr{F}$

$V = [V_1 V_2 \cdots V_{N+1}]^T$

$\mathscr{F} = [L_0 0 \cdots 0 V_F L_F 0 \cdots 0]^T$

$r_j = \dfrac{L_j}{V_j} = \theta_j \left(\dfrac{L_j}{V_j} \right)_a \quad (j = 1, 2, \cdots, N + 1)$

3. $J_n(F/\theta)\, \Delta\theta_n = -F_n$

[Same form as Equation (4-49)]

4. $A_i \dfrac{\partial v_i}{\partial \theta_k} = C_{ki}$

[Same form as Equation (4-51)]

5. $R \dfrac{\partial V}{\partial \theta_k} = E_k$

[Same form as Equation (4-52)]

6. $\mathscr{g}_n(G/T)\, \Delta T_n = -G_n$

[Same form as Equation (4-53); the G_j's are of the same form as given by Equation (4-39) except for $j = f - 1, f$, and $N + 1$.]

7. $J_n(F/\theta) \dfrac{\partial\theta}{\partial T_k} = -\dfrac{\partial F}{\partial T_k}$

[Same form as Equation (4-57)]

to the fact that for conventional and complex columns as well as systems of these columns, there exist total reflux problems for which the θ method constitutes an exact solution (11).

A brief summary of other methods which have been proposed follows. Greenstadt *et al.* (9) proposed a procedure for solving distillation problems in which the Newton-Raphson equations for each stage were applied in succession. Other applications of the Newton-Raphson method which have been proposed by Boynton (2), Naphtali *et al.* (14), Newman (16), and Tierney *et al.* (19). Of these proposed applications, the one proposed by Boynton (2) is the closest related to Method I of the multi-θ method, Boynton took the liquid rates $\{L_j\}$ to be the independent variables and for each choice of these variables, the temperatures required to satisfy the component-material balances and equilibrium relationships were found by use of the Newton-Raphson method. The results so obtained were then used in the enthalpy balances to compute a new set of liquid rates by use of one application of the Newton-Raphson method. Tierney's method (19) is very similar to Procedure 4 in that the same choice of independent variables (the vapor rates and temperatures) is used in each of these methods.

PROBLEMS

4-1. Verify the result given by Equation (4-11).

4-2. Show that

$$\lim_{A_i \to 0} \frac{A_i^{N+1} - A_i}{A_i^{N+1} - 1} \cong A_i \qquad \text{(for } N + 1 > 1\text{)}$$

$$\lim_{A_i \to \infty} \frac{A_i^{N+1} - A_i}{A_i^{N+1} - 1} = 1 \qquad \text{(for } N + 1 > 1\text{)}$$

$$\lim_{N \to \infty} \frac{A_i^{N+1} - A_i}{A_i^{N+1} - 1} = 1 \qquad \text{(for } A_i > 1\text{)}$$

$$\lim_{N \to \infty} \frac{A_i^{N+1} - A_i}{A_i^{N+1} - 1} = A_i \qquad \text{(for } A_i < 1\text{)}$$

$$\lim_{A_i \to 1} \frac{A_i^{N+1} - A_i}{A_i^{N+1} - 1} = \frac{N}{N + 1} \qquad (N + 1 > 1)$$

4-3. Plot the curves (σ_{j+1} versus η_j) given by Equations (4-32) and (4-33). Locate the solution set $\{\sigma_{j+1}, \eta_j\}$ for any trial relative to the solution set $\{\sigma_{j+1} = 1, \eta_j = 1\}$ at convergence of the problem.

4-4. If the dependencies of the component flow rates on the temperatures are neglected in the partial differentiation of the G_j functions given by Equation (4-38), then show that the Jacobian matrix given by Equation (4-41) reduces to the following tridiagonal matrix:

$$
J_n(G/T) = \begin{bmatrix}
\dfrac{\partial G_1}{\partial T_1} & \dfrac{\partial G_1}{\partial T_2} & 0 & 0 & \cdots & 0 \\[2mm]
\dfrac{\partial G_2}{\partial T_1} & \dfrac{\partial G_2}{\partial T_2} & \dfrac{\partial G_2}{\partial T_3} & 0 & \cdots & 0 \\[2mm]
\cdots\cdots\cdots\cdots\cdots\cdots\cdots\cdots \\[1mm]
0\cdots\cdots\cdots 0 & \dfrac{\partial G_N}{\partial T_{N-1}} & \dfrac{\partial G_N}{\partial T_N}
\end{bmatrix}
$$

where

$$
\frac{\partial G_j}{\partial T_{j-1}} = \sum_{i=1}^{c} l_{j-1,i}\frac{\partial h_{j-1,i}}{\partial T_{j-1}} = L_{j-1}\sum_{i=1}^{c} x_{j-1,i}\frac{\partial h_{j-1,i}}{\partial T_{j-1}};
$$

$$
\frac{\partial G_j}{\partial T_j} = -\sum_{i=1}^{c} v_{ji}\frac{\partial H_{ji}}{\partial T_j} - \sum_{i=1}^{c} l_{ji}\frac{\partial h_{ji}}{\partial T_j};
$$

$$
\frac{\partial G_j}{\partial T_{j+1}} = \sum_{i=1}^{c} v_{j+1,i}\frac{\partial H_{j+1,i}}{\partial T_{j+1}}.
$$

4-5. For the case in which a system contains both single-phase lights and single-phase heavy components, show that the appropriate forms of the expressions given by Equations (4-48) and (4-50) are given by the following expressions:

$$
F_j = \frac{1}{V_j}\left[\sum_{\substack{i=1 \\ i\neq H,L}}^{c}\left(\frac{1}{K_{ji}}-1\right)v_{ji} + \frac{l_H}{\left(\dfrac{L_j}{V_j}\right)_a\theta_j} - v_L\right] \tag{A}
$$

$$
\frac{\partial F_j}{\partial\theta_k} = \frac{1}{V_j}\left\{\left[\sum_{\substack{i=1 \\ i\neq H,L}}^{c}\left(\frac{1}{K_{ji}}-1\right)\frac{\partial v_{ji}}{\partial\theta_k}\right] - \left[\frac{l_H}{\left(\dfrac{L_j}{V_j}\right)_a\theta_j^2}\right]\frac{\partial\theta_j}{\partial\theta_k} - F_j\frac{\partial V_j}{\partial\theta_k}\right\} \tag{B}
$$

where

$$
\frac{\partial\theta_j}{\partial\theta_k} = 0, \text{ for } j \neq k.
$$

4-6. Verify the formulas given by Equations (4-51), (4-52), and (4-55).

4-7. Construct the graphs of the functions given by Equations (4-35) and (4-48) for a column with one plate and for $\theta_1 \geqq 0$.

4-8(a). Verify the result given by Equation (4-58).

4-8(b). Show that for the case of a single plate, Equation (4-57) reduces to the same result which may be obtained by use of Equation (4-16) and (4-21) where it is understood that for each choice of T_1, the corresponding value of θ_1 is that θ_1 that makes $F_1 = 0$, that is, show that both methods give

$$
\frac{\partial\theta_1}{\partial T_1} = \frac{\displaystyle\sum_{i=1}^{c}\left[\left(\frac{1}{K_{1i}}-1\right)\frac{A_{1i}}{1+A_{1i}} - \frac{1}{K_{1i}}\right]\frac{v_{1i}}{K_{1i}}\frac{dK_{1i}}{dT_1}}{\displaystyle\sum_{i=1}^{c}\left(\frac{1}{K_{1i}}-1\right)\left(\frac{A_{1i}}{1+A_{1i}}\right)\left(\frac{v_{1i}}{\theta_1}\right)}
$$

4-9. Verify the relationships given by Equation (4-60).

Hints: (1) First recall that any one of the equations represented by Equation

(4-53) is of the form

$$0 = G_j + \frac{DG_j}{\partial T_1}\Delta T_1 + \frac{DG_j}{\partial T_2}\Delta T_2 + \cdots + \frac{DG_j}{\partial T_N}\Delta T_N$$

(2) Next observe that

$$\frac{DG_j}{\partial T_k} = \frac{\partial G_j}{\partial \theta_1}\frac{\partial \theta_1}{\partial T_k} + \frac{\partial G_j}{\partial \theta_2}\frac{\partial \theta_2}{\partial T_k} + \cdots + \frac{\partial G_j}{\partial \theta_N}\frac{\partial \theta_N}{\partial T_k} + \frac{\partial G_j}{\partial T_k}$$

$$= \left[\frac{\partial G_j}{\partial \theta_1}\frac{\partial G_j}{\partial \theta_2}\cdots\frac{\partial G_j}{\partial \theta_N}\right]\frac{\partial \theta}{\partial T_k} + \frac{\partial G_j}{\partial T_k}$$

(3) Show that

$$\begin{bmatrix} \left(\left[\dfrac{\partial G_1}{\partial \theta}\right]\dfrac{\partial \theta}{\partial T_1}\right) & \cdots & \left(\left[\dfrac{\partial G_1}{\partial \theta}\right]\dfrac{\partial \theta}{\partial T_N}\right) \\ \vdots & & \vdots \\ \left(\left[\dfrac{\partial G_N}{\partial \theta}\right]\dfrac{\partial \theta}{\partial T_1}\right) & \cdots & \left(\left[\dfrac{\partial G_N}{\partial \theta}\right]\dfrac{\partial \theta}{\partial T_N}\right) \end{bmatrix}$$

$$= \begin{bmatrix} \dfrac{\partial G_1}{\partial \theta_1} & \cdots & \dfrac{\partial G_1}{\partial \theta_N} \\ \vdots & & \vdots \\ \dfrac{\partial G_N}{\partial \theta_1} & \cdots & \dfrac{\partial G_N}{\partial \theta_N} \end{bmatrix}\begin{bmatrix} \dfrac{\partial \theta_1}{\partial T_1} & \cdots & \dfrac{\partial \theta_1}{\partial T_N} \\ \vdots & & \vdots \\ \dfrac{\partial \theta_N}{\partial T_1} & \cdots & \dfrac{\partial \theta_N}{\partial T_N} \end{bmatrix}$$

where

$$\left[\frac{\partial G_j}{\partial \theta}\right] = \left[\frac{\partial G_j}{\partial \theta_1}\frac{\partial G_j}{\partial \theta_2}\cdots\frac{\theta G_j}{\partial \theta_N}\right].$$

(4) Make use of the relationship given by Equation (4-57).

NOTATION*

A_{ji} = absorption factor for component i on plate j; $A_{ji} = L_j/K_{ji}V_j$ for the single-θ method and $A_{ji} = \theta_j(L_j/V_j)_a/K_{ji}$ for the multi-θ method of convergence.

A_i = a square matrix for each component i; defined by Equation (4-26).

$\mathcal{B}$ = a difference in flow rates; see Equation (4-32).

C = column vector defined below Equations (4-52) and (4-55).

E = column vector defined below Equation (4-52).

$\mathpzc{f}_i$ = a feed vector in the component-material balances; defined by Equation (4-26).

f_j = a bubble point function for plate j; see Equations (4-35) and (4-36).

F_j = a dew point function for plate j; see Equations (4-47) and (4-48).

*See also Chapter 3 notations.

F = a column vector that contains the values of the dew point functions; defined below Equation (4-49).

$\mathfrak{F}$ = the feed vector that appears in the total material balances; defined below Equation (4-44).

G_j = enthalpy function; defined by Equation (4-38) or (4-39).

G = a column vector that contains the values of the G_j's; defined by Equation (4-41).

h_{ji} = enthalpy of one mole of pure component i in the liquid phase at temperature of plate j.

$h_j = \sum\limits_{i=1}^{c} h_{ji}x_{ji}$, the enthalpy per mole of liquid leaving plate j.

H_{ji} = enthalpy of one mole of pure component i in the vapor phase at the temperature of plate j.

$H_j = \sum\limits_{i=1}^{c} H_{ji}y_{ji}$, the enthalpy per mole of vapor leaving plate j.

J = Jacobian matrix that appears in the matrix equations for the Newton-Raphson method.

K_{ji} = the K value for component i; evaluated at the temperature and pressure of plate j.

l_{ji} = flow rate at which component i in the liquid phase leaves plate j mole/time.

$\mathfrak{L}_j$ = a total flow rate that approaches L_j as convergence to the problem is approached; see Equation (4-30).

L_j = total flow rate at which the liquid leaves plate j mole/time.

N = total number of plates; also, N is the number of the bottom plate, since the plates are numbered down from the top of the column.

P = column pressure.

R = a tridiagonal matrix; defined by Equation (4-45).

S_{ji} = stripping factor for component i and plate j; $S_{ji} = K_{ji}V_j/L_j$ for the single-θ method of convergence, and $S_{ji} = K_{ji}(V_j/L_j)_a/\theta_j$ for the multi-θ method of convergence.

T_j = temperature of plate j.

ΔT = a column vector; defined by Equation (4-41).

v_{ji} = flow rate at which component i leaves plate j in the vapor phase mole/time.

$\mathfrak{V}_j$ = a total flow rate that approaches V_j as convergence to the problem is approached; see Equation (4-30).

V_j = total flow rate at which the vapor leaves plate j mole/time.

V = column vector of the vapor rates V_j; defined by Equation (4-45).

$$\frac{\partial V}{\partial \theta_k} = \left[\frac{\partial V_1}{\partial \theta_k} \frac{\partial V_2}{\partial \theta_k} \cdots \frac{\partial V_N}{\partial \theta_k} \right]^T.$$

v_i = column vector of the component flow rates in the vapor phase; defined by Equation (4-26).

$$\frac{\partial \boldsymbol{v}_i}{\partial \theta_k} = \left[\frac{\partial v_{1i}}{\partial \theta_k} \frac{\partial v_{2i}}{\partial \theta_k} \quad \cdots \quad \frac{\partial v_{Ni}}{\partial \theta_k} \right]^T.$$

Subscripts

    a = assumed value.

    ca = calculated value.

    H = single-phase heavy component.

    L = single-phase light component.

Greek Letters

    η_j = a multiplier; see Equation (4-29).

    θ = a multiplier; defined by Equations (4-15) and (4-22).

    θ_j = a multiplier associated with each plate j; defined by Equations (4-23) and (4-43).

    $\Delta\boldsymbol{\theta}$ = a column vector defined below Equation (4-49).

    σ_j = a multiplier; see Equations (4-29).

Mathematical Symbols

    $\sum\limits_{\substack{i=1 \\ i \neq H, L}}^{c}$ = sum over all components from 1 to c except H and L.

    $\{T_j\}$ = set of all T_j's for plates $j = 1$ through $j = N$.

$[x_1 x_2 x_3]^T$ = transpose of the row vector $[x_1 x_2 x_3]$ which is equal to the column vector

$$\begin{bmatrix} x_1 \\ x_2 \\ x_3 \end{bmatrix}$$

REFERENCES

1. Billingsley, D. S., and G. W. Boynton, "Iterative Methods for Solving Problems in Multicomponent Distillation at the Steady State,' *A. I. Ch. E. Journal*, **17**, (1971), 65.

2. Boynton, G. W., "Iteration Solves Distillation," *Hydrocarbon Processing*, **49**, (1970), 153.

3. Boyum, A. A., "A New Methodology for Rigorous Solution of Equilibrium Stage Separation Problems," Ph. D. Dissertation, Polytechnic Institute of Brooklyn, 1966.

4. Brown, G. G., and M. Souders, Jr. "Fundamental Design of Absorbing and Stripping Columns for Complex Vapors," *Ind. End. Chem.*, **24**, (1932), 519.

5. Broyden, C. G., "A Class of Methods for Solving Nonlinear Simultaneous Equations," *Mathematics of Computation*, **19**, (1965), 577.

6. Carnahan, B., H. A. Luther, and J. O. Wilkes, *Applied Numerical Methods.* New York: John Wiley & Sons, Inc., 1969.

7. Chao, K. C., and J. D. Seader, "A General Correlation of Vapor-Liquid Equilibria in Hydrocarbon Mixtures," *A. I. Ch. E. Journal*, **7**, (1961), 598.

8. Friday, J. R., and B. D. Smith, "An Analysis of the Equilibrium Stage Separations Problem—Formulation and Convergence," *A. I. Ch. E. Journal*, **10**, (1964), 698.

9. Greenstadt, John, Yonathan Bard, and Burt Morse, "Multicomponent Distillation Calculation on the IBM 704," *Ind. Eng. Chem.*, **50**, (1958) 1644.

10. Holland, C. D., *Multicomponent Distillation*, Englewood Cliffs, N. J.: Prentice-Hall, Inc., 1963.

11. ———, and G. P. Pendon, "Exact Solutions Given by the θ-Method of Convergence for Complex and Conventional Distillation Columns," *Chemical Engineering Journal*, Loughborough University of Technology **7**, (1972), 10.

12. ———, G. P. Pendon, A. E. Hutton, R. McDaniel, and I. Yamada, "Solution of Absorber, Stripper, and Liquid-Liquid Extraction Problems by Use of the θ-Method of Convergence," *J. K. I. Ch. E.*, **11**, (1973) 8.

13. Kremser, A., "Theoretical Analysis of Adsorption Process," *National Petro. News*, **22**, (May 21, 1930), p. 43.

14. Naphtali, L. M., and D. S. Sandholm, "Multicomponent Separation Calculations by Linearization," *A. I. Ch. E. Journal*, **17**, (1971), 148.

15. Nartker, T. A., J. M. Srygley, and C.D. Holland, "Solution of Problems Involving Systems of Distillation Columns," *Can. J. Chem. Engr.*, (Aug., 1966), 217.

16. Newman, J. S., "Temperature Computed for Distillation," *Petroleum Refiner*, **42**, (1963), 141.

17. Pendon, G. P., A.E. Hutton, C.D. Holland, and R. McDaniel, "Solution of Absorber, Stripper, and Liquid-Liquid Extraction Problems by Use of the Multiple Theta Method of Convergence," Presented at the International Congress on the Use of Electronic Computers in Chemical Engineering, 123rd Event of the European Federation of Chemical Engineering, Paris, France, April 25-27, 1973.

18. Sujata, A. D., "Absorber Stripper Calculations Made Easier," *Hydrocarbon Processing & Petroleum Refiner*, **40**, (1961), 137.

19. Tierney, J. W., and J. L. Yanosik, "Simultaneously Flow and Temperature Correction in the Equilibrium Stage Problem," *A. I. Ch. E. Journal*, **6**, (1969), 897.

20. Tomich, J. F., "A New Simulation Method for Equilibrium Stage Processes," *A. I. Ch. E. Journal*, **16**, (1970), 229.

21. Wilde, J. W., and C. S. Beightler, *Foundations of Optimization.* Englewood Cliffs, N. J.: Prentice-Hall, Inc., 1967.

Graphical and Numerical Methods for Solving Extraction Problems

5

Two types of extraction processes are considered in this chapter, liquid-liquid extraction and liquid-solid extraction. The class of separation processes called liquid-liquid extraction depends on the relative distribution of one or more components between two liquid phases.

Unlike distillation processes which depend on the variation of vapor pressures with temperature, liquid-liquid extraction separations depend on the nonideal behavior of the distribution of the components between the two liquid phases. Also, relative to the distillation process, the equilibrium data available for making design calculations for extraction processes are relatively meager, particularly for systems with four or more components. Equilibrium data are available, however, for several ternary and a few quaternary systems. An abbreviated presentation of a well-known graphical method is given. This method is applicable for the solution of problems involving ternary mixtures. Numerical methods are presented which are applicable to problems involving the separation of multicomponent mixtures, provided, of course, two and only two liquid phases exist. The graphical method is presented in Part 1 and the numerical methods in Part 3. In Part 2, it is demonstrated that solid-liquid extraction problems may be solved graphically by use of the general method described for liquid-liquid extraction problems.

In Chapter 12, the results of several field tests are presented and analyzed. These tests were made on a column located at the Baytown Refinery of the Humble Oil and Refining Company. In this column, liquid sulfur dioxide was used to extract aromatics from a kerosene feed containing paraffinic, naphthenic, and aromatic compounds. This extraction process is recognized as the well-known Edeleanu process (5, 6. 7) which was the first successful application of liquid-liquid extraction in petroleum processing. There follows a brief description of the column employed in the plant tests and a brief summary of other types of equipment employed in liquid-liquid extraction.

Equipment

Equipment used to carry out separations by the liquid-liquid process has the objective of creating a large amount of surface area between the two phases. The reason for promoting the formation of interfacial area is that the rates of mass transfer between the two liquid phases are proportional to the amount of interfacial area. Packing, baffles, sieve plates, mechanical mixers, and centrifugal force are commonly used to enhance the formation of interfacial area. The column used to make the plant tests was a packed column. A diagram of this column is shown in Figure 5-1. The column consisted of a carbon steel vessel which was 67 ft in height and 5 ft in diameter (internal). The column was packed with $1\frac{1}{4}$-in. Raschig rings, and the height of the packed section was 44.5 ft. Temperatures of all entering and leaving streams were measured. In addition, temperatures were measured at five other locations within the column as indicated in Figure 5-1. The solvent or extract phase consisted of liquid sulfur dioxide which entered near the top of the column and was dispersed by means of a spray ring. The kerosene feed, the lighter of the two liquids, entered near the bottom of the column and flowed up through the packing countercurrently to the extract phase and left at the top of the column as the raffinate. The column was operated such that throughout the packed section, the extract phase was dispersed in the continuous raffinate phase. The interface between the raffinate and extract phases at the bottom of the column was maintained at the desired level as indicated in Figure 5-1 by means of a liquid level controller. A further discussion of this column is presented in Chapter 12 in connection with its use in making field tests.

Instead of packing, other types of contacting devices, such as sieve plates, baffles, spray columns, wetted wall columns, and pulse columns, may be employed as described in considerable detail by others (2, 23). A variety of mechanically agitated extractors are available commercially. Typical of these are the rotary annular extractors (4, 20), rotary-disk contactors (13), the Mixco Lightnin CM Contactor [also known as the Oldshue-Rushton (11)

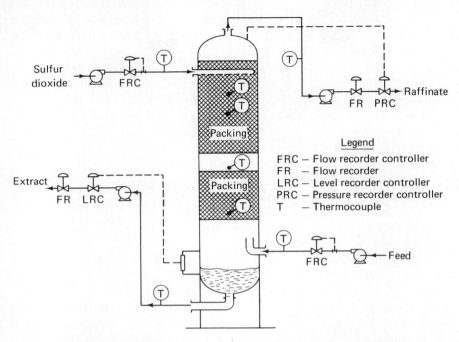

Figure 5-1. Simplified flow diagram of extractor. [Taken from A. E. Hutton and C. D. Holland, "Use of Field Tests in the Modeling of a Liquid-liquid Extraction Column," *Chem. Eng. Sci.*, **27**, (1972), 919.]

Extractor], Scheibel Extractors (15, 16, 17), and Pulsed Extractors (24). Centrifugal extractors are also available (12).

Operational of continuous extraction column

Calculational procedures are presented in subsequent sections of this chapter. These procedures may be applied to problems involving any one of the four modes of operation displayed in Figures 5-2 through 5-4.

A schematic diagram of a typical countercurrent extractor is shown in Figure 5-2. When the feed or raffinate phase is heavier than the solvent or extract phase, it is introduced at the top of the column at the rate L_0 and leaves at the bottom of the column as the raffinate at the rate L_N. The solvent phase enters at the bottom of the column at the rate V_{N+1} and leaves at the top of the column as the extract at the rate V_1. When the solvent is heavier than the feed, it is introduced at the top of the column at the rate L_0 and leaves at the bottom of the column at the rate L_N while the feed enters at the bottom of the column at the rate V_{N+1} and leaves at the top of the column at the rate V_1. The plates or stages are numbered from the top of the column down; the first stage is assigned the number 1 and the last stage the number N.

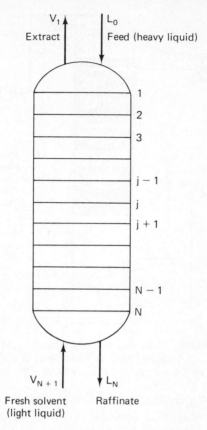

Figure 5-2. Sketch of a countercurrent extractor with multiple stages for the case in which the feed is heavier than the solvent.

Suppose two feeds are available at say rates F and L_0 and further suppose that the feed L_0 is leaner in the component to be recovered by the solvent than is the feed F. Rather than mix the two feeds before introducing them to the column, the feeds may be introduced separately as shown in Figure 5-3.

Instead of L_0 being a second feed in Figure 5-3, it may consists of a second solvent. Occasions arise when the feed F consists principally of two components, say A and B, and it is desired to separate these components by solvent extraction. If two immiscible solvents exist, one of which is selective for A and the other for B, then the separation may be effected by an arrangement such as the one shown in Figure 5-3, where L_0 represents a second solvent.

The use of extract reflux as indicated in Figure 5-4 aids in effecting separations in the same manner that reflux aids in making separations in conventional distillation columns. The solvent removal element represents the process required to separate the extract stream V_1 into a relatively pure

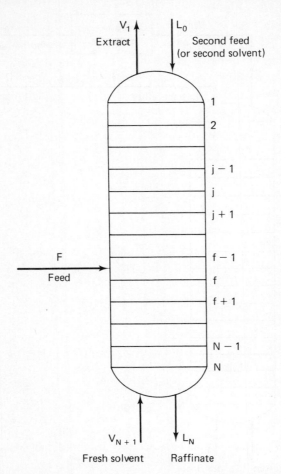

Figure 5-3. Sketch of an extractor operated with either two feeds or two solvents.

solvent stream V_0 and two streams (reflux L_0, product D) which are relatively rich in the component (or components) removed by the solvent. The product and reflux streams have the same composition. The "solvent removal" element may consist of a conventional distillation column or alternately the separation may be achieved through the use of a decanter.

Part 1 : Graphical Solution of Separation Problems Involving Ternary Mixtures

For the case of two liquid phases, the Gibbs' phase rule [(number of phases) + (number of degrees of freedom) = (number of components) + 2] asserts that the number of degrees of freedom is equal to the number of components.

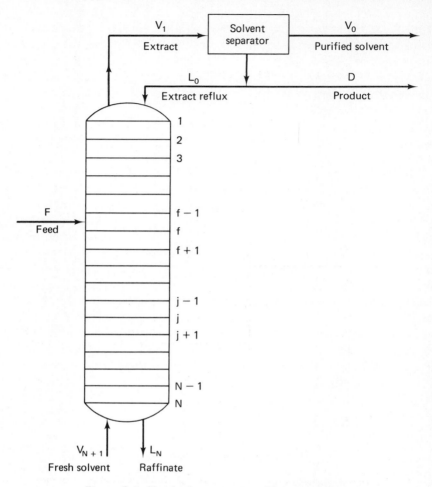

Figure 5-4. Sketch of an extractor with extract reflux.

Then if the temperature and pressure of a ternary system composed of two phases are fixed, the fixing of one mole fraction in either phase completely determines the state of equilibrium.

The characteristics of such a system may be illustrated by use of the data for the benzene-acetone-water system (1). It is customary to characterize the mutual solubilities of the constituents of ternary systems in terms of the mutual solubilities of the three binary pairs. At 30°C, acetone and water are completely miscible liquids and so is the liquid pair acetone and benzene, but water and benzene are for all practical purposes completely immiscible liquids. Thus, acetone is conveniently characterized as the distributed component between the two immiscible liquids, water and benzene. Equilibrium data at 30°C and at the vapor pressure of the system are presented in Table 5-1. These data are commonly represented graphically on either rectangular or equilateral triangular coordinate systems. A graphical representation of the data in Table 5-1 on rectangular coordinates is presented in Figure 5-5. The region bounded by the curve $\overline{ACFEDB}$, the abscissa, and the hypotenuse represents the totality of all single-phase mixtures. The curve $\overline{ACFEDB}$ was determined by plotting the equilibrium sets of compositions given in Table 5-1. The composition sets of the benzene phase lie along the curve $\overline{ACFP}$ and those for the water phase lie along curve $\overline{BDFP}$. Selected sets of compositions of the benzene phase and the corresponding sets of composition of the water phase which are in equilibrium with them are presented in Table 5-2. In order to retain the correspondence of the sets of equilibrium compositions, tie lines such as lines $\overline{CD}$ and $\overline{EF}$ in Figure 5-5 are employed. To any point, say G in Figure 5-5, lying in the two-phase region, the following physical significance may be attached. Point G may be regarded as some initial mixture whose total mass fractions (mass fractions computed on the basis of the total mass of each constituent present regardless of state) are the coordinates of point G. Such a mixture will separate into two equilibrium phases with the compositions E and F, which are given by the tie line passing through point G as shown in Figure 5-5. As the mass fraction of acetone in the initial mixture is increased, it is seen that the length of the tie lines decreases. Point P, called the "plait point," represents the limit as the lengths of the tie lines approach zero. This point is also called a "critical point" because the compositions of the two liquid phases approach each other at this point and become indistinguishable.

TABLE 5-1

SOLUBILITY DATA FOR THE BENZENE-ACETONE-WATER SYSTEM
AT 30°C, PRESSURES LESS THAN 1 ATM.*†

Benzene wt. %	Acetone wt. %	Water wt. %
0.1	0.0	99.9
0.2	10.0	89.8
0.4	20.0	79.6
0.9	30.0	69.1
1.8	40.0	58.2
4.1	50.0	45.9
11.2	60.0	28.8
23.8	64.1	12.1
99.9	0.0	0.1
89.8	10.0	0.1
79.4	20.0	0.6
69.1	30.0	0.9
58.0	40.0	2.0
46.3	50.0	3.7
32.8	60.0	7.2

*The effect of pressure on the distribution was estimated to be in order of 0.01% per atmosphere. The pressures encountered here were less than one atmosphere gauge.

†These data were read from Figure 4, S.W. Briggs and E.W. Comings, "Effect of Temperature on Liquid-liquid Equilibrium," *Ind. Eng. Chem.*, **35**, (1943), 413.

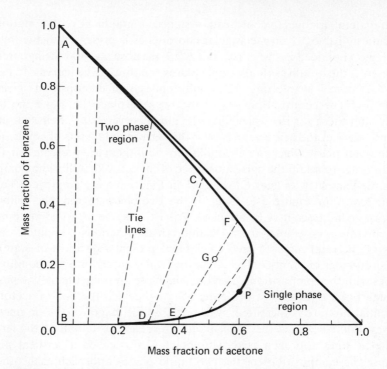

Figure 5-5. Equilibrium phase diagram for the Benzene-acetone-water system at 30° C. [Taken from S. W. Briggs and E. W. Comings, "Effect of Temperature on Liquid-liquid Equilibrium," *Ind. Eng. Chem.*, **35**, (1943), 413.]

TABLE 5-2
TIE LINES FOR THE BENZENE-ACETONE-WATER SYSTEM
AT 30°C, PRESSURES LESS THAN 1 ATM.*†

BENZENE PHASE wt. %			WATER PHASE wt. %		
Benzene	Acetone	Water	Benzene	Acetone	Water
94.0	5.8	0.2	0.1	5.0	94.9
86.7	13.1	0.2	0.2	10.0	89.8
68.7	30.4	0.9	0.4	20.0	79.6
49.8	47.2	3.0	0.9	30.0	69.1
34.5	58.9	6.6	1.8	40.0	58.2
23.9	64.1	12.0	4.1	50.0	45.9

*The effect of pressure on the distribution was estimated to be in order of 0.01 % per atmosphere. The pressures encountered here were less than one atmosphere gauge.

†Taken from Briggs and Comings, "Effect of Temperature on Liquid-liquid Equilibrium," p. 414.

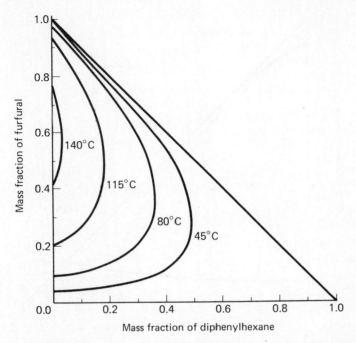

Figure 5-6. Effect of temperature on the phase behavior of the Docosane-diphenylexane-furfural system. (Taken from Briggs and Comings, "Effect of Temperature on Liquid-liquid Equilibrium," p. 417.)

The influence of pressure on the shape of the equilibrium curve of the phase diagram is generally negligible. However, the effect of temperature on the phase equilibria may be quite pronounced as demonstrated by Briggs and Comings (1) for the docosane-diphenylhexane-furfural system (see Figure 5-6).

Systems having solubility curves with two branches such as the one shown in Figure 5-7 are of particular interest because for systems of this type extract reflux may be used advantageously in effecting separations. In Figure 5-7, the two-phase region lies inside the two branches, that is, it is the region bounded by the two branches, the ordinate and the hypotenuse. The single-phase region lies outside this bounded region.

Graphical solution of separation problems involving ternary mixtures at steady-state operation

The graphical procedure presented below rests upon the validity of several relationships that are established in succeeding paragraphs. Let the stages of an extractor be numbered as depicted in Figure 5-2. Let the total flow rate

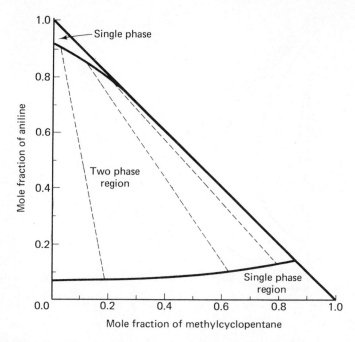

Figure 5-7. Example of a system having a solubility curve with two branches: The aniline-methylcyclopentane-n-hexane system at 25° C. (Taken from Darwent, B. de B., and C. A. Winkler, "The System n-Hexane-Methylcyclopentane-Aniline,"*J. Phys. Chem.*, **47**, (1943), 447.

in pounds per hour (or moles per hour) of the extract phase be denoted by V and the total flow rate of the raffinate phase by the symbol L. Let the mass fraction (or mole fraction) of component i in the extract phase be denoted by y_i and the mass fraction (or mole fraction) of component i in the raffinate phase by the symbol x_i. Also, in the developments the follow, the solute is denoted by the subscript A, the solvent by the subscript S, and the third component (the principal constituent of the raffinate phase) by the subscript C.

The *addition point* $(z_{J,A}, z_{J,S})$ in Figure 5-8 is denoted by z_J and defined by the total material balance

$$V_{N+1} + L_0 = L_N + V_1 = J \qquad (5\text{-}1)$$

and the component-material balances

$$V_{N+1}y_{N+1,i} + L_0 x_{0,i} = V_1 y_{1i} + L_N x_{Ni} = J z_{Ji} \qquad (i = A, C, S) \qquad (5\text{-}2)$$

PROPOSITION I. The point z_J $(z_{J,A}, z_{J,S})$ lies on the straight line connecting the points $y_{N+1}(y_{N+1,A}, y_{N+1,S})$ and $x_0(x_{0,A}, x_{0,S})$ (see Figure 5-8).

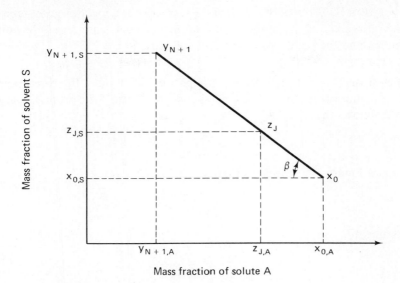

Figure 5-8. Graph of the points y_{N+1}, z_J, and x_0.

Proof:

For any component i, Equations (5-1) and (5-2) may be combined to give

$$V_{N+1}y_{N+1,i} + L_0 x_{0,i} = (V_{N+1} + L_0)z_{Ji} \tag{5-3}$$

which is readily rearranged to give

$$\frac{V_{N+1}}{L_0} = \frac{z_{J,i} - x_{0,i}}{y_{N+1,i} - z_{J,i}} \tag{5-4}$$

Now let z_J be arbitarily selected as any point which is not coincident with y_{N+1} and x_0. Then, the slopes of the two line segments $\overline{z_J y_{N+1}}$ and $\overline{z_J x_0}$ are as follows:

$$\text{slope of } \overline{z_J y_{N+1}} = \frac{y_{N+1,S} - z_{J,S}}{y_{N+1,A} - z_{J,A}}$$

$$\text{slope of } \overline{z_J x_0} = \frac{x_{0,S} - z_{J,S}}{x_{0,A} - z_{J,A}}$$

Equation (5-4) may be stated for components A and S and rearranged to give

$$\frac{y_{N+1,S} - z_{J,S}}{y_{N+1,A} - z_{J,A}} = \frac{x_{0,S} - z_{J,S}}{x_{0,A} - z_{J,A}}$$

Consequently, the slopes of lines $\overline{z_J y_{N+1}}$ and $\overline{z_J x_0}$ are equal. Since the point z_J is common to these two line segments, it follows that the points y_{N+1}, z_J, and x_0 all lie on the same straight line.

PROPOSITION II. The ratio V_{N+1}/L_0 of the respective external flow rates is equal to the ratio $\overline{z_J x_0}/\overline{y_{N+1} z_J}$ of the lengths of line segments connecting the points z_J and x_0 and the points y_{N+1} and z_J (see Figure 5-8).

Proof:

Based on Proposition I, let the straight line containing the three points y_{N+1}, z_J, and x_0 be constructed as shown in Figure 5-8. From Equation (5-4) it follows that

$$\frac{V_{N+1}}{L_0} = \frac{x_{0,A} - z_{J,A}}{z_{J,A} - y_{N+1,A}}$$

and from Figure 5-8, it follows that

$$\frac{x_{0,A} - z_{J,A}}{z_{J,A} - y_{N+1,A}} = \frac{(\cos \beta)(\overline{z_J x_0})}{(\cos \beta)(\overline{y_{N+1} z_J})} = \frac{\overline{z_J x_0}}{\overline{y_{N+1} z_J}}$$

Consequently,

$$\frac{V_{N+1}}{L_0} = \frac{\overline{z_J x_0}}{\overline{y_{N+1} z_J}} \tag{5-5}$$

PROPOSITION III. The point z_J lies on the straight line connecting the points x_N and y_1 (see Figure 5-9).

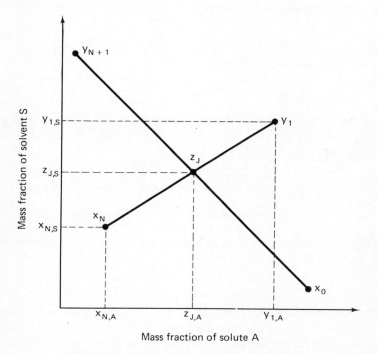

Figure 5-9. Graph of the points x_N, z_J, y_1, x_0, and y_{N+1}.

Proof:

Since a proof of this proposition may be obtained in a manner analogous to that shown for Proposition I, only an outline of the proof is presented. Equation (5-2) may be solved with Equation (5-1) to give

$$\frac{L_N}{V_1} = \frac{y_{1i} - z_{J,i}}{z_{J,i} - x_{Ni}} \tag{5-6}$$

Next, formulate expressions for the slopes of the line segments $\overline{x_N z_J}$ and $\overline{z_J y_1}$ (see Figure 5-9). To show that the slopes of these line segments are equal, make use of Equation (5-6).

PROPOSITION IV. The ratio L_N/V_1 of the respective external flow rates is equal to the ratio $\overline{z_J x_0}/\overline{y_{N+1} z_J}$ of the lengths of the respective line segments, that is,

$$\frac{L_N}{V_1} = \frac{\overline{z_J x_0}}{\overline{y_{N+1} z_J}} \tag{5-7}$$

Proof:

The proof of this relationship follows by use of an argument analogous to that presented for Proposition II.

PROPOSITION V. If the difference point $x_\Delta(x_{\Delta,A}, x_{\Delta,S})$ is defined by

$$x_{\Delta i} = \frac{L_0 x_{0i} - V_1 y_{1i}}{L_0 - V_1} \qquad (i = A, C, S) \tag{5-8}$$

show that the ratio remains constant for all j, that is, show that

$$x_{\Delta i} = \frac{L_j x_{ji} - V_{j+1} y_{j+1,i}}{L_j - V_{j+1}} \qquad (i = A, C, S; 1 \leqq j \leqq N) \tag{5-9}$$

Proof:

The total material balance enclosing the first j extraction stages (see Figure 5-2) is given by

$$L_0 - V_1 = L_j - V_{j+1} \qquad (1 \leqq j \leqq N) \tag{5-10}$$

Similarly, the component-material balance over the first j stages is given by

$$L_0 x_{0i} - V_1 y_{1i} = L_j x_{ji} - V_{j+1} y_{j+1,i} \qquad (i = A, C, S; 1 \leqq j \leqq N) \tag{5-11}$$

Thus, it follows that

$$\frac{L_0 x_{0i} - V_1 y_{1i}}{L_0 - V_1} = \frac{L_j x_{ji} - V_{j+1} y_{j+1,i}}{L_j - V_{j+1}}$$

and, consequently, the two expressions given by Equations (5-8) and (5-9) for $x_{\Delta i}$ are equivalent.

PROPOSITION VI. The points x_Δ, x_0, and y_1 lie on the same straight line (see Figure 5-10).

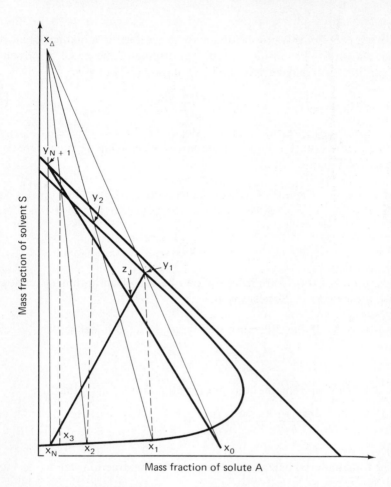

Figure 5-10. Relationship between the difference point x_Δ and the pairs of points (x_0, y_1), (x_1, y_2), . . . , (x_N, y_{N+1}).

Proof:

That the slopes of the line segments $\overline{x_0 y_1}$ and $\overline{x_0 x_\Delta}$ are equal is shown as follows:

$$\text{slope of } \overline{x_0 y_1} = \frac{y_{1,S} - x_{0,S}}{y_{1,A} - x_{0,A}}$$

$$\text{slope of } \overline{x_0 x_\Delta} = \frac{x_{0,S} - x_{\Delta,S}}{x_{0,A} - x_{\Delta,A}}$$

Let $x_{\Delta,S}$ and $x_{\Delta,A}$ be replaced by their equivalents as given by Equation (5-8). Then,

$$\text{slope of } \overline{x_0 x_\Delta} = \frac{x_{0,S} - \left(\dfrac{L_0 x_{0,S} - V_1 y_{1,S}}{L_0 - V_1} \right)}{x_{0,A} - \left(\dfrac{L_0 x_{0,A} - V_1 y_{1,A}}{L_0 - V_1} \right)}$$

When the expression on the right-hand side is cleared of fractions, and the result so obtained is compared with the above formula for $\overline{x_0 y_1}$, it follows that

$$\text{slope of } \overline{x_0 x_\Delta} = \frac{y_{1,S} - x_{0,S}}{y_{1,A} - x_{0,A}} = \text{slope of } \overline{x_0 y_1}$$

Since the slopes of $\overline{x_0 y_1}$ and $\overline{x_0 x_\Delta}$ are equal and have the point x_0 in common, it follows that x_0, x_Δ, and y_1 all lie on the same straight line. The next proposition is a generalization of Proposition VI.

PROPOSITION VII. For each $j(j = 0, 1, 2, \ldots, N)$, the points x_Δ, x_j, and y_{j+1} all lie on the same straight line (see Figure 5-10).

Proof:

For each choice of $j(j = 0, 1, 2, \ldots, N)$, the proof is carried out in the same manner as that shown above for the case of $j = 0$.

The alternate use of Proposition VII and the equilibrium relationship given by the tie lines, for the purpose of determining the number of equilibrium stages required to effect a specified separation, is illustrated as follows. Suppose that the line passing through x_Δ, x_0, and y_1 has been located. Then the composition of the raffinate phase that is in equilibrium with the extract of composition y_1 leaving the first stage is given by following the tie line from y_1 to x_1 as shown in Figure 5-10. Then by Proposition VII, the points x_Δ, x_1, and y_2 all lie on the same straight line. Next x_2 is located by the tie line passing through y_2. Again, by Proposition VII the points x_Δ, x_2, and y_3 all lie on the same straight line. Continuation of this procedure yields that number of stages required to reduce the mass fraction of A in the solute to value equal or less than some specified value $x_{N,A}$. Also, it should be noted that the graphical procedure may be initiated by starting with the Nth rather than the first stage.

The graphical method developed above is easily extended to include the solution of problems involving columns with two feeds (see Figure 5-3) and to columns which use extract reflux (see Figure 5-4) as outlined in Problems 5-1 and 5-2.

To demonstrate the application of the graphical method described and developed above, the following example is presented.

ILLUSTRATIVE EXAMPLE 5-1

Acetone is to be recovered from a feed (L_0) that is 50% acetone and 50% water (by weight) by extraction with benzene. The raffinate (L_N) is to contain no more than 1% acetone by weight. The solvent (V_{N+1}) may be assumed to be pure benzene. The extraction is to be carried out at 30°C under the pressure exerted by the system at this temperature. The solubility and tie line data presented in Tables 5-1 and 5-2 may be used in the solution of the problem. On the basis of a solvent to feed ratio (V_{N+1}/L_0) of 1.5, find the number of theoretical stages required to effect the specified separation.

Solution: The number of theoretical stages may be determined as follows:

Step 1. From the composition given for the feed mixture, locate the point x_0 as shown in Figure 5-11.

Step 2. Since the solvent is pure benzene $y_{N+1,S} = 1.0$ and $y_{N+1,A} = 0.0$, the point y_{N+1} is located as shown in Figure 5-11.

Step 3. Pass a straight line through the points x_0 and y_{N+1}.

Step 4. By Proposition I the addition point z_J lies on the straight line connecting x_0 and y_{N+1}, and it may be located by use of Proposition II. By this proposition the lengths of the line segments $\overline{z_J x_0}$ and $\overline{y_{N+1} z_J}$ are related to the specified value of V_{N+1}/L_0 as follows:

$$\frac{V_{N+1}}{L_0} = \frac{\overline{z_J x_0}}{\overline{y_{N+1} z_J}} = 1.5$$

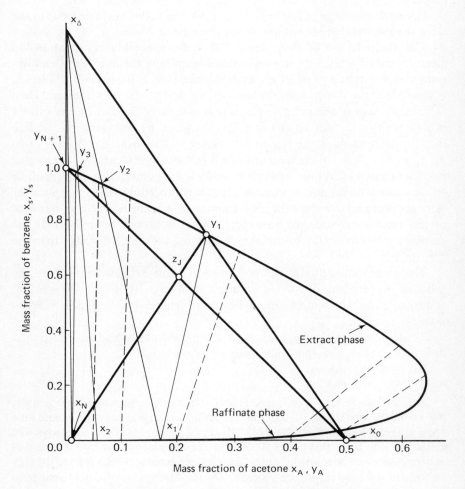

Figure 5-11. Graphical solution of Illustrative Example 5-1.

If the length of the line $\overline{y_{N+1}x_0}$ is represented by 17.5 arbitrary but equal units of length, then the length of the line segment $\overline{y_{N+1}z_J}$ may be computed as follows:

$$\overline{z_Jx_0} + \overline{y_{N+1}z_J} = \overline{y_{N+1}x_0} = 17.5$$

from which it follows that

$$\overline{y_{N+1}z_J} = \frac{17.5}{1 + \dfrac{\overline{z_Jx_0}}{\overline{y_{N+1}z_J}}} = \frac{17.5}{1 + 1.5} = 7.0$$

The point z_J is then located by assigning a length of 7.0 units to the line segment $\overline{y_{N+1}z_J}$ and a length of 10.5 units to $\overline{z_Jx_0}$ as shown.

Step 5. By Proposition III, the points z_J, x_N, and y_1 lie on the same straight line. Since x_N is an equilibrium point, it lies on the lower curve (the raffinate phase) and since y_1 is an equilibrium point, it lies on the upper curve (the extract phase). Since it is given that $x_{N,A} = 0.01$, the point x_N may be located because it lies on the lower equilibrium curve, which coincides with the abscissa at these low mass fractions. Thus, the intersection of the line passing through x_N and z_J with the upper equilibrium curve locates the point y_1 as shown in Figure 5-11.

Step 6. The difference point x_Δ is located as follows. Since the points x_Δ, x_0, y_1 all lie on the same straight line (Proposition VI) and the points x_Δ, x_N, y_{N+1} also lie on the same straight line (Proposition VII), it follows that x_Δ is the intersection of these two straight lines as demonstrated in Figure 5-11.

Step 7. Next, the number of equilibrium stages is determined by carrying out the following construction. Since y_1 and x_1 are in equilibrium, the point x_1 is located on the lower equilibrium curve by construction of a tie line that intersects the lower equilibrium curve at the same relative position between the two adjacent tie lines as that of the point y_1.

Step 8. Since the points x_1, y_2, and x_Δ all lie on the same straight line (by Proposition VII), the point y_2 is located by connecting the points x_1 and x_Δ as shown. Since y_2 is in equilibrium with x_2, the point x_2 is located in a manner analogous to that described for x_1. Next, y_3 is located by connecting the points x_2 and x_Δ with a straight line as shown (by Proposition VII). Finally, since y_3 is in equilibrium with x_3 and x_3 coincides with x_N, it follows that three equilibrium stages are required to effect the specified separation.

From the standpoint of analysis, the following type of solid-liquid extraction is seen to be a special case of liquid-liquid extraction.

Part 2. Solid-Liquid Extraction

In this separation process, a solute A is removed from a solid C by dissolving the solute A in an appropriate solvent S. Solid-liquid extraction which is characterized by the following conditions is considered below.

1. It is supposed that the solid is dissolved by neither the solute nor the solvent.
2. The solute and solvent are miscible in all proportions.
3. The mass ratio of the solute to solvent in the solvent or overflow phase is equal to the mass ratio of the solute to solvent in the solution retained by the solid.

In the following analysis it is supposed that a given solute A is removed from a solid C by a series of equilibrium stages. Countercurrent flow of the solvent phase relative to the solid phase is supposed, and thus the material balances and corresponding relationship developed above are applicable to solid-liquid extraction. In view of the three conditions stated above, unique relationships do exist, however, for solid-liquid extraction. By the first two conditions, it follows that for the underflow stream (composed of the solid and the solution retained by it)

$$x_{jA} + x_{jS} + x_{jC} = 1 \tag{5-12}$$

and for the overflow stream

$$y_{jA} + y_{jS} = 1 \tag{5-13}$$

Since the overflow phase does not contain any solid (condition 1), it follows that $y_{jC} = 0$. The line $y_{jC} = 0$ is represented by the hypotenuse in Figure 5-12. Consequently, the points corresponding to overflow composition lie on the hypotenuse of Figure 5-12.

By the third condition, it follows that

$$\frac{y_{jA}}{y_{jS}} = \frac{x_{jA}}{x_{jS}} \tag{5-14}$$

The curve containing all possible underflow compositions x_j such as curve D in Figure 5-12 must be determined experimentally. It remains, however, to show that the equilibrium compositions corresponding to the points x_j and y_j for any stage j lie on straight lines that pass through the origin.

PROPOSITION VIII. For any stage j, the straight line passing through x_j and y_j also passes through the origin (see Figure 5-12).

Proof:

Let the origin be denoted by O. Then

$$\text{slope of the line segment } \overline{Ox_j} = \frac{x_{jS}}{x_{jA}}$$

$$\text{slope of the line segment } \overline{Oy_j} = \frac{y_{jS}}{y_{jA}}$$

By Equation (5-14), it follows that the slopes of the line segments are equal. Since the origin O is common to both line segments, it follows that they coincide. Therefore, the straight line passing through x_j and y_j also passes through the origin.

For countercurrent, multiple stage, solid-liquid extraction, a graphical solution may be obtained for problems by use of Propositions I through VIII. In order to demonstrate the similarities between liquid-liquid extraction and solid-liquid extraction, the following example is presented.

ILLUSTRATIVE EXAMPLE 5-2

It is desired to extract a solute A which is retained by a solid C by means of a continuous countercurrent, multiple-stage extractor with a solvent S. The composition of the feed L_0 is $x_{0,A} = 0.25$ and $x_{0,C} = 0.75$. The mass fraction of the solute retained by the solid is to be reduced to 0.05 or less. If a ratio of solvent to feed (V_{N+1}/L_0) equal to 0.3 is to be used, find the number of equilibrium stages required to effect the separation. It may be supposed that the underflow curve D in Figure 5-12 was determined at experimental conditions similar to those of the proposed extraction. It may also be supposed that the three conditions listed at the beginning of this section on solid-liquid extraction are satisfied by the proposed extraction.

Solution: This problem is solved by following the same set of steps outlined for Illustrative Example 5-1.

Step 1. The point $x_0(x_{0,A} = 0.25, x_{0,C} = 0.75)$ is plotted as shown in Figure 5-12.

Step 2. The point y_{N+1} ($y_{N+1,S} = 1.0$) is plotted as shown in Figure 5-12.

Step 3. Points x_0 and y_{N+1} are connected by a straight line as shown.

Step 4. By Proposition I, z_J lies on the line constructed in Step 3. By Proposition II,

$$\frac{V_{N+1}}{L_0} = \frac{\overline{z_J x_0}}{\overline{y_{N+1} z_J}} = 0.3$$

Let the line $\overline{x_0 y_{N+1}}$ be represented by 10.3 arbitrary but equal units of length. Then

$$\overline{z_J x_0} + \overline{y_{N+1} z_J} = \overline{y_{N+1} x_0} = 10.3$$

from which it follows that

$$\overline{y_{N+1} z_J} = \frac{10.3}{1 + \dfrac{\overline{z_J x_0}}{\overline{y_{N+1} z_J}}} = \frac{10.3}{1.3} = 7.92$$

and $\overline{z_J x_0} = 10.3 - 7.92 = 2.38$, which locates the point z_J as shown.

Step 5. Since the point x_N lies on the underflow curve D and it is given that $x_{N,A} = 0.05$, this point may be plotted as shown in Figure 5-12. The straight line passing through x_N and z_J intersects the hypotenuse at y_1 (by Proposition III), which locates y_1 as shown.

Step 6. The intersection of the straight line passing through y_{N+1} and x_N with the straight line passing through y_1 and x_0 is the difference point x_Δ (by Propositions VI and VII) (see Figure 5-12).

Step 7. The number of equilibrium stages is determined by the following construction. The straight line passing through the origin and point y_1 intersects the underflow curve D at the equilibrium composition x_1 (by Proposition VIII).

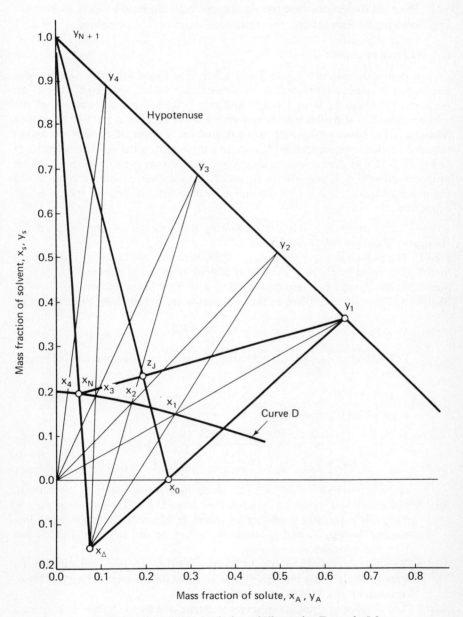

Figure 5-12. Graphical solution of Illustrative Example 5-2.

Step 8. Since the points x_1, y_2, and x_Δ all lie on the same straight line (by Proposition VII), the point y_2 is located by connecting the points x_1 and x_Δ as shown. Next, x_2 is located by passing a straight line through the origin and y_2 as shown (by Proposition VIII). Next, y_3 is located by connecting the points x_2 and x_Δ with a straight line (by Proposition VII). By the continuation of this procedure it is seen that four equilibrium stages are required to satisfy the specified separation requirement that $x_{N,A} \leq 0.05$.

Part 3. Solution of Liquid-liquid Extraction Problems by Use of Numerical Methods

The numerical methods presented are applicable for systems containing any number of components distributed between the two liquid phases. Five methods [the single-θ method, direct iteration, the multi-θ method, the Newton-Raphson method as proposed by Tierney et al. (21, 22), and Hanson's method (9)] are described and compared by use of numerical examples.

Fortunately, liquid-liquid extraction columns are customarily operated at conditions that closely approximate isothermal operation because temperature dependent equilibrium and enthalpy data for these systems are seldom available. Thus, in the developments that follow, only isothermal operation is considered. When temperature dependent data are available, the effect of temperature may be included by use of calculational procedures analogous to those presented in Chapter 4.

Isothermal operation of a countercurrent extractor (see Figure 5–2)

The separation process of liquid-liquid extraction is based on the fact that there exist certain liquid pairs that are so highly nonideal relative to one another that they are only partially miscible. This separation process consists of the transfer of one or more components from one of the liquid phases to the other one.

In the development that follows, perfect plates are assumed, that is,

$$\bar{f}_i^V = \bar{f}_i^L$$

Through the use of the concepts of classical thermodynamics, this expression may be stated in the customary form for vapor-liquid equilibria, namely,

$$y_i = K_i x_i \tag{5-15}$$

In this case, each K_i is defined by

$$K_i = \frac{\gamma_i^L}{\gamma_i^V}$$

The modeling of columns with plates as well as packed columns is considered in Chapter 12.

Application of the single-θ method of convergence to liquid-liquid extraction problems

Consider first the class of problems involving the countercurrent extraction unit shown in Figure 5-2. Suppose the problem to be solved consists of the determination of the product distribution that can be expected when the fresh solvent rate V_{N+1}, the feed rate L_0, the number of stages, the composition sets $\{x_{0i}\}$ and $\{y_{N+1,i}\}$ and the temperature and pressure of the column are specified.

To initiate the proposed calculational procedure, sets of solvent and raffinate rates $\{(V_j)_a\}$ and $\{(L_j)_a\}$ as well as sets of compositions, $\{(x_{ji})_a\}$ and $\{(y_{ji})_a\}$ are assumed, and these are used to compute the corresponding set of K values, $\{(K_{ji})_a\}$. On the basis of these assumed values, the absorption factors,

$$(A_{ji})_a = \left(\frac{L_j}{K_{ji}V_j}\right)_a$$

are computed. After the A_{ji}'s have been computed, the component flow rates may be determined by use of the same matrix equations employed for absorbers [see Equation (4-25)]. After the component flow rates $\{(v_{ji})_{ca}\}$ have been determined by solving the matrix equation given [Equation (4-25)] for each component i, the corresponding set of component flow rates $\{(l_{ji})_{ca}\}$ in the other phase are calculated by use of the well-known relationship:

$$(l_{ji})_{ca} = (A_{ji})_a(v_{ji})_{ca}$$

Next, the single-θ method of convergence is applied in precisely the same manner described in Chapter 4. After the desired $\theta > 0$ that minimizes $g(\theta)$ has been found, the corresponding sets of corrected values for the total flow rates and compositions $[\{V_j\}, \{L_j\}, \{x_{ji}\}, \{y_{ji}\}]$ are found by use of Equations (4-28) through (4-34). These sets of corrected compositions are used to compute a set of activity coefficients γ_{ji}^V and γ_{ji}^L [where γ_{ji}^V is a function of $\{y_{ji}\}$ and γ_{ji}^L is a function of $\{x_{ji}\}$ for each component i on each plate.] The corrected activity coefficients are next employed to compute a new set of K values $\{K_{ji}\}$ by use of the definition given below Equation (5-15).

On the basis of these improved sets of values of the variables, $\{V_j\}, \{L_j\}, \{K_{ji}\}$, the procedure described is repeated by first solving the component-material balances [Equation (4-25)]. The procedure described is repeated until the tolerance on θ and/or the V_j's has been satisfied.

A summary of the steps of the proposed calculational procedure follows:

1. Assume values for the following sets of variables $\{V_j\}, \{y_{ji}\}$, and $\{x_{ji}\}$. On the basis of the assumed set $\{(V_j)_a\}$, compute the corre-

sponding set of total flow rates $\{(L_j)_a\}$ by use of the total material balances. From the assumed sets of compositions $\{(y_{ji})_a\}$ and $\{(x_{ji})_a\}$, compute the activity coefficients and then the corresponding K values $\{(K_{ji})_a\}$ by use of the definition given below Equation (5-15).

2. Use the results of Step 1 to solve the component-material balances [Equation (4-25)] for each component i and denote the flow rates so obtained by $\{(v_{ji})_{ca}\}$. Then compute the corresponding set of flow rates $\{(l_{ji})_{ca}\}$ for each component in the other phase by use of the relationship: $(l_{ji})_{ca} = (A_{ji})_a(v_{ji})_{ca}$.

3. On the basis of the calculated flow rates $\{(v_{ji})_{ca}\}$ and $\{(l_{ji})_{ca}\}$ found in Step 2 and the total flow rates $\{(V_j)_a\}$ and $\{(L_j)_a\}$ assumed in Step 1, find the θ that minimizes $g(\theta)$. On the basis of this θ, compute the corresponding sets of compositions and total flow rates by use of Equations (4-28) through (4-34). Use the sets of corrected compositions to compute a set of corrected K values, (K_{ji}). (Hutton 10) used the search method known as the "golden section" (25) to find the θ that minimized $g(\theta)$.)

4. If θ and/or other variables such as the flow rates satisfy preassigned tolerances, convergence has been achieved; otherwise, the procedure is repeated. The most recent sets of values of the variables found in Step 3 become the assumed values of the variables in Step 1.

USE OF THE SINGLE-θ METHOD FOR SOLVING OTHER TYPES OF EXTRACTION PROBLEMS. If the feed is lighter than the solvent stream, the feed is identified as V_{N+1} rather than L_0 (see Figure 5-2). The calculational procedure for the case in which the feed is identified as V_{N+1} is, however, identically the same as the one described above.

When one feed is introduced on an intermediate plate f in the column as shown in Figure 5-3, the calculational procedure is the same as the one described above except that the material balances must reflect the addition of the feed on plate f (see Problem 5-3). In the treatment of all feed plates as well as the plates upon which the solvents enter, Model 1 of Chapter 3 is employed. In this model, it is supposed that all of the feed intimately contacts the liquid on and leaving the feed plate as well as the lighter liquid passing up through the liquid on the feed plate and that the feed distributes itself between the two liquid phases such that the lighter stream V_f leaving the feed plate f is in equilibrium with the heavier stream leaving L_f. The component-material balance enclosing the feed plate f is then given by

$$A_{f-1,i}v_{f-1,i} - (A_{fi} + 1)v_{fi} + v_{f+1,i} = -FX_i \qquad (5\text{-}16)$$

where

$$A_{f-1,i} = \frac{L_{f-1}}{V_{f-1}K_{f-1,i}};$$

$$A_{fi} = \frac{L_f}{V_f K_{fi}}.$$

The formula for v_{1i} for the column shown in Figure 5-3 is obtained by replacing $(v_{N+1,i} + l_{0i})$ in Equation (4-28) by $(v_{N+1,i} + l_{0i} + FX_i)$. The total flow rates are computed in a manner similar to that demonstrated in Chapter 4 for absorbers (see Problem 5-4).

When extract reflux is used to aid in effecting a given separation as shown in Figure 5-4, the calculational procedure and convergence method are the same as the one described above except for the material-balance matrices that must be revised as required to reflect the column configuration. The "solvent removal" stage shown in Figure 5-4 is similar to a condenser on a distillation column. Unlike a condenser, however, the solvent removal stage seldom consists of a single equilibrium stage. Instead, the "solvent removal" generally consists of some other type of separation process, which should be solved simultaneously with the extractor problem. In order to focus attention on the solution of extractor problems, however, it will be supposed that the separations effected by the solvent removal stage are independent of the flow rate and composition of the feed V_1 to the solvent removal stage. That is, it is supposed that the separation factor s_i for each component i is constant, where

$$s_i = \frac{L_0 x_{0i} + DX_{Di}}{V_0 y_{0i}} = \left(\frac{L_0}{D} + 1\right)\frac{d_i}{v_{0i}} \tag{5-17}$$

The material balance enclosing the solvent removal stage is given by

$$v_{1i} = v_{0i} + l_{0i} + d_i = v_{0i} + \left(\frac{L_0}{D} + 1\right)d_i \tag{5-18}$$

Elimination of d_i from Equations (5-17) and (5-18) gives

$$-(1 + s_i)v_{0i} + v_{1i} = 0 \tag{5-19}$$

For any component i, the material balance enclosing plate 1 is given by

$$v_{2i} + l_{0i} - v_{1i} - l_{1i} = 0 \tag{5-20}$$

The ratio l_{0i}/v_{0i} may be stated in terms of s_i through the use of Equation (5-17).

$$\frac{l_{0i}}{v_{0i}} = \left(\frac{\frac{L_0}{D}}{\frac{L_0}{D} + 1}\right)s_i \tag{5-21}$$

By use of this relationship and the equilibrium relationship $l_{1i} = A_{1i}v_{1i}$, it is possible to restate Equation (5-20) in the following form:

$$\left(\frac{\frac{L_0}{D}}{1 + \frac{L_0}{D}}\right)s_i v_{0i} - (A_{1i} + 1)v_{1i} + v_{2i} = 0 \tag{5-22}$$

The remaining component-material balances are developed in the usual way, and the complete set of material balances for any component i in a column having the configuration shown in Figure 5-4 is represented by a matrix equation of the same form as Equation (4-25), $A_i \boldsymbol{v}_i = -\boldsymbol{f}_i$. In this case the matrices A_i, $\boldsymbol{v}_i$, and $\boldsymbol{f}_i$ have the following definitions:

$$
A_i = \begin{bmatrix}
-(1 + s_i) & 1 & 0 & 0 & 0 & \ldots & 0 \\
\dfrac{s_i}{1 + D/L_0} & -(1 + A_{1i}) & 1 & 0 & 0 & \ldots & 0 \\
0 & A_{1i} & -(1 + A_{2i}) & 1 & 0 & \ldots & 0 \\
\hdotsfor{7} \\
0 & \hdotsfor{4} & 0 & A_{N-1,i} - (1 + A_{Ni})
\end{bmatrix}
$$

$$\boldsymbol{v}_i = [v_{0i} v_{1i} \ldots v_{fi} \ldots v_{Ni}]^T$$

$$\boldsymbol{f}_i = [00 \ldots 0 F X_i 0 \ldots 0 v_{N+1,i}]^T \tag{5-23}$$

(The element FX_i occurs in row f as indicated by Equation (5-16).)

For an extractor with reflux, the single-θ method differs slightly from the procedure shown for the column in Figure 5-2; the multiplier θ is defined as follows:

$$\frac{l_{Ni}}{v_{0i}} = \theta \left(\frac{l_{Ni}}{v_{0i}}\right)_{ca} \tag{5-24}$$

where the subscript "co" used in Chapter 3 to denote the corrected values have been omitted from the left-hand side in the interest of simplicity. Again, it is required that the corrected rates satisfy the overall component-material balance,

$$FX_i + v_{N+1,i} = v_{0i} + d_i + l_{Ni} \tag{5-25}$$

Since both L_0/D and the set of s_i's are known and remain fixed for all trials, it follows that d_i/v_{0i} is uniquely determined for all trials by Equation (5-17). Then, by use of Equation (5-17), the definition of θ as given by Equation (5-24), and Equation (5-25), the following formula for corrected values for v_{0i} is obtained.

$$v_{0i} = \frac{FX_i + v_{N+1,i}}{1 + \left(\dfrac{s_i}{\dfrac{L_0}{D} + 1}\right) + \theta \left(\dfrac{l_{Ni}}{v_{0i}}\right)_{ca}} \tag{5-26}$$

For a column having extract reflux (Figure 5-4), the defining equations for the corrected values of l_{ji}, v_{ji}, L_j, V_j, $\mathcal{L}_j$, $\mathcal{U}_j$, x_{ji}, and y_{ji} are of the same form as those given in Chapter 4, and they may be obtained from the corresponding expressions in Chapter 4 by replacing v_{1i} by v_{0i}. For example, the expression for l_{ji} as given by Equation (4-29) becomes

$$l_{ji} = \eta_j \left(\frac{l_{ji}}{v_{0i}}\right)_{ca} v_{0i} \tag{5-27}$$

The calculation of the corrected values of D and L_0 merits some attention because it differs from the remainder of the calculational procedure that follows closely that described in Chapter 4 for absorbers. Since each member of the set $\{s_i\}$ as well as L_0/D remains fixed, it follows that the ratio of the corrected values of d_i and v_{0i} are given by Equation (5-17), that is,

$$\frac{d_i}{v_{0i}} = \frac{s_i}{\dfrac{L_0}{D} + 1} \tag{5-28}$$

When this equation is solved for d_i and both sides of the expression so obtained are summed over all components i, the following expression for the corrected value of D is obtained:

$$D = \sum_{i=1}^{c} d_i = \left(\frac{1}{\dfrac{L_0}{D} + 1}\right) \sum_{i=1}^{c} s_i v_{0i} \tag{5-29}$$

Likewise, the following expression for the corrected value of L_0 is obtained from Equation (5-21), namely,

$$L_0 = \sum_{i=1}^{c} l_{0i} = \left(\frac{1}{1 + \dfrac{D}{L_0}}\right) \sum_{i=1}^{c} s_i v_{0i} \tag{5-30}$$

Next it will be shown that σ_1, which is defined by

$$V_1 = \sigma_1 \mho_1 = \sigma_1 \sum_{i=1}^{c} \left(\frac{v_{1i}}{v_{0i}}\right)_{ca} v_{0i}$$

is equal to unity. Elimination of $(v_{1i}/v_{0i})_{ca}$ by use of Equation (5-19) gives

$$V_1 = \sigma_1 \sum_{i=1}^{c} (1 + s_i) v_{0i} \tag{5-31}$$

When the results given by Equations (5-29), (5-30), and (5-31) are substituted into the total-material balance

$$V_1 - L_0 - D - V_0 = 0 \tag{5-32}$$

the following expression is obtained upon rearrangement

$$\sigma_1 \sum_{i=1}^{c} (1 + s_i) v_{0i} - \sum_{i=1}^{c} (1 + s_i) v_{0i} = 0$$

which yields the result

$$\sigma_1 = 1. \tag{5-33}$$

The calculation of the remaining members of the sets $\{\eta_j\}$ and $\{\sigma_j\}$ is carried out in a manner analogous to that described in Chapter 4 for absorbers (see Problem 5-4).

Application of the method of direct iteration to liquid-liquid extraction problems

If θ is set equal to unity throughout the calculational procedures described for the single-θ method, that is, if the calculated values of the variables at the end of each trial are taken to be the assumed values of the variables for making the next trial, the method of direct iteration is obtained. Although this method converged relatively rapidly for the numerical examples presented herein, Hutton (10) produced other examples for which this method either failed to converge or converged very slowly.

Application of the multi-θ method of convergence to liquid-liquid extraction problems

When it is supposed that the temperatures are known throughout the column, Methods I and II of the multi-θ method described in Chapter 4 become identical and are referred to herein as simply the multi-θ method of convergence. For liquid-liquid extraction problems involving columns of the type illustrated in Figure 5-2, the multi-θ method of convergence is applied as described in Chapter 4 for absorbers except for the manner in which the dependency of the activity coefficients (or K values) on the compositions is taken into account. When the dependency of the K values on composition is neglected in the development of the formulas for the partial derivatives, all of the equations stated in Chapter 4 for absorbers are applicable to liquid-liquid extraction columns of the type described by Figure 5-2. To account for the dependency of the K values on the compositions, however, a minor modification of the calculational procedure described for absorbers is necessary.

In one modification of the calculational procedure for absorbers, the K values were held fixed until the corresponding solution set of θ_j's had been determined. Then, on the basis of these compositions, a new set of K values was computed, and this set was used to find a new solution set of θ_j's. Continuation of this process led to convergence in relatively few trials and a relatively small amount of computer time (see Example 5-3 in Tables 5-3 and 5-4). The multi-θ method was the fastest of those investigated; only 0.41 minutes of computer time (IBM 360/65) were required to make the 11 trials for Example 5-3 (see Part III of Table 5-4).

In a second modification of the procedure described in Chapter 4 for absorbers, the K values were reevaluated at the end of each trial through the material balances by use of sets of normalized compositions. When this procedure was used to solve Example 5-1, a total of 21 trials and 0.35 minutes of computer time (IBM 360/65) were required to obtain a solution of the same accuracy as stated at the bottom of Table 5-4. In the solution of this problem, it was, of course, necessary to modify the equations stated in Chapter 4 for absorbers to reflect the introduction of an additional feed F on plate f as shown in Figure 5-3.

Either one of the above modifications of the general procedure described in Chapter 4 could lead to convergence problems because the dependency of the K_{ji}'s on the θ_j's was neglected in the partial differentiation of the functions F_j. Alternately, one could choose the L_j/V_j's, the x_{ji}'s and y_{ji}'s as the

TABLE 5-3
STATEMENT OF EXAMPLE 5-3

Component	Component No.	l_{0i} (solvent) moles/hr	$v_{N+1,i}$ (solvent) moles/hr	FX_i (feed) moles/hr
Acetone	1	0.0	0.0	0.1
Ethanol	2	0.0	0.0	0.1
Chloroform	3	0.8	0.0	0.0
Water	4	0.0	1.0	0.0

Other Specifications:
Total number of plates $N = 15$, feed plate for F is $f = 6$, and feed plate for V_{N+1} is $N = 15$ (see Figure 5-3). The activity coefficients are computed by use of the following three-suffix Margules equation for component I in a mixture of L components

$$\gamma_I = \exp\left\{\left[2x_I \sum_{J=1}^{L} x_J A_{IJ}\right] + \left[\sum_{J=1}^{L} (x_J)^2 A_{IJ}\right] + \right.$$

$$\left[\sum_{\substack{J=1 \\ I\neq J \\ J\neq K \\ J<K}}^{L} \sum_{K=2}^{L} x_J x_K A^*_{IJK}\right] - \left[2 \sum_{I=1}^{L} \left\{(x_I)^2 \sum_{J=1}^{L} x_J A_{JI}\right\}\right] - $$

$$\left. 2\left[\sum_{\substack{I=1 \\ I\neq J \\ I\neq K \\ J<K}}^{L} \sum_{J=2}^{L} \sum_{K=3}^{L} x_I x_J x_K A^*_{IJK}\right]\right\}$$

where

$$A^*_{IJK} = \frac{1}{2}(A_{IJ} + A_{JI} + A_{IK} + A_{KI} + A_{JK} + A_{KJ})$$

[This equation consists of a corrected form of the expression presented in Reference (9).]
The values of A_{IJ} are as follows:

$A_{11} = 0$	$A_{12} = 0.5446$	$A_{13} = -0.9417$	$A_{14} = 1.872$
$A_{21} = 0.599$	$A_{22} = 0.0$	$A_{23} = 1.61$	$A_{24} = 1.46$
$A_{31} = -0.674$	$A_{32} = 0.501$	$A_{33} = 0.0$	$A_{34} = 5.91$
$A_{41} = 1.338$	$A_{42} = 0.877$	$A_{43} = 4.76$	$A_{44} = 0.0$

independent variables in a manner similar to that suggested by Roche and Staffin (14). However, such an approach leads to extremely large matrices.

The application of the multi-θ method to a column with reflux of the type described above for the column shown in Figure 5-4 merits additional consideration. For problems involving columns of this type, there are $N + 1$ unknown values of L_j/V_j, namely, $L_0/V_0, L_1/V_1, \ldots, L_N/V_N$, and consequently there exist $N + 1$ unknown multipliers, $\theta_0, \theta_1, \ldots, \theta_{N+1}, \theta_N$. The component-material balances and equilibrium relationships may be represented by Equation (5-23), provided, of course, that the A_{ji}'s in the matrix A_i are taken to have the meaning

$$A_{ji} = \frac{\theta_j}{K_{ji}}\left(\frac{L_j}{V_j}\right)_a \tag{5-34}$$

The total material balances are again given by an equation of the same form as Equation (4-44), $RV = -\mathfrak{F}$, except that in this case, the tridiagonal

TABLE 5-4
SOLUTION OF EXAMPLE 5-3

I. Initial Assumptions and the Solution Set of Flow Rates $\{V_j\}$

Plate No.	V_j (initial) lb-moles/hr	V_j (final) lb-moles/hr	Other Initial Specifications
1	1.4	1.048	All streams V_j were assumed to
2	1.4	1.102	have the same initial composition
3	1.4	1.142	as V_{N+1}. All streams L_j were as-
4	1.4	1.180	sumed to have the initial composi-
5	1.4	1.226	tion of the combined streams L_0
6	1.4	1.307	and F. The initial set of activity
7	1.0	1.327	coefficients was computed on the
8	1.0	1.322	basis of these sets of assumed com-
9	1.0	1.313	positions for V_j and L_j.
10	1.0	1.299	
11	1.0	1.275	
12	1.0	1.239	
13	1.0	1.190	
14	1.0	1.131	
15	1.0	1.069	

II. Final Compositions of Streams L_N and V_1

Component No.	x_{Ni}	y_{1i}
1	0.10506	0.23973×10^{-6}
2	0.04531	0.54253×10^{-1}
3	0.83613	0.39686×10^{-2}
4	0.01350	0.94178

III. Comparison of the Proposed Calculational Procedures for Example 5-3

Calculational Procedure	No. of Trials*
Tierney et al. (21)	24
Hanson et al. (9)	42
Single-θ method	17
Direct iteration ($\theta = 1$)	23
Multi-θ method	11

*The convergence criterion used was $\left|\dfrac{\Delta V}{V}\right| \leq 10^{-3}$, where $\Delta V = V_{correct} - V_{calculated}$. The correct set was obtained by making a large number of trials to converge within the accuracy of the IBM 360/65 computer.

matrix R and the conformable column vectors V and $\mathfrak{F}$ have the following definitions:

$$
R = \begin{bmatrix}
-\rho_0 & 1 & 0 & 0 & 0 & \dots & 0 \\
r_0 & -(1+r_1) & 1 & 0 & 0 & \dots & 0 \\
0 & r_1 & -(1+r_2) & 1 & 0 & \dots & 0 \\
\multicolumn{7}{c}{\dotfill} \\
0 & \multicolumn{4}{c}{\dotfill} 0 & r_{N-1} & -(1+r_N)
\end{bmatrix}
$$

$$
V = [V_0 V_1 \dots V_f \dots V_{N-1} V_N]^T
$$

$$
\mathfrak{F} = [00 \dots 0 F 0 \dots 0 V_{N+1}]^T \tag{5-35}
$$

where

$$
\rho_0 = 1 + \left(1 + \frac{D}{L_0}\right) r_0 ;
$$

$$
r_j = \theta_j \left(\frac{L_j}{V_j}\right)_a \qquad (j = 0, 1, 2, \dots, N).
$$

(The element F is located in row f of the column vector $\mathfrak{F}$.)

The dew point functions F_j for $j = 1$ through $j = N$ are the same as those given by Equation (4-48). The function F_0 corresponding to the solvent separator is developed by taking an expression of the same form as Equation (4-47) as a starting point, that is,

$$
F_0 = \frac{\sum\limits_{i=1}^{c} l_{0i}}{L_0} - \frac{\sum\limits_{i=1}^{c} v_{0i}}{V_0} \tag{5-36}
$$

The component flow rate l_{0i} may be expressed in terms of v_{0i} by use of Equation (5-21), and since the total flow rate L_0 may be stated in terms of V_0,

$$
L_0 = r_0 V_0 = \theta_0 \left(\frac{L_0}{V_0}\right)_a V_0 \tag{5-37}
$$

it is possible to restate Equation (5-36) in terms of v_{0i} and V_0 as follows:

$$
F_0 = \frac{1}{V_0} \left[\sum_{i=1}^{c} \left(\frac{\dfrac{s_i}{r_0}}{1 + \dfrac{D}{L_0}} - 1 \right) v_{0i} \right] \tag{5-38}
$$

The partial derivatives of the F_j's ($j = 1, 2, \dots, N$) with respect to the θ_k's ($k = 0, 1, 2, \dots, N$) are given by Equation (4-50). The partial derivatives of F_0 with respect to θ_k ($k = 0, 1, 2, \dots, N$) are given by

$$
\frac{\partial F_0}{\partial \theta_k} = \frac{1}{V_0} \left\{ \sum_{i=1}^{c} \left[\left(\frac{\dfrac{s_i}{r_0}}{1 + \dfrac{D}{L_0}} - 1 \right) \frac{\partial v_{0i}}{\partial \theta_k} - \left(\frac{s_i}{1 + \dfrac{D}{L_0}} \right) \frac{v_{0i}}{r_0^2} \frac{\partial r_0}{\partial \theta_k} \right] - F_0 \frac{\partial V_0}{\partial \theta_k} \right\}
$$

$$
\tag{5-39}
$$

where

$$\frac{\partial r_0}{\partial \theta_k} = 0, \text{ for } k \neq 0;$$

$$\frac{\partial r_0}{\partial \theta_k} = \left(\frac{L_0}{V_0}\right)_a, \text{ for } k = 0.$$

The partial derivatives of the F_j's with respect to the θ_k's are needed to compute an improved set of θ_j's by use of Equation (4-49). $J_n \Delta \theta_n = -F_n$. In this case, the matrices J_n, $\Delta \theta_n$, and F_n have the following definitions:

$$J_n(F/\theta) = \begin{bmatrix} \dfrac{\partial F_0}{\partial \theta_0} & \cdots & \dfrac{\partial F_0}{\partial \theta_N} \\ \cdots\cdots\cdots\cdots \\ \dfrac{\partial F_N}{\partial \theta_0} & \cdots & \dfrac{\partial F_N}{\partial \theta_N} \end{bmatrix}$$

$$\Delta \theta_n = [\Delta \theta_0 \quad \cdots \quad \Delta \theta_N]^T$$

$$F_n = [F_0 F_1 \quad \cdots \quad F_N]^T$$

(5-40)

To evaluate the elements of J_n, the partial derivatives of the v_{ji}'s and V_j's with respect to the θ_k's are needed. The values of these partial derivatives are computed by use of Equations (4-51) and (4-52), $A_i \, \partial v_i / \partial \theta_k = C_{ki}$ and $R \, \partial V / \partial \theta_k = E_k$, respectively. The matrices A_i, v_i, R, and V are defined by Equations (5-23) and (5-35) while C_{ki} and E_k have the following definitions:

$$C_{0i} = [00\ldots 0]^T; \qquad C_{1i} = [0C_{1i}(-C_{1i})0\ldots 0]^T:$$

$$C_{N-1,i} = [0\ldots 0C_{N-1,i}(-C_{N-1,i})]^T; \qquad C_{Ni} = [0\ldots 0C_{Ni}]^T:$$

$$C_{ki} = v_{ki}\frac{\partial A_{ki}}{\partial \theta_k} = \frac{v_{ki}}{K_{ki}}\left(\frac{L_k}{V_k}\right)_a \qquad (k = 1, 2, \ldots, N)$$

$$E_0 = [e_0(-E_0)0\ldots 0)]^T; \qquad E_1 = [0E_1(-E_1)0\ldots 0)]^T;$$

$$E_{N-1} = [0\ldots 0E_{N-1}(-E_{N-1})]^T; \qquad E_N = [0\ldots 0E_N]^T;$$

$$E_k = \left(\frac{L_k}{V_k}\right)_a V_k \qquad (k = 0, 1, 2, \ldots, N); \qquad e_0 = \left(1 + \frac{D}{L_0}\right)E_0$$

Again, the composition effects on the K values may be taken into account by use of either of the two methods described above in the discussion of Example 5-3. (The results shown in Table 5-4 were obtained by reevaluation of the K values after each solution set of θ_j's had been obtained.)

When the temperatures vary throughout the column and enthalpy data are available, the multi-θ method as well as the single-θ method are applied to extraction problems in the same manner as described in Chapter 4 for absorbers.

Use of the Newton-Raphson method as proposed by Tierney and Bruno

As discussed in Chapter 4, both the multi-θ method as well as Tierney's method are based on the use of the Newton-Raphson method. In the method of Tierney et al. (21, 22), the vapor flow rates $\{V_j\}$ and the variables $(\{L_j\}, \{x_{ji}\}, \{y_{ji}\})$ are taken to be dependent variables. A development of the equations for the case of isothermal operation follows.

ISOTHERMAL OPERATION OF A COUNTERCURRENT EXTRACTOR. The development that follows is applicable to extractors of the type shown in Figure 5-2. For such an extractor, there exist N values of the independent variables, namely, $V_1, V_2, \ldots, V_N$. Now let N functional relationships be selected so that when the solution set of the V's has been selected, all functions take on the value of zero simultaneously. Tierney and Bruno selected the following N functional relationships

$$F_j = \sum_{i=1}^{c} x_{ji} - \sum_{i=1}^{c} y_{ji}, \qquad (1 \leq j \leq N) \tag{5-41}$$

When the equilibrium relationship given by Equation (5-15) is used to eliminate the y_{ji}'s from Equation (5-41), the following bubble-point form of the functions is obtained.

$$F_j = \sum_{i=1}^{c} (1 - K_{ji}) x_{ji}, \qquad (1 \leq j \leq N) \tag{5-42}$$

The Newton-Raphson equation is given by

$$J_n(F/V) \, \Delta V_n = -F_n \tag{5-43}$$

where

$$J_n(x/V) = \begin{bmatrix} \sum_{i=1}^{c} (1 - K_{1i})\dfrac{\partial x_{1i}}{\partial V_1} & \cdots & \sum_{i=1}^{c} (1 - K_{1i})\dfrac{\partial x_{1i}}{\partial V_N} \\ \cdots\cdots\cdots\cdots\cdots\cdots\cdots\cdots\cdots\cdots\cdots\cdots \\ \sum_{i=1}^{c} (1 - K_{Ni})\dfrac{\partial x_{Ni}}{\partial V_1} & \cdots & \sum_{i=1}^{c} (1 - K_{Ni})\dfrac{\partial x_{Ni}}{\partial V_N} \end{bmatrix};$$

$$\Delta V_n = [\Delta V_1 \Delta V_2 \ldots \Delta V_N]^T;$$

$$F_n = [F_1 F_2 \ldots F_N]^T;$$

$$\Delta V_j = V_{j,n+1} - V_{j,n}.$$

The elements of J_n and F_n are, of course, evaluated on the basis of the assumed set of V_j's used to make the nth trial.

The development of the formulas for the evaluation of the elements of J_n follows. In order to evaluate the terms that appear in the expressions for the partial derivatives of the x_{ji}'s with respect to the V_j's, it is necessary to develop expressions for the calculation of the L_j's, x_{ji}'s, and the partial derivatives of the L_j's with respect to the V_j's for an assumed set of V_j's.

First, formulas for the calculation of the total flow rates $\{L_j\}$ for an assumed set $\{V_j\}$ are obtained by use of the total material balances, where the input rates L_0 and V_{N+1} are known. For any assumed set of V_j's (V_1, V_2, V_3, ..., V_N), the corresponding set of L_j's (L_1, L_2, ..., L_N) follows immediately from the total material balances.

The partial derivatives of the L_j's with respect to the V_j's may be found by taking the partial derivatives of the total material balances. The results obtained are as follows:

$$\frac{\partial L_j}{\partial V_j} = -1 \qquad (1 \leq j \leq N)$$

$$\frac{\partial L_j}{\partial V_{j+1}} = 1 \qquad (1 \leq j \leq N - 1)$$

$$\frac{\partial L_j}{\partial V_m} = 0 \qquad (m \neq 1, m \neq j+1, 1 \leq j \leq N)$$

$$\frac{\partial V_j}{\partial V_m} = 0 \qquad (m \neq j, 1 \leq j \leq N)$$

$$\frac{\partial V_j}{\partial V_m} = 1 \qquad (m = j, 1 \leq j \leq N) \tag{5-44}$$

This same set of results may be obtained by use of the calculus and algebra of matrices as demonstrated by Tierney et al. (21). When the equilibrium relationship [Equation (5-15)] is used to eliminate the y_{ji}'s from the component-material balances enclosing each stage, the resulting set of equations for an extractor of the type shown in Figure 5-2 may be stated in matrix form for any component i as follows:

$$[L + VK_i]x_i = -\boldsymbol{\mathcal{f}}_i \tag{5-45}$$

where

$$\boldsymbol{\mathcal{f}}_i = [L_0 x_{0i} 0 \ldots 0 V_{N+1} y_{N+1,i}]^T;$$

$$x_i = [x_{1i} x_{2i} \ldots x_{N-1,i} x_{N,i}]^T;$$

$$K_i = \begin{bmatrix} K_{1i} & 0 & 0 & \ldots & 0 \\ 0 & K_{2i} & 0 & \ldots & 0 \\ \multicolumn{5}{c}{\ldots\ldots\ldots\ldots\ldots} \\ 0 & \ldots\ldots & 0 & & K_{Ni} \end{bmatrix};$$

$$V = \begin{bmatrix} -V_1 & V_2 & 0 & 0 & \ldots & 0 \\ 0 & -V_2 & V_3 & 0 & \ldots & 0 \\ \multicolumn{6}{c}{\ldots\ldots\ldots\ldots\ldots\ldots\ldots\ldots} \\ 0 & \ldots\ldots & 0 & 0 & \ldots & -V_N \end{bmatrix};$$

$$L = \begin{bmatrix} -L_1 & 0 & 0 & 0 & \dots & 0 \\ L_1 & -L_2 & 0 & 0 & \dots & 0 \\ \multicolumn{6}{c}{\dotfill} \\ 0 \dots & 0 & 0 & L_{N-1} & & -L_N \end{bmatrix}.$$

When Equation (5-45) is solved for x_i, the following result is obtained:

$$x_i = -Z_i^{-1} \mathcal{f}_i \tag{5-46}$$

where the matrix Z_i is defined by

$$Z_i = [L + VK_i] \tag{5-47}$$

Matrix Equation (5-46) is used to calculate the x_{ji}'s for each component i as follows. On the basis of an assumed set of V_j's, the corresponding values of the L_j's are found by using the total material balances. The K_{ji}'s are evaluated on the basis of the most recent sets of compositions. Since the elements of $\mathcal{f}_i$ are known, it is evident that Equation (5-46) may be solved for the x_{ji}'s for each component i.

Next, the formulas for the evaluation of the partial derivatives of the x_{ji}'s with respect to the V_j's are developed. Differentiation of the members of matrix Equation (5-45) with respect to a particular V_j yields

$$\left[\frac{\partial L}{\partial V_j} + \frac{\partial V}{\partial V_j} K_i \right] x_i + Z_i \frac{\partial x_i}{\partial V_j} = 0 \tag{5-48}$$

which may be solved by matrix algebra to give

$$\frac{\partial x_i}{\partial V_i} = -Z_i^{-1} \left[\frac{\partial L}{\partial V_j} + \frac{\partial V}{\partial V_j} K_i \right] x_i \tag{5-49}$$

Elements of the matrices $\partial L / \partial V_j$ and $\partial V / \partial V_j$ are given by Equation (5-44), and the elements of x_i are given by Equation (5-46). On the basis of the x_{ji}'s so obtained, the corresponding y_{ji}'s are computed by use of the original set of K_{ji}'s. If it is desired [as pointed out by Tierney et al. (21)], the most recent set of x_{ji}'s and y_{ji}'s may be normalized before the subsequent calculation of the elements of ΔV_n as given by Equation (5-43). Once the elements of ΔV_n have been obtained, the V_j's for the next trial are readily computed by use of the definition of ΔV_j that follows Equation (5-43). The V_j's obtained and the total material balances are used to determine the L_j's to be used for the next trial.

When the variation of temperatures is taken into account, Tierney et al. (21, 22) proposed a matrix method in which the vapor rates $\{V_j\}$ and temperatures $\{T_j\}$ were taken to be the independent variables. Except for the fact that the derivatives were determined by use of matrix equations, the procedure is analogous to Procedure 4 of Chapter 4.

Application of method II of Hanson et al. to liquid-liquid extraction problems

This procedure, called "Method II" by Hanson *et al.* (9), is in many respects similar to another procedure that Hanson *et al.* called Method I. Method II was recommended by Hanson *et al.* as being more stable for problems involving highly nonideal solutions such as liquid-liquid extraction problems. As pointed out by Hanson *et al.*, Smith *et al.* (18, 19) proposed a procedure that was essentially the same as Hanson's Method I.

The first step of this procedure (Method II) is the assumption of sets of compositions $\{y_{ji}\}$ and $\{x_{ji}\}$ for each component i and plate j as well as sets of the total flow rates $\{L_j\}$ and $\{V_j\}$. It is, of course, supposed that the number of plates N, the temperature of the column, and the complete definitions of the streams V_{N+1} and L_0 have been specified for a column of the type shown in Figure 5-2.

On the basis of the assumed sets of compositions, the activity coefficients γ_{ji}^V and γ_{ji}^L are calculated for each component i and for each plate j.

Next, improved sets of compositions are calculated by use of component-material balances as follows. The component-material balance enclosing plate 1 of the column in Figure 5-2 is obtained by setting $j = 1$ in the following expression:

$$V_{j+1}y_{j+1,i} + L_{j-1}x_{j-1,i} - V_jy_{ji} - L_jx_{ji} = 0 \tag{5-50}$$

The equilibrium relationship

$$y_{ji} = \frac{\gamma_{ji}^L}{\gamma_{ji}^V}x_{ji} = K_{ji}x_{ji} \tag{5-51}$$

may be used to eliminate x_{ji} from Equation (5-50) to give the following expression for y_{ji}, namely,

$$y_{ji} = \frac{V_{j+1}y_{j+1,i} + L_{j-1}x_{j-1,i}}{V_j + \dfrac{L_j}{K_{ji}}} \tag{5-52}$$

To evaluate the right-hand side of Equation (5-52) for $j = 1$, the values of the variables initially assumed are employed. After the y_{1i}'s have been found, the corresponding set of x_{1i}'s are found by use of Equation (5-51). This set of x_{1i}'s is then used to compute the y_{2i}'s. In this case [$j = 2$ in Equation (5-52)], the remaining values of the variables V_3, y_{3i}, L_1, and L_2 on the right-hand side of Equation (5-52) are taken equal to those initially assumed. After the y_{2i}'s have been computed, the corresponding x_{2i}'s are computed by use of Equation (5-51). The procedure is continued by using these x_{2i}'s in the calculation of the y_{3i}'s. Finally, one obtains a new set of values for the y_{Ni}'s and the x_{Ni}'s for the bottom plate N.

Next, further improved sets of compositions are found by calculating back up through the column. First, one computes the $y_{N-1,i}$'s by use of Equation (5-52) and the most recent sets of compositions, the initially assumed total flow rates, and the initial set of K_{ji}'s (those computed at the beginning of the calculation down through the column). The calculation is continued back up through the column by use of the most recent sets of compositions in a manner analogous to that described for calculating down from the top. The compositions obtained are used to predict improved sets of total flow rates and K values that are used to make the next trial through the column. In the interest of brevity, the details of this predictive procedure

TABLE 5-5
CONVERGENCE CHARACTERISTICS OF THE FOUR PROCEDURES USED
TO SOLVE EXAMPLE 5-3

Trial No.	Single-θ method θ	Single-θ method $V_7{}^*$	Multi-θ method $V_7{}^*$	Direct Iteration $V_7{}^*$	Tierney et al. (21) $V_7{}^*$	Hanson et al. (9) V_7
1	0.62438	1.2943	1.1967	1.2943	1.2943	1.1355
2	1.01094	1.4180	1.2569	1.3606	1.2114	1.1952
3	1.00531	1.3850	1.2828	1.3895	1.3523	1.2340
4	1.00625	1.3564	1.2984	1.4038	1.3350	1.2647
5	1.00436	1.3383	1.3091	1.4075	1.3807	1.2908
6	1.00301	1.3281	1.3167	1.4030	1.4014	1.3134
7	1.00216	1.3231	1.3219	1.3933	1.4107	1.3325
8	1.00147	1.3211	1.3251	1.3811	1.4076	1.3477
9	1.00096	1.3209	1.3269	1.3685	1.3971	1.3595
10	1.00067	1.3216	1.3277	1.3569	1.3831	1.3682
11	1.00037	1.3227	1.3280	1.3472	1.3688	1.3740
12	1.00021	1.3239		1.3396	1.3560	1.3773
13	1.00010	1.3249		1.3339	1.3423	1.3784
14	1.00002	1.3258		1.3299	1.3378	1.3778
15	0.99997	1.3265		1.3274	1.3322	1.3757
16	0.99997	1.3269		1.3259	1.3286	1.3726
17	1.00000	1.3272			1.3252	1.3687
18					1.3250	1.3643
19					1.3251	1.3597
20					1.3253	1.3551
21					1.3257	1.3506
22					1.3260	1.3464
23						1.3426
24						1.3392
25						1.3362
30						1.3274
35						1.3255
40						1.3260
42						1.3263

*These were the values obtained at the end of the component-material balances.

were omitted. [They are given not only in the original presentation by Hanson *et al.* (9) but also by Hutton (10) who stated them in terms of the notation used herein.]

Comparison of the proposed numerical methods

Example 5-3 (Tables 5-3 and 5-4) was used to compare the calculational procedures proposed by Hanson *et al.* (9) and Tierney *et al.* (21) with the single-θ method of convergence (10), the multi-θ method of convergence, and the method of direct iteration. This example involves a column having the same configuration as the one shown in Figure 5-3, and it consists of a minor modification of the one presented by Hanson *et al.* (9).

The convergence characteristics of the five methods used to solve Example 5-3 are presented in Table 5-5. The variation of the flow rate V_7 that was used for comparison of the five methods was fairly typical of the variations exhibited by the other flow rates.

Example 5-4 (Tables 5-6 and 5-7) is presented to demonstrate the use of the single-θ and multi-θ methods of convergence as well as the method of direct iteration for solving liquid-liquid extraction problems that involve the use of a solvent separator (see Figure 5-4). The configuration of the column involved in Example 5-2 is the same as the one shown in Figure 5-9 except that the feed F_1 enters on plate 1 and the feed F_2 enters on plate f. This example is similar to a problem considered by Roche *et al.* (14), who applied the Newton-Raphson method with the $\{V_j\}, \{L_j\}, \{x_{ji}\}$ as the inde-

TABLE 5-6
STATEMENT OF EXAMPLE 5-4

Component	Component No.	$F_1 X_{1i}$ (solvent) lb-moles/hr	$F_2 X_{2i}$ (feed) lb-moles/hr	$v_{N+1, i}$ (solvent) lb-moles/hr	Separator Factors*
Acetone	1	0.0	0.1	0.0	200
Ethanol	2	0.0	0.1	0.0	570
Chloroform	3	0.8	0.0	0.0	350
Water	4	0.0	0.0	1.0	7×10^{-4}

Other Specifications

This column has the same general configuration as the one shown in Figure 5-4 except that instead of one stream F, this column has two such streams, F_1 and F_2. Solvent F_1 enters on plate 1, feed F_2 enters on plate 6, and solvent V_{N+1} enters on plate 15. The total number of plate $N = 15$. The reflux ratio L_0/D is fixed at 0.25, and the activity coefficients are to be computed by use of the expressions and constants A_{IJ} stated for Example 5-3.

*The operation of the solvent separator is described by the specification of the separator factors s_i, where

$$s_i = \frac{L_0 x_{0i} + D X_{Di}}{V_0 y_{0i}}$$

pendent variables. The statement of Example 5-4 is presented in Table 5-6, and the results obtained by the two θ methods and direct iteration are presented in Table 5-7.

TABLE 5-7

SOLUTION OF EXAMPLE 5-4

I. *Initial Assumptions and the Solution Set of Flow Rates* $\{V_j\}$

Plate No.	V_j (initial) lb-moles/hr	V_j (final) lb-moles/hr	Other Initial Assumptions
0	1.36	0.987	All streams V_j were assumed to
1	1.4	1.060	have the same initial composi-
2	1.4	1.111	tion as V_{N+1}. All streams L_j
3	1.4	1.148	were assumed to have the initial
4	1.4	1.185	composition of the combined
5	1.4	1.231	streams F_1 and F_2. The initial
6	1.4	1.312	set of activity coefficients were
7	1.0	1.332	computed on the basis of these
8	1.0	1.328	sets of assumed compositions
9	1.0	1.320	for V_j and L_j.
10	1.0	1.305	
11	1.0	1.281	
12	1.0	1.246	
13	1.0	1.196	
14	1.0	1.136	
15	1.0	1.071	

II. *Final Compositions* * *of Streams* L_N, V_0, *and* D

Component No.	x_{Ni}	y_{0i}	X_{Di}
1	0.10479	0.14034×10^{-8}	0.37540×10^{-5}
2	0.04700	0.12196×10^{-3}	0.92976
3	0.83454	0.13005×10^{-4}	0.60879×10^{-1}
4	0.01359	0.99987	0.93610×10^{-2}

III. *Convergence Characteristics of the Single-θ method, Direct Iteration, and the Multi-θ method*

Calculational Procedure	No. of Trials	Computational Time (min)*
Single-θ method	15	0.82
Direct iteration ($\theta = 1$)	21	0.65
Multi-θ method	11	0.60

*The convergence criterion used was $\left| \dfrac{\Delta V}{V} \right| \leqq 10^{-3}$ (see Item III of Table 5-4).

PROBLEMS

5-1. The graphical procedure described in the text is easily extended to handle columns with two feeds as shown in Figure 5-3. First, let the difference points x_Δ and x'_Δ be defined as follows:

$$x_\Delta = \frac{L_0 x_{0i} - V_1 y_{1i}}{L_0 - V_1} \qquad (i = A, C, S) \tag{A}$$

and

$$x'_\Delta = \frac{L_N x_{Ni} - V_{N+1} y_{N+1,i}}{L_N - V_{N+1}} \qquad (i = A, C, S) \tag{B}$$

Next, let the addition stream J and the addition point z_J be defined by

$$L_0 + F = J \tag{C}$$

$$L_0 x_{0i} + F X_i = J z_{Ji} \qquad (i = A, C, S) \tag{D}$$

where F is the flow rate of the feed that enters the column on plate f and X_i is its composition. Finally, let the addition stream J' and the addition point z'_J be defined by

$$L_0 + F_1 + V_{N+1} = L_N + V_1 = J' \tag{E}$$

$$L_0 x_{0i} + F X_i + V_{N+1} y_{N+1,i} = L_N x_{Ni} + V_1 y_{1i} = J' z'_{Ji} \tag{F}$$

By use of these definitions and the required material balances for the column shown in Figure 5-3, prove the following propositions for the case in which the two feeds have a totality of three components and the column is to be operated isothermally in the two-phase region.

1. $x_{\Delta i} = \dfrac{L_j x_{ji} - V_{j+1} y_{j+1,i}}{L_j - V_{j+1}}$ $\qquad (0 \leq j \leq f - 1, i = A, C, S)$

2. $x'_{\Delta i} = \dfrac{L_j x_{ji} - V_{j+1} y_{j+1,i}}{L_j - V_{j+1}}$ $\qquad (f \leq j \leq N, i = A, C, S)$

3. The points x_Δ, x'_Δ, and X all lie on the same straight line.
4. The points z_J, y_1, and x'_Δ all lie on the same straight line.
5. The points x_N, y_{N+1}, and x'_Δ all lie on the same straight line.
6. The points x_0, y_1, and x_Δ all lie on the same straight line.
7. The points x_N, z'_J, and y_1 all lie on the same straight line.
8. The points y_{N+1}, z'_J, and z_J all lie on the same straight line.

9. $\dfrac{L_N}{V_1} = \dfrac{z'_{Ji} - y_{1i}}{x_{Ni} - z'_{Ji}}$ $\qquad (i = A, C, S)$

10. $\dfrac{L_N}{V_1} = \dfrac{\overline{z'_J y_1}}{x_N z'_J}$

11. $\dfrac{V_{N+1}}{J} = \dfrac{z'_{Ji} - z_{Ji}}{y_{N+1,i} - z'_{Ji}}$ $\qquad (i = A, C, S)$

12. $\dfrac{V_{N+1}}{J} = \dfrac{\overline{z'_J z_J}}{y_{N+1} z'_J}$

The relationships developed above may be used in the graphical determination of the minimum number of stages required to effect a specified separation. In particular, the following procedure for the location of the feed plate f minimizes the total number of stages required to effect the separation. Let the graphical solution be commenced at stage 1. Use the difference point x_Δ until a stage is reached whose x_A is less than X_A and then use x'_Δ for the remainder of the graphical procedure.

5-2. When the solubility curve has two branches as shown in Figure 5-7, extract reflux may be used advantageously to improve the separation. In this operation, the solvent stream V_0 is commonly returned to the system as "fresh solvent." Thus, the compositions of V_0 and V_{N+1} may be taken to be the same, that is, $y_{0i} = y_{N+1\,i}$ $(i = A, C, S)$. Also, the compositions of the streams L_0 and D will be taken to be the same, that is, $y_{0i} = y_{N+1,i}$ $(i = A, C, S)$. Also, the compositions of the streams L_0 and D will be taken to be the same since these streams leave the solvent removal unit as a single stream $(L_0 + D)$, that is, it will be supposed that $x_{0i} = X_{Di}$ $(i = A, C, S)$.

Except for the additional relationships that exist as a consequence of the solvent removal unit, the remaining relationships are the same as those stated for the column in Problem 5-1.

(a) Show that the following relationships represent the total and component-material balances for the solvent removal unit (see Figure 5-4):

$$V_1 = L_0 + V_0 + D \tag{A}$$

$$V_1 y_{1i} = L_0 x_{0i} + V_0 y_{0i} + DX_{Di} \qquad (i = A, C, S) \tag{B}$$

(Note: y_{0i} may be replaced by its equivalent $y_{N+1,i}$.)

(b) By use of the relationships in Part (a), show that the expression given by Proposition 1 of Problem 5-1 for $x_{\Delta i}$ reduces to

$$x_{\Delta i} = \frac{V_0 y_{N+1,i} + DX_{Di}}{V_0 + D} \qquad (i = A, C, S) \tag{C}$$

(c) Show that the points X_D, y_{N+1}, and x_Δ all lie on the same straight line.

(d) Verify the following relationships for the solvent recovery unit:

1. $\dfrac{V_0}{D} = \dfrac{X_{Di} - x_{\Delta i}}{x_{\Delta i} - y_{N+1,i}}$ $\qquad (i = A, C, S)$

2. $\dfrac{V_0}{D} = \dfrac{\overline{X_D x_\Delta}}{\overline{x_\Delta y_{N+1}}}$

3. $\dfrac{L_0}{V_0 + D} = \dfrac{x_{\Delta i} - y_{1i}}{y_{1i} - X_{Di}}$ $\qquad (i = A, C, S)$

4. $\dfrac{L_0}{V_0 + D} = \dfrac{\overline{x_\Delta y_1}}{\overline{y_1 X_D}}$

5-3. Show that the component-material balances for any component i for the extractor shown in Figure 5-3 may be represented by the matrix equation given Equation (4-25), provided that the zero element of $\mathcal{f}_i$ corresponding to the equation for plate f is set equal to FX_i.

5-4(a). Show that Equation (4-34) may be used to compute the η_j's of the single-θ method for the column shown in Figure 5-3, provided the following values of $\mathcal{B}$ are used.

$$\mathcal{B} = V_1 - L_0, \qquad (1 \leqq j \leqq f - 1)$$

$$\mathcal{B} = V_1 - L_0 - F, \quad (1 \leqq j \leqq N - 1)$$

(b). Repeat Part (a) for the column shown in Figure 5-4 except that for this column show that the values of $\mathfrak{B}$ to be used in Equation (4-34) in the calculation of the η_j's are given by

$$\mathfrak{B} = V_0 + D, \qquad (1 \leqq j \leqq f - 1)$$
$$\mathfrak{B} = V_0 + D - F, \qquad (f \leqq j \leqq N - 1)$$

5-5. Verify the expressions given below Equation (5-40) for C_{ki} and E_k.

5-6. Barley malt is contacted with water in a countercurrent multiple-stage extraction unit at a popular brewery. Assume that the solvent water is pure (as advertised) and that the barley malt is composed of an active ingredient schmaltz (A) and inert solid malt (C).

Equilibrium data has determined that each pound of inert solid will retain 1.4 lbs of solution regardless of the relative amounts of water and schmaltz in the solution. The barley malt to solvent ratio is 1 to 2.5 and the feed is composed of 75% inert malt, 20% water (all by weight). Find the number of equilibrium stages necessary to produce a final overhead product of 9.0% schmaltz.

5-7. Repeat Illustrative Example 5-1 for the case in which it is desired to find the solvent to feed ratio (V_{N+1}/L_0) required to effect the specified separation with two equilibrium stages.

5-8. Repeat Illustrative Example 5-1 for the case in which the feed L_0 contains 68% acetone and 32% water (by weight), and an additional feed F equal in total flow rate to L_0 and having the composition of 22% acetone and 78% water (by weight) is to be treated in the column for the purpose of recovering the acetone. A solvent to total feed ratio $V_{N+1}/(L_0 + F)$ of 0.75 of equilibrium stages N required to reduce the acetone in the raffinate to 1% when the feed F is introduced at the optimum location.

NOTATION*

A_i = a square matrix; defined by Equations (4-26) and (5-23).

F_0 = a function defined by Equation (5-36).

F = flow rate of the feed introduced on the intermediate plate f; see Figures 5-3 and 5-4.

F_j = functions defined by (4-48).

F_n = column vector of the functions F_j; defined by Equations (5-40) and (5-43).

$\mathcal{L}_i, \mathfrak{F}$ = feed vectors defined by Equation (5-23) and Equation (5-35), respectively.

J = a square matrix in the formula for the Newton-Raphson equations; defined by Equations (5-40) and (5-43).

J = sum of two streams; defined by Equation (5-1).

K_i = a square matrix of K values for component i; defined below Equation (5-45).

*See also notations in Chapters 3 and 4.

L = a square matrix that contains the L_j's as elements; defined below Equation (5-45).

L_j = total flow rate of the underflow leaving stage j in moles per hour or pounds per hour.

s_i = separation factor for each component i; defined by Equation (5-17).

V_j = total flow rate of the overflow leaving stage j in moles per hour or pounds per hour.

V = a column matrix which contains the V_j's as elements, defined by Equation (5-35); also used to denote a square matrix that contains the V_j's as elements defined below (5-45).

ΔV_n = a column vector having the elements ΔV_j ($1 \leq j \leq N$); defined by Equation (5-43).

X_i = mole fraction of component i in the feed F; see Figure 5-3 and 5-4.

x_{ji} = mole fraction (or mass fraction) of component i in the underflow leaving stage j.

y_{ji} = mole fraction (or mass fraction) of component i in the overflow leaving stage j.

x_Δ = difference point.

x_i = a column vector whose elements are $x_{1i}, x_{2i}, \ldots, x_{N-1,i}, x_{Ni}$.

z_J = addition point.

Z_i = a square matrix defined by Equation (5-47).

Greek letters

γ_i^L, γ_i^V = activity coefficients of streams L and V, respectively.

REFERENCES

1. Briggs, S. W., and E. W. Comings, "Effect of Temperature on Liquid-Liquid Equilibrium," *Ind. Eng. Chem.*, **35**, (1943), 411.

2. Brown, G. G., and Associates, *Unit Operations.* New York: John Wiley & Sons, Inc., 1953.

3. Darwent, B. de B., and C. A. Winkler, "The System n-Hexane-Methycyclopentane-Aniline," *J. Phys. Chem.*, **47**, (1943) 447.

4. Davis, M. W., and E. J. Weber, "Liquid-Liquid Extraction . . . Between Rotating Concentric Cylinders," *Ind. Eng. Chem.*, **52**, (1960), 929.

5. Edeleanu, L., "Refining with Liquid Sulphur Dioxide," *Jour. Inst. Petroleum Technologists*, **18**, (1932), 900.

6. Edeleanu, L. and G. Gane, "Hydrocarbures Estraits des Goudrons Acides du Petrole," *Report of the 3rd International Petroleum Congress-Bucarest*, **2** (1907) 665

7. Engler, C., and L. Ubbelohde, "Über das Edeleanusche Verfahren der Raffination von Erdöl mit Schwefeldioxyd," *Z. angew. Chem.*, **26**, (1913), 177.

8. Greenstadt, John, Yonathan Bard, and Burt Morse, "Multicomponent Distillation on the IBM 704," *Ind. Eng. Chem.*, **50**, (1958), 1644.

9. Hanson, D. N., J. H. Duffin, and G. F. Somerville, *Computation of Multistage Processes.* New York: Reinhold Publishing Corporation, 1962.

10. Hutton, A. E., and C. D. Holland, "Use of Field Tests in the Modeling of a Liquid-Liquid Extraction Column," *Chem. Eng. Sci.*, **27**, (1972), 919. See also Liquid-Liquid Extraction in Packed Columns at Steady-State Operation," Ph. D. Dissertation by A. E. Hutton, Texas A & M University, College Station, Texas, 1971.

11. Oldshue, J. Y., and J. H. Rushton, "Continuous-Extraction in a Multistage Mixer Column," *Chem. Eng. Progr.*, **48**, (1952), 297.

12. Podbielniak, W. J., "Method of Securing Countercurrent Contact of Fluids by Centrifugal Action," U.S. Patents 2,044,996 (1936); "Centrifugal Countercurrent Contact Apparatus," U. S. Patents 2,670,132 (1954).

13. Reman, G. H., "A New Efficient Extraction Apparatus: The Rotating Disc Contactor," Proc. 3rd World Petroleum Congress, Hague, Sect. III, (1951) 121. U. S. Patents 2,601,674 (1952); 2,729,545 (1956); 2,912,310 (1959).

14. Roche, E. C., and H. K. Staffin, "Rigorous Solution of Multicomponent Multistage Liquid-Liquid Extraction Problems," paper presented at 61st Annual A. I. Ch. E. Meeting. Los Angeles, California, December (1968).

15. Schiebel, E. G., "Fractional Liquid Extraction—Part I," *Chem. Eng. Progr.*, **44**, (1948), 681. "Extraction Apparatus," U. S. Patent 2,493,310 (1950).

16. ———, "Performance of an Internally Baffled Multistage Extraction Column," *A. I. Ch. E. Journal*, **2**, (1956), 74.

17. ———, U. S. Patent 2,850,362 (1958).

18. Smith, B. D., *Design of Equilibrium Stage Processes.* New York: McGraw-Hill Book Company, 1963.

19. Smith, B. D., and W. K. Brinkley, "Rigorous Solution of Multicomponent, Multistage Extraction Problems," *A. I. Ch. E. Journal*, **6**, (1960) 451.

20. Spense, R. W., and R. J. W. Streeton, "Fluid Contactor Apparatus," U. S. Patent 2,742,348 (1956).

21. Tierney, J. W., and J. A. Bruno, "Equilibrium Stage Calculations," *A. I. Ch. E. Journal*, **13**, (1967), 556.

22. Tierney, J. W., and J. L. Yanosik, "Simultaneous Flow and Temperature Correction in the Equilibrium Stage Problem," *A. I. Ch. E. Journal*, **15**, (1969), 897.

23. Treybal, R. E., *Liquid Extraction*, 2nd ed. New York: McGraw-Hill Book Company, 1963.

24. Van Dÿck, W. J. D., "Process and Apparatus for Intimately Contacting Fluids," U. S. Patent 2,011,186 (1935).

25. Wilde, D. J., and C. S. Beightler, *Foundations of Optimization.* Englewood Cliffs, N. J.: Prentice-Hall, Inc., 1967.

Section II:
Fundamentals and Modeling
of Rate Processes

Fundamentals of Modeling   **6**

In the second section of this book, the rates of simultaneous mass and heat transfer are taken into account in the modeling of separation processes. The fundamental rate expressions which are commonly accepted for heat and mass transfer are presented in this chapter. To demonstrate the use of these expressions in the formulation of the equations required to describe a process, a wide variety of relatively simple processes are considered in the first two chapters of this section. Techniques for the formulation of the equations for the relative simple processes are demonstrated for both steady state and unsteady state operation.

The remaining chapters in this section (Chapters 8-12) are devoted to accounting for the rates of mass and heat transfer in separation processes at steady state operation. The techniques demonstrated may be applied, however, in the modeling of separation processes at unsteady state operation as shown in references (4, 6, 8).

In this chapter, techniques for setting up the material and energy ba___ for systems at unsteady-state operation are considered first, and ___ of the various rate expressions for simultaneous mass and h___ presented. In the methods presented, the material and ___ represented by integral or integral-difference equati___ the time and space domains are used in the fo___ energy balances. Since this method constit___ may check a given balance by subst___ problem of signs takes care of it___

Section II:
Fundamentals and Modeling
of Rate Processes

Fundamentals of Modeling   **6**

In the second section of this book, the rates of simultaneous mass and heat transfer are taken into account in the modeling of separation processes. The fundamental rate expressions which are commonly accepted for heat and mass transfer are presented in this chapter. To demonstrate the use of these expressions in the formulation of the equations required to describe a process, a wide variety of relatively simple processes are considered in the first two chapters of this section. Techniques for the formulation of the equations for the relative simple processes are demonstrated for both steady state and unsteady state operation.

The remaining chapters in this section (Chapters 8-12) are devoted to accounting for the rates of mass and heat transfer in separation processes at steady state operation. The techniques demonstrated may be applied, however, in the modeling of separation processes at unsteady state operation as shown in references (4, 6, 8).

In this chapter, techniques for setting up the material and energy balances for systems at unsteady-state operation are considered first, and then some of the various rate expressions for simultaneous mass and heat transfer are presented. In the methods presented, the material and energy balances are represented by integral or integral-difference equations. Fixed increments in the time and space domains are used in the formulation of the material and energy balances. Since this method constitutes an actual balance, the student may check a given balance by substituting numbers in the equation. The problem of signs takes care of itself. All one has to do is to select a coordinate system to represent the boundary and the inputs or outputs relative to the positive directions of the selected coordinate system.

The integral or integral-difference forms of the balances have two other significant advantages over the differential or partial differential forms of these balances. Unlike the differential or partial differential equations for these balances, the integral or integral-difference forms may contain any finite number of point discontinuities caused by the input of heat Q or the output of work W at points along the boundary of the system. No difficulty at all is encountered in including these terms in the integral or integral-difference balances. In contrast, the differential or partial differential equations for these balances do not even exist at these points of discontinuity. In a sense, then, the differential or partial differential equations for the balances could be regarded as a subset of the more general integral or integral-difference equations.

Another advantage of the integral or integral-difference equations is the fact that they are suitable for direct solution by numerical methods. If, however, it is desired to obtain the differential or partial differential equation corresponding to the given integral or integral-difference equations, this can be done by applying at most two fundamental theorems of calculus, the *mean value theorem of differential calculus* (B-2)* and the *mean value theorem of integral calculus* (B-3). The final form of the differential or partial differential equation is obtained by allowing the increments in the space and/or the time domains to go to zero.

The techniques for formulating the integral or integral-difference forms of the material and energy balances and the corresponding differential or partial differential equations are introduced in Part 1. They are further demonstrated by the solution of a wide variety of problems in Chapter 7.

In Part 2, generally accepted rate expressions for various types of mass and heat transfer are introduced. These expressions could just as well have been called *models*. Although these rate expressions are not generally exact, they have proven useful over the years for the description of these rate processes, and they are customarily included in that body of information commonly referred to as fundamentals.

Part 1. Material Balances and Energy Balances

Material balances constitute an application of the *law of conservation of mass*. For purposes of application, a convenient statement of this law follows: Except for the conversion of mass to energy and conversely, mass can neither be created nor destroyed. Consequently, for a system in which

*See Appendix B, Theorem 2.

the conversion of mass to energy and conversely is not involved, it follows that during the time period from $t = t_n$ to $t = t_n + \Delta t$

$$\begin{pmatrix} \text{input of mass} \\ \text{to the system} \\ \text{during the} \\ \text{time period } \Delta t \end{pmatrix} - \begin{pmatrix} \text{output of mass} \\ \text{from the system} \\ \text{during the time} \\ \text{period } \Delta t \end{pmatrix} = \begin{pmatrix} \text{accumulation of} \\ \text{mass within the} \\ \text{system during the} \\ \text{time period } \Delta t \end{pmatrix} \qquad (6\text{-}1)$$

The accumulation term is defined as follows:

$$\begin{pmatrix} \text{accumulation of} \\ \text{mass within the} \\ \text{system during the} \\ \text{time period } \Delta t \end{pmatrix} = \begin{pmatrix} \text{amount of} \\ \text{mass in the} \\ \text{system at} \\ \text{time } t_n + \Delta t \end{pmatrix} - \begin{pmatrix} \text{amount of} \\ \text{mass in the} \\ \text{system at} \\ \text{time } t_n \end{pmatrix} \qquad (6\text{-}2)$$

The particular part of the universe under consideration is called the "system." The transformation of Equation (6-1) into a more convenient form for use in the description of industrial processes is perhaps best understood through the consideration of two relatively simple examples.

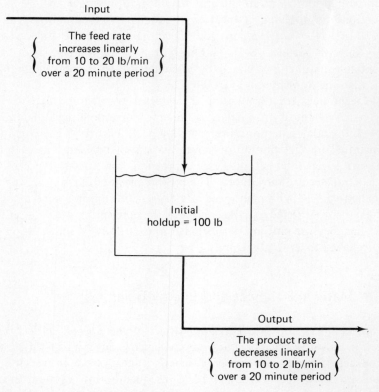

Input

{ The feed rate increases linearly from 10 to 20 lb/min over a 20 minute period }

Initial holdup = 100 lb

Output

{ The product rate decreases linearly from 10 to 2 lb/min over a 20 minute period }

Figure 6-1. An unsteady-state process with input and output rates that vary linearly with time, Illustrative Example 6-1.

ILLUSTRATIVE EXAMPLE 6-1

Initially, the mixer shown in Figure 6-1 is at steady-state operation with a feed rate F of 10 lb*/min, a product rate B of 10 lb/min, and a holdup U of 100 lb. At time $t = 0$, the feed rate commences to increase linearly and attains a rate of 20 lb/min at the end of 20 min. At time $t = 0$, the product rate commences to decrease linearly and attains a value of 2 lb/min at the end of 20 min. Calculate the holdup at the end of 20 min.

Solution: The variations of the input and output rates are demonstrated graphically in Figure 6-2. The input and outputs over the 20-min period are seen to be equal

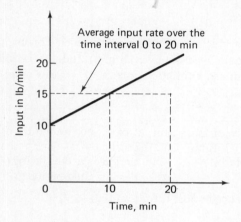

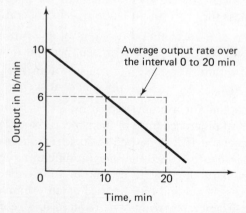

Figure 6-2. Variations of the inputs and outputs for Illustrative Example 6-2.

*Note that where there is no chance for confusion to arise between pounds force lb_f and pounds mass lb_m, the symbol lb is used.

to the areas under the respective curves over this time period. For the case of a linear variation of a function over a given time period, the area under the curve is equal to the arithmetic average of the values of the function at the beginning and at the end of the time period times the time period. Thus, from Equation (6-1), it follows that

$$\underbrace{\left(15\frac{lb}{min}\right)(20 \text{ min})}_{\substack{\text{area under the}\\\text{input curve}}} - \underbrace{\left(6\frac{lb}{min}\right)(20 \text{ min})}_{\substack{\text{area under the}\\\text{output curve}}} = U \text{ (lb)} - 100 \text{ (lb)} \qquad (6\text{-}3)$$

and

$$U = 280 \text{ lb}$$

Example 6-1 is a special case of the more general problem in which each of the rates [the input feed rate F (lb/min) and the output product rate B (lb/min)] varies in any continuous manner. For this case,

$$\underbrace{\int_{t_n}^{t_n+\Delta t} F\, dt}_{\substack{\text{input of mass}\\\text{during time}\\\text{period } \Delta t}} - \underbrace{\int_{t_n}^{t_n+\Delta t} B\, dt}_{\substack{\text{output of}\\\text{mass during}\\\text{time period}\\\Delta t}} = \underbrace{U\Big|_{t_n+\Delta t} - U\Big|_{t_n}}_{\substack{\text{accumulation of}\\\text{mass within the}\\\text{system during}\\\text{time period } \Delta t}} \qquad (6\text{-}4)$$

Since F and B are continuous, it follows that

$$\int_{t_n}^{t_n+\Delta t} (F - B)\, dt = U\Big|_{t_n+\Delta t} - U\Big|_{t_n} \qquad (6\text{-}5)$$

These examples suggest the following alternate statement of the law of conservation of mass for processes at unsteady-state operation:

$$\int_{t_n}^{t_n+\Delta t} \left[\left(\begin{array}{l}\text{input of mass to the}\\\text{system per unit time}\end{array}\right) - \left(\begin{array}{l}\text{output of mass from the}\\\text{system per unit time}\end{array}\right)\right] dt$$
$$= \left(\begin{array}{l}\text{mass within the system}\\\text{at time } t_n + \Delta t\end{array}\right) - \left(\begin{array}{l}\text{mass within the}\\\text{system at time } t_n\end{array}\right) \qquad (6\text{-}6)$$

When the equations for a model are to be solved by use of numerical methods, the integral-difference form of the material balance given by Equation (6-6) is satisfactory. If it is desired to look for an analytical solution, it is generally desirable to reduce the integral-difference equations to their corresponding ordinary differential or partial differential equations. To illustrate this process, Equation (6-5) is used as an example. Consider first the left-hand side. These integrals may be reduced to algebraic form by use of the *mean value theorem of integral calculus* (B-3).

$$\int_{t_n}^{t_n+\Delta t} (F - B)\, dt = (F - B)\Big|_{t_n+\alpha\Delta t}^{\Delta t} \qquad (6\text{-}7)$$

where $0 \leq \alpha \leq 1$. The term $(F - B)\Big|_{t_n+\alpha\Delta t}$ represents the mean value of the

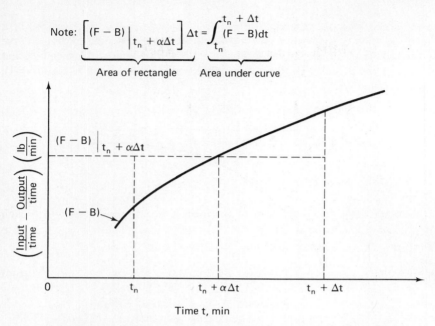

Note: $\left[(F - B) \Big|_{t_n + \alpha \Delta t} \right] \Delta t = \int_{t_n}^{t_n + \Delta t} (F - B) dt$

$\underbrace{\qquad\qquad}_{\text{Area of rectangle}}$ $\underbrace{\qquad\qquad}_{\text{Area under curve}}$

Figure 6-3. Geometrical interpretation of the mean value theorem of integral calculus.

function $(F - B)$ over the interval of time from t_n to $t_n + \Delta t$. The geometrical interpretation of the term $(F - B)\Big|_{t_n + \alpha\Delta t}$ is shown in Figure 6-3. The mean value of $(F - B)$ is that particular value of $(F - B)$ such that the rectangle of height $(F - B)\Big|_{t_n + \alpha\Delta t}$ and base Δt has an area equal to that represented by the integral on the left-hand side of Equation (6-7).

Application of the *mean value theorem of differential calculus* (B-2) to the right-hand side of Equation (6-5) gives

$$U\Big|_{t_n + \Delta t} - U\Big|_{t_n} = \Delta t \frac{dU}{dt}\Big|_{t_n + \beta\Delta t} \qquad (6\text{-}8)$$

where $0 < \beta < 1$. The time $t_n + \beta\Delta t$ is that particular value of t $(t_n \leq t \leq t_n + \Delta t)$ at which the slope of the tangent to the curve of U versus t is equal to the slope of the chord drawn from t_n to $t_n + \Delta t$, as illustrated in Figure 6-4.

From Equations (6-5), (6-7), and (6-8), it follows that

$$(F - B)\Big|_{t_n + \alpha\Delta t} = \frac{dU}{dt}\Big|_{t_n + \beta\Delta t} \qquad (6\text{-}9)$$

In the limit, as Δt goes to zero, Equation (6-9) reduces to

$$(F - B)\Big|_{t_n} = \frac{dU}{dt}\Big|_{t_n} \qquad (6\text{-}10)$$

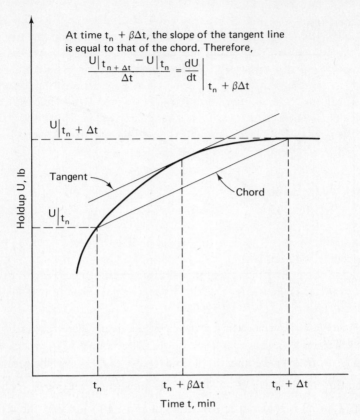

Figure 6-4. Geometrical interpretation of the mean value theorem of differential calculus.

Since t_n was arbitrarily selected as any t in the time domain other than $t = 0$ (the time of upset), Equation (6-10) holds for all $t > 0$, or

$$F - B = \frac{dU}{dt} \qquad (t > 0) \tag{6-11}$$

Energy balances

The first law of thermodynamics asserts that the energy of the universe is constant. Consequently, the total amount of energy entering minus that leaving the system must be equal to the accumulation of energy within the system. The following form of the energy balance is readily applied to systems at unsteady-state operation:

$$\int_{t_n}^{t_n + \Delta t} \left[\binom{\text{input of energy to the}}{\text{system per unit time}} - \binom{\text{output of energy from the}}{\text{system per unit time}} \right] dt$$

$$= \binom{\text{energy within}}{\text{the system}} \bigg|_{t_n + \Delta t} - \binom{\text{energy within}}{\text{the system}} \bigg|_{t_n} \tag{6-12}$$

In accounting for all of the energy entering and leaving a system, the energy equivalents of the net heat absorbed by the system and the net work done on the system must be taken into account. The terms *heat* and *work* represent energy in the state of transition between the system and its surroundings. A system that has work done on it experiences the conversion of mechanical energy into internal energy. Throughout the following analysis, one pound-mass (one lb_m) of fluid is taken as a basis. Thus, *KE*, *PE*, and *E* denote energies in Btu per lb_m of fluid. The total energy possessed by a unit mass of fluid is denoted by E_T,

$$E_T = E + KE + PE \qquad (6\text{-}13)$$

The enthalpy *H* is defined by

$$H = E + Pv \qquad (6\text{-}14)$$

where

P = pressure in lb_f/ft^2, where lb_f means pounds force;

v = specific volume, ft^3/lb_m.

In the interest of simplicity, the mechanical equivalent of heat (778 ft lb_f/Btu) has been omitted as the divisor of the *Pv* product in Equation (6-14). For convenience, let

$$H_T = E_T + Pv \qquad (6\text{-}15)$$

For a flow system at steady-state operation, Equation (6-12) reduces to the well-known expression $\Delta H = Q$, where no work is done by the system on the surroundings, and where the kinetic and potential energy changes are negligible. For an unsteady-state process, however, the expressions are not quite so simple.

There follows an application of Equation (6-12) to several systems of interest.

Fluids flowing in pipes

Throughout the developments that follow it is supposed that the pipe is flowing full and that perfect mixing occurs in the radial direction and no mixing occurs in the horizontal direction z (see Figure 6-5). Let z_j, z_{j+1}, t_n, and t_{n+1} be arbitrarily selected in the space and time domains of interest $(0 < z < z_T, t > 0)$. Then Δz and Δt are fixed as follows:

$$\Delta z = z_{j+1} - z_j$$

$$\Delta t = t_{n+1} - t_n$$

The formulation of the energy balance for the element of volume from z_j to z_{j+1} over the time period from t_n to t_{n+1} follows. The energy entering the element per unit time in the fluid at z_j is given by

$$\begin{pmatrix} \text{input of energy per} \\ \text{unit time by flow} \end{pmatrix} = (w\,E_T)\Big|_{z_j} \qquad (6\text{-}16)$$

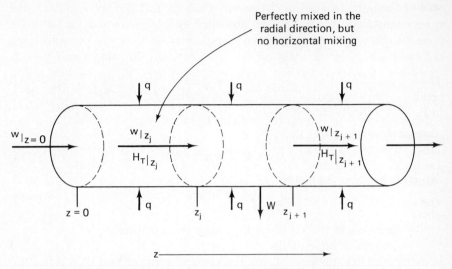

Figure 6-5. Energy balance on the element of volume from z_j to z_{j+1} for a flow system.

At any time $t(t_n \leq t \leq t_{n+1})$, the work required to force one pound mass of fluid into the element of volume is given by

$$\binom{\text{work per}}{\text{unit mass}} = \int_0^v P \, dv \bigg|_{z_j} = Pv \bigg|_{z_j} \tag{6-17}$$

Note that this work, Pv, may vary with time throughout the time period Δt. At any time t,

$$\begin{pmatrix} \text{rate at which work is done} \\ \text{on the element of volume} \\ \text{by the entering fluid} \end{pmatrix} = (wPv) \bigg|_{z_j} \tag{6-18}$$

Let q [Btu/(hr ft)] denote the rate of heat transfer at each $z(z_j \leq z \leq z_{j+1})$, where it is understood that q varies in a continuous manner with respect to z at any t. Then

$$\begin{pmatrix} \text{heat transferred across the} \\ \text{boundary of the element} \\ \text{of volume per hr} \end{pmatrix} = \int_{z_j}^{z_{j+1}} q \, dz \tag{6-19}$$

With regard to the shaftwork done by the system on the surroundings, two cases are considered. In the first case, the work is done (energy leaves the system) at a point z lying between z_j and z_{j+1}, as shown in Figure 6-5, and the rate at which the system does work on the surroundings is denoted by W(Btu/hr). In the second case, it is supposed that work is done by the system in a continuous manner at each point on the boundary, and the rate at which work is done per ft of length is denoted by $\mathcal{W}$[Btu/(hr ft)]. For the second case,

$$\begin{pmatrix} \text{shaftwork done} \\ \text{by the element of} \\ \text{volume per hr} \end{pmatrix} = \int_{z_j}^{z_{j+1}} \mathcal{W}\, dz \qquad (6\text{-}20)$$

The integral-difference equation is set up for the first case, and then the final result for the second case is obtained therefrom. The input terms that appear in Equation (6-12) are as follows:

$$\begin{pmatrix} \text{input of energy to} \\ \text{the element of} \\ \text{volume during the} \\ \text{time period } \Delta t \end{pmatrix} = \int_{t_n}^{t_{n+1}} \left[(wE_T) \Big|_{z_j,t}^* + (wPv) \Big|_{z_j,t} + \int_{z_j}^{z_{j+1}} q\, dz \right] dt \qquad (6\text{-}21)$$

The output terms are

$$\begin{pmatrix} \text{output of energy} \\ \text{from the element} \\ \text{of volume during} \\ \text{the time period } \Delta t \end{pmatrix} = \int_{t_n}^{t_{n+1}} \left[(wE_T) \Big|_{z_{j+1},t} + (wPv) \Big|_{z_{j+1},t} + W \right] dt \qquad (6\text{-}22)$$

The accumulation of energy (the right-hand side of Equation (6-12)) is given by

$$\begin{pmatrix} \text{accumulation of energy} \\ \text{within the element of} \\ \text{volume during the time} \\ \text{period } \Delta t \end{pmatrix} = \int_{z_j}^{z_{j+1}} (\rho S E_T) \Big|_{t_{n+1},z}\, dz - \int_{z_j}^{z_{j+1}} (\rho S E_T) \Big|_{t_n,z}\, dz$$

$$(6\text{-}23)$$

where ρ is the density (lb_m/ft^3) of the fluid and S is the cross-sectional area of the element of volume as shown in Figure 6-5. The cross-sectional area S is generally independent of z, and it will be considered constant throughout the remainder of this development. Since $\rho = 1/v$, it follows from Equation (6-15) that

$$\rho E_T = \rho H_T - \rho(Pv) = \rho H_T - P \qquad (6\text{-}24)$$

Then, Equation (6-23) may be restated in terms of H_T as follows:

$$\begin{pmatrix} \text{accumulation of} \\ \text{energy during} \\ \text{time } \Delta t \end{pmatrix} = \int_{z_j}^{z_{j+1}} \left[\rho H_T \Big|_{t_{n+1},z} - \rho H_T \Big|_{t_n,z} \right] S\, dz$$

$$- \int_{z_j}^{z_{j+1}} \left[P \Big|_{t_{n+1},z} - P \Big|_{t_n,z} \right] S\, dz \qquad (6\text{-}25)$$

Also, when Equation (6-15) is used to state the inputs and outputs in terms of H_T, the final result for the energy balance is given by

*The notation $(wE_T) \Big|_{z_j,t}$ is used to denote the value of wE_T at $z = z_j$ for any t in the closed interval $t_n \leq t \leq t_{n+1}$.

$$\int_{t_n}^{t_{n+1}} \left[(wH_T) \bigg|_{z_j, t} - (wH_T) \bigg|_{z_{j+1}, t} + \int_{z_j}^{z_{j+1}} q \, dz - W \right] dt$$

$$= \int_{z_j}^{z_{j+1}} \left[(\rho H_T) \bigg|_{n+1, z} - (\rho H_T) \bigg|_{t_n, z} \right] S \, dz - \int_{z_j}^{z_{j+1}} \left[P \bigg|_{t_{n+1}, z} - P \bigg|_{t_n, z} \right] S \, dz \qquad (6\text{-}26)$$

Examination of the second integral on the right-hand side of Equation (6-26) shows that it has the physical significance of being the difference between the amount of work required to sweep out the element of volume at times t_{n+1} and t_n.

If the element of volume does shaftwork on the surroundings at each point along the boundary, then W in Equation (6-26) is replaced by the expression given by Equation (6-20).

Equation (6-26) may be solved by use of numerical methods whereby the integrals are replaced by algebraic approximations. Alternately, it may be converted to a partial differential equation (provided W does not occur at z_j) and then solved numerically or, under certain conditions, an analytical solution may be obtained.

Liquid flowing through a perfect mixer with an open boundary

For the mixer shown in Figure 6-6, the energy balance over the time period Δt is given by

$$\int_{t_n}^{t_{n+1}} [w_i H_{Ti} - w_0 H_{T0} - (w_i - w_0) P_s v_s + Q - W] \, dt = \mathfrak{M} E_{Ts} \bigg|_{t_{n+1}} - \mathfrak{M} E_{Ts} \bigg|_{t_n}$$

$$(6\text{-}27)$$

Since

$$w_i - w_0 = \frac{d\mathfrak{M}}{dt}$$

$$\mathfrak{M} E_{Ts} \bigg|_{t_{n+1}} - \mathfrak{M} E_{Ts} \bigg|_{t_n} = \int_{t_n}^{t_{n+1}} \frac{d(\mathfrak{M} E_{Ts})}{dt} \, dt \qquad (6\text{-}28)$$

and the relationship $H_{Ts} = E_{Ts} + P_s v_s$ exists, it is possible to restate Equation (6-27) in the following equivalent form:

$$\int_{t_n}^{t_{n+1}} [w_i H_{Ti} - w_0 H_{T0} + Q - W] \, dt$$

$$= \mathfrak{M} H_{Ts} \bigg|_{t_{n+1}} - \mathfrak{M} H_{Ts} \bigg|_{t_n} - \int_{t_n}^{t_{n+1}} \mathfrak{M} \frac{d(P_s v_s)}{dt} \, dt \qquad (6\text{-}29)$$

The corresponding differential equation is

$$w_i H_{Ti} - w_0 H_{T0} + Q - W = \frac{d(\mathfrak{M} H_{Ts})}{dt} - \frac{\mathfrak{M} d(P_s v_s)}{dt} \qquad (t > 0) \qquad (6\text{-}30)$$

In many processes, the second term on the right-hand side of Equation (6-30) is small and may be neglected.

When $w_i = w_0 = 0$, the system shown in Figure 6-6 reduces to a batch system. A brief consideration of batch systems follows.

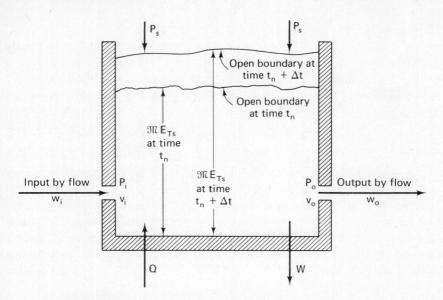

Figure 6-6. Sketch of a variable-mass, variable-energy system with an open boundary.

Batch system

A typical batch system is shown in Figure 6-7. In the event that the system possesses an expanding boundary, the work done by the expanding boundary on the surroundings is included in the term W. For the batch system shown in Figure 6-7, the energy balance over the time period from t_n to t_{n+1} is given by

$$\int_{t_n}^{t_{n+1}} \left[\int_0^{z_T} q\, dz - W \right] dt = \rho V E \Big|_{t_{n+1}} - \rho V E \Big|_{t_n} \tag{6-31}$$

The corresponding integral-differential equation is given by

$$\int_0^{z_T} q\, dz - W = \frac{d(\rho V E)}{dt} \tag{6-32}$$

In the complete description of processes, rate expressions not only for heat transfer but also for mass transfer and chemical reactions are required. An enumeration of some of these follows.

Part 2. Rate Expressions

The rate expressions for heat transfer are presented first. Heat transfer by conduction, conduction plus transfer by turbulence, convection, and radia-

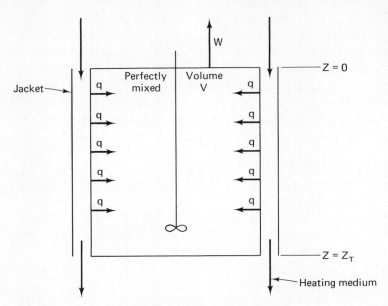

Figure 6-7. A typical batch system.

tion are considered. The rate of heat transfer by conduction is given by the well-known Fourier equation

$$Q_{cd} = -kA\frac{dT}{dz} \qquad (6\text{-}33)$$

where

Q_{cd} = rate of heat transfer in the positive direction of z, Btu/hr;
k = thermal conductivity, Btu/(hr ft °F);
A = area perpendicular to the direction of transfer, ft²;
T = temperature, °F;
z = distance, ft, measured in a specified direction.

Instead of the system of units used in these definitions, any consistent set may be employed. *Note that if heat is actually transferred in the positive direction of* z, *then* Q$_{cd}$ *will be equal to a positive number, which results from the fact that for this case* T *decreases with* z *and consequently* dT/dz *is negative. If* Q$_{cd}$ *is a negative number, then the flow of heat is in the negative direction of* z.

For the transfer of heat through a fluid in turbulent motion, it is customary to represent the rate of heat transfer through the fluid by conduction plus the rate of heat transfer caused by the turbulent motion by the following expression:

$$Q_{cd} = -\kappa A\frac{dT}{dz} \qquad (6\text{-}34)$$

where

κ = effective thermal conductivity that is used to account for heat transfer by both conduction and turbulent motion effects, Btu/(hr ft °F).

The rate of heat transfer by forced convection such as the transfer of heat from a metal surface at temperature T_A to a fluid flowing past the surface with bulk temperature T_B is given by Newton's law of cooling as

$$q_{cv} = ha(T_A - T_B) \qquad (6-35)$$

at each point along the path of flow, where

q_{cv} = rate of heat transfer by forced convection per ft of length of flow path, Btu/(hr ft);

h = coefficient for heat transfer by forced convection, Btu/(hr ft² °F);

a = heating surface perpendicular to the direction of heat transfer per ft of length, ft²/ft.

For an exchanger at steady-state operation, the bulk temperature T_B is defined as that temperature which the fluid stream would possess at a given position if it were withdrawn at this position and mixed. This same definition suffices to define the bulk temperature at unsteady-state operation provided that it is recognized that a limiting process is involved wherein the withdrawal time approaches zero. From a mathematical point of view, a rate expression of this form $q \propto (T_A - T_B)$ rather than the form $q \propto dT/dz$ is needed to describe the rate of heat transfer across the interface because the derivative dT/dz is generally regarded as being discontinuous at the interface.

When radiant heat transmission is involved, the expressions proposed by others (2, 7) for the net rate of heat transfer by radiation are recommended.

For steady-state heat transfer, the expression

$$q = Ua(T_A - T_B) \qquad (6-36)$$

is commonly used to express the rate of heat transfer from the hot stream A at the bulk temperature T_A through, say, a metal wall to the cold stream B at the bulk temperature T_B. The quantity U[Btu/(hr ft² °F)] is called the overall coefficient of heat transfer, and its formula for the case mentioned follows Equation (6-44), which is also valid for unsteady-state operation subject to the restrictions enumerated in the following development.

Equation (6-36) is a valid rate expression at unsteady-state operation provided that (1) the holdup of energy by the metal wall separating the two streams is negligible and (2) the transfer of heat along the wall by conduction is negligible. On the basis of these conditions, it is shown as follows that Equation (6-36) is applicable at unsteady-state operation. Consider Figure 6-8, a sketch of a double-pipe heat exchanger in which heat flows from stream A through a metal wall to stream B. Note that T_1, T_2, T_3, and T_4 vary both

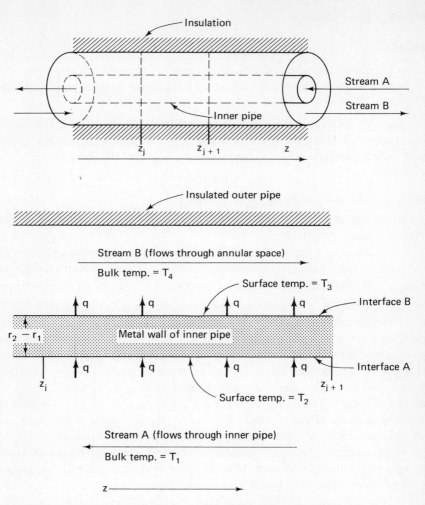

Figure 6-8. Section from z_j to z_{j+1} of the metal wall of the inner pipe of a double-pipe heat exchanger.

with z and with time t. In this case, it is convenient to express the conduction rate in Btu/hr per ft of length z. Then, at a given z and at a given time t,

$$q_{cd}\bigg|_{r=r_1} = -k2\pi r_1 \frac{dT}{dr}\bigg|_{r_1} \quad \text{and} \quad q_{cd}\bigg|_{r=r_2} = -k2\pi r_2 \frac{dT}{dr}\bigg|_{r_2}$$

where r_1 and r_2 are the radii corresponding, respectively, to the internal and external diameters, of the inner pipe in Figure 6-8. Consider first the energy balance enclosing interface A, where it is supposed that the interface encloses the metal surface at $r = r_1$ and the liquid at the surface of the wall at $r = r_1$. Energy enters the interface by forced convection and leaves by conduction. Since the holdup of interface A is zero, it follows that

$$\int_{t_n}^{t_{n+1}} \left[\int_{z_j}^{z_{j+1}} \left[q_{cv}\bigg|_{r_1} - q_{cd}\bigg|_{r_1} \right] dz \right] dt = 0 \tag{6-37}$$

208

where

$$q_{cv}\bigg|_{r_1} = h_1 a_1(T_1 - T_2).^*$$

Similarly, the energy balance enclosing interface B is given by

$$\int_{t_n}^{t_{n+1}} \left[\int_{z_j}^{z_{j+1}} \left[q_{cd}\bigg|_{r_2} - q_{cv}\bigg|_{r_2} \right] dz \right] dt = 0 \qquad (6\text{-}38)$$

where

$$q_{cv}\bigg|_{r_2} = h_2 a_2(T_3 - T_4)$$

Next consider the energy balance enclosing the metal wall. If the holdup of energy by the wall as well as the conduction of heat in the direction z is neglected, then

$$\int_{t_n}^{t_{n+1}} \left[\int_{z_j}^{z_{j+1}} \left[q_{cd}\bigg|_{r_1} - q_{cd}\bigg|_{r_2} \right] dz \right] dt = 0 \qquad (6\text{-}39)$$

Since each of the integrals given by Equations (6-37) through (6-39) is equal to zero for all choices of z_j, z_{j+1}, t_n, and t_{n+1} in the space and time domains of interest, it follows that each of the respective integrands is identically equal to zero for all z and t in these domains, that is,

$$q_{cv}\bigg|_{r_1} - q_{cd}\bigg|_{r_1} = 0, \quad q_{cd}\bigg|_{r_2} - q_{cv}\bigg|_{r_2} = 0 \quad \text{and} \quad q_{cd}\bigg|_{r_1} - q_{cd}\bigg|_{r_2} = 0 \qquad (6\text{-}40)$$

Since $q_{cd}\bigg|_{r_1} = q_{cd}\bigg|_{r_2} = q_{cd}$ at any z, it follows that q_{cd} is independent of r at any time t. Since at any z and t

$$q_{cd} = -k2\pi r_1 \frac{dT}{dr}\bigg|_{r_1} = -k2\pi r_2 \frac{dT}{dr}\bigg|_{r_2} \qquad (6\text{-}41)$$

it follows that dT/dr must vary with r. Thus,

$$\int_{r_1}^{r_2} \frac{dr}{r} = -\int_{T_2}^{T_3} \frac{2\pi k}{q_{cd}} dT \qquad (6\text{-}42)$$

which may be integrated to give

$$q_{cd} = \frac{a_{lm} k}{\Delta r}(T_2 - T_3) \qquad (6\text{-}43)$$

where k is independent of T and where

$a_{lm} = 2\pi r_{lm};$

$$r_{lm} = \frac{r_2 - r_1}{\log_e \dfrac{r_2}{r_1}};$$

$\Delta r = r_2 - r_1.$

---

*Actually, the notation $q_{cv}\bigg|_{r_1,t}$ should have been used, but the t was omitted in this and the other integrands in the interest of simplicity.

From Equations (6-40), (6-43), and the definitions of the q_{cv}'s, it follows that

$$\frac{q}{h_1 a_1} + \frac{q}{h_2 a_2} + \frac{q\Delta r}{ka_{lm}} = (T_1 - T_2) + (T_2 - T_3) + (T_3 - T_4)$$

which yields the desired result

$$q = Ua(T_1 - T_4) \tag{6-44}$$

where

$$q = q_{cd}\Big|_{r_1} = q_{cv}\Big|_{r_1} = q_{cd}\Big|_{r_2} = q_{cv}\Big|_{r_2}$$

$$U = \frac{1}{\dfrac{a}{h_1 a_1} + \dfrac{a}{h_2 a_2} + \dfrac{a\Delta r}{ka_{lm}}}$$

(Note, as in steady-state heat transfer, a may be selected arbitrarily. The value of U, however, depends on the particular choice made for a.) Consequently, under the conditions enumerated above, Equation (6-36) is a valid rate expression at unsteady-state operation. In the modeling of large systems at unsteady-state operation, the holdup of energy by the films can be neglected, but the holdup of energy by the metal walls cannot always be neglected (4).

Rates of mass transfer

As pointed out by others (2, 3), the rates of mass transfer have analogies in heat transfer. Various rate expressions have been treated in detail by Bird (2), among others. Only two rate expressions are presented herein. The first of these, *Fick's first law of diffusion* for binary mixtures, is given by

$$J_A = -D_V A \frac{dC_A}{dz} \tag{6-45}$$

where
  A = area perpendicular to the direction of diffusion, ft²;
  C_A = concentration of component A (1b mole)/ft³;
  D_V = volumetric diffusion coefficient for a binary mixture, ft²/hr;
  J_A = rate at which component A diffuses (or moves relative to the mixture) (lb moles)/hr.

In the application of this rate expression, bulk flow induced either by the molecular diffusion or by any other mechanism must be taken into account in the formulation of material balances (see Problem 7-4).

If a gas phase is in turbulent motion in the positive direction of z, there is also transfer of component i in the direction of z because of the turbulent motion of the flowing stream. To account for this transfer, the rate expression given by Equation (6-45) is commonly modified in a manner analogous to that for heat transfer by conduction, that is,

$$\mathcal{J}_A = -\mathcal{D} A \frac{dC_A}{dz} \tag{6-46}$$

where

$\mathcal{J}_A$ = net rate of transfer of component A in the positive direction of z (1b moles)/hr;

$\mathcal{D}$ = effective diffusion coefficient that is used to account for both molecular diffusion and mixing effects; defined by Equation (6-46).

Over the years, the transfer of mass from the bulk conditions of flowing streams to interfaces and more recently the transfer of mass across the interfaces of different phases have received considerable attention. The rate expressions commonly employed are analogous in form to those used for heat transfer by forced convection except that either partial pressure, concentration, or fugacity replaces temperature as the potential for transfer.

Although thermodynamics does not give any information about the speed with which a transfer of mass occurs, it does give information on the forces that tend to drive a mass transfer process toward equilibrium, and in this respect it is helpful in the formulation of the rate expression for the transfer of mass across interfaces (3).

Consider first the transfer of any component i from a liquid phase L to a vapor phase V in contact with it. A net transfer of component i from the liquid to the vapor phase will occur until the two phases reach a state of equilibrium. The two phases are said to be at a state of equilibrium provided that the temperature and pressure in the respective phases are equal and the fugacity of each component in the vapor phase is equal to its fugacity in the liquid phase, that is, a state of equilibrium exists provided that

$$T^V = T^L, \quad P^V = P^L, \quad \text{and} \quad \bar{f}_i^V = \bar{f}_i^L \text{ (for each component } i) \qquad (6\text{-}47)$$

In order for a net transfer of any component i from the liquid phase L to the vapor phase V to occur, it is necessary that

$$\bar{f}_i^L > \bar{f}_i^V \qquad (6\text{-}48)$$

Thus, so long as

$$\bar{f}_i^L - \bar{f}_i^V > 0 \qquad (6\text{-}49)$$

there exists an "escaping tendency" or "driving force" for the transfer of component i from phase L to phase V; see Figure 6-9. Equation (6-49) suggests that the rate of transfer of component i across the interface should be proportional to the difference in its fugacities $(\bar{f}_i^L - \bar{f}_i^V)$. Thus, it is consistent from the standpoint of thermodynamic considerations to postulate that the rate of mass transfer N_i^I of any component $i(1 \leq i \leq c)$ across the liquid-vapor interface is given by

$$N_i^I = k_I a_i S(\bar{f}_i^\circ - \bar{f}_i^v) \qquad (6\text{-}50)$$

where

N_i^I = rate of mass transfer of any component across the vapor-liquid interface (1b moles)/(hr ft);

$k_I a_i$ = product of the mass transfer coefficient for the interface; (1b-

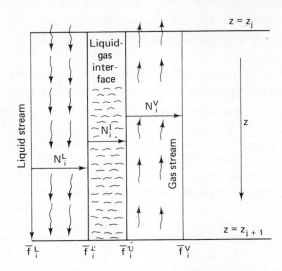

Figure 6-9. Mass transfer from a liquid to a gas stream.

moles)/(hr atm ft²) and the interfacial area per cubic foot of empty column;

a = interfacial area per cubic foot of empty column;

S = cross-sectional area of the empty column.

The superscript I refers to the vapor-liquid interface. The fugacities $\bar{f}_i^{\varrho}$ and $\bar{f}^{v}$ are evaluated at the conditions of the liquid and vapor phases at the interface, respectively, for each component i.

Also, note that the convention has been adopted that when a appears as a multiplier of a heat transfer coefficient, it has the dimensions of ft² of interfacial area for heat transfer per ft of length, and that when a appears as a multiplier of a mass transfer coefficient, it has the dimensions of ft² of interfacial area for mass transfer per cubic ft of empty column. It should be observed that Newton's law of cooling could have been formulated in a manner analogous to that shown for mass transfer.

When a liquid and vapor phase are in contact, such as in a wetted wall column, the rate of mass transfer of any component i from the liquid phase to the liquid side of the interface and from the vapor side of the interface to the vapor stream are represented, respectively, by expressions of the same form as Equation (6-50), namely,

$$N_i^L = k_L a_i S(\bar{f}_i^L - \bar{f}_i^{\varrho}) \qquad (6\text{-}51)$$

$$N_i^V = k_G a_i S(\bar{f}_i^{v} - \bar{f}_i^V) \qquad (6\text{-}52)$$

where

N_i^L = rate of mass transfer of component i from the flowing liquid stream to the liquid side of the interface (lb moles)/(hr ft);

N_i^V = rate of mass transfer of component i from the vapor side of the interface to the flowing stream (lb moles)/(hr ft);

$\bar{f}_i^L, \bar{f}_i^V$ = fugacities of component i evaluated at the bulk conditions of the liquid and vapor streams, respectively, atm.

Since these rate processes are in series, it follows that at steady state

$$N_i^L = N_i^I = N_i^V = N_i \qquad (6\text{-}53)$$

When these three rate expressions are added and rearranged, one obtains

$$N_i\left[\frac{1}{k_L a_i} + \frac{1}{k_I a_i} + \frac{1}{k_G a_i}\right] = S[(\bar{f}_i^L - \bar{f}_i^\mathcal{L}) + (\bar{f}_i^\mathcal{L} - \bar{f}_i^\upsilon) + (\bar{f}_i^\upsilon - \bar{f}_i^V)] \qquad (6\text{-}54)$$

This expression may be restated in the following form:

$$N_i = K_G a_i S(\bar{f}_i^L - \bar{f}_i^V) \qquad (6\text{-}55)$$

where the overall mass transfer coefficient $K_G a_i$ is defined as follows:

$$\frac{1}{K_G a_i} = \frac{1}{k_G a_i} + \frac{1}{k_I a_i} + \frac{1}{k_L a_i} \qquad (6\text{-}56)$$

When the resistance to transfer across the interface is negligible,

$$\lim_{k_I \to \infty} K_G a = \frac{1}{\dfrac{1}{k_L a} + \dfrac{1}{k_G a}} \qquad (6\text{-}57)$$

This limiting case is sometimes referred to as the *two-film theory* for mass transfer. The resistances to the mass transfer in the vapor and liquid phases are concentrated in the vapor and liquid films. The potential for transfer of a component across the liquid film is the fugacity difference $(\bar{f}_i^L - \bar{f}_i^\mathcal{L})$, and the potential for transfer across the vapor film is $(\bar{f}_i^\upsilon - \bar{f}_i^V)$.

Experimental evidence presented by Tung et al. (9) and Emmert et al. (5) suggests that k_I is very large except at very high rates of mass transfer. Thus, for all practical purposes $\bar{f}^\upsilon \cong \bar{f}^\mathcal{L}$, or "equilibrium" at the interface exists. This approximation is used in subsequent applications.

Since the rate expression given by Equation (6-55) involves the fugacities evaluated at the bulk conditions, it would be desirable to employ such an expression at unsteady-state operation. That Equation (6-55) is also valid at unsteady-state operation is shown as follows. A graphical representation of the mass transfer process under consideration is given in Figure 6-9. Since the holdup of component i in the liquid is zero at the plane of discontinuity of the liquid phase (the liquid side of the interface), a material balance for any component i at the plane of discontinuity over the time period Δt and the height of column Δz is given by

$$\int_{t_n}^{t_{n+1}} \left[\int_{z_j}^{z_{j+1}} (N_i^L - N_i^I)\, dz\right] dt = 0 \qquad (6\text{-}58)$$

Similarly, a material balance for any component i at the plane of discontinuity of the gas phase yields

$$\int_{t_n}^{t_{n+1}} \left[\int_{z_j}^{z_{j+1}} (N_i^I - N_i^V) \, dz \right] dt = 0 \tag{6-59}$$

Since each of the integrals given by Equations (6-58) and (6-59) is equal to zero for all choices of z_j, z_{j+1}, t_n, and t_{n+1} in the space and time domains of interest, it follows that each of the respective integrands is identically equal to zero for all z and t in these domains, that is,

$$N_i^L - N_i^I = 0, \qquad N_i^I - N_i^V = 0 \tag{6-60}$$

from which Equation (6-53) follows. Consequently, Equation (6-55) is also applicable to systems at unsteady-state operation.

PROBLEMS

6-1. Verify the result given by (a) the first expression of Equation (6-28), and (b) the expression given by Equation (6-29).

6-2. Verify the result given by Equation (6-30).

6-3. Verify the result given by Equation (6-32).

NOTATION

a = square feet of surface area (perpendicular to the direction of heat or mass transfer) per foot of pipe (ft²/ft) for heat transfer and per cubic foot of empty tower (ft²/ft³) for mass transfer. (Although the interfacial areas for heat transfer may differ from those for mass transfer, the same symbol is used for both.)

A = total area perpendicular to the direction of heat or mass transfer; ft².

C_A, C_i = concentration of component A and i, respectively (lb moles)/ft³. (In some problems it is convenient to express the concentration in the units lb/ft³.)

D_V = volumetric diffusion coefficient for mass transfer by molecular diffusion; ft²/hr.

E = internal energy per unit mass (or per mole of material).

E_T = total energy per unit mass of material; defined by Equation (6-13).

h = coefficient of heat transfer by convection, Btu/(hr ft² °F).

$\tilde{f}_i$ = fugacity of component i in a mixture, atm.

H = enthalpy per unit mass (or per mole) of material; $H = E + Pv$.

$H_T = E_T + Pv$.

k = thermal conductivity, Btu/(hr ft °F).

k_G, k_I, k_L = mass transfer rate constants; (lb moles)/(hr ft² atm).

J_A = rate of mass transfer of component A by molecular diffusion in the positive direction of z (lb moles)/hr.

$\mathcal{J}_A$ = rate of mass transfer of component A by molecular diffusion and by turbulent motion in the positive direction of z (lb moles)/hr.

K_G = overall coefficient for mass transfer (lb moles)/(hr ft^2 atm).

KE = kinetic energy per unit mass of fluid.

$\mathfrak{M}$ = total mass of the system.

N_i^I, N_i^V, N_i^L = rate of mass transfer of a given component from the liquid to the vapor side of the interface, from the vapor of the interface to the bulk conditions of the vapor phase, and from the bulk conditions of the liquid phase to the liquid side of the interface, respectively (lb moles)/(hr ft °F).

P = pressure, lb$_f$/ft^2.

PE = potential energy per unit mass of material.

q = rate of heat transfer per unit length of pipe, Btu/(hr ft).

Q = rate of heat transfer, Btu/hr.

Q_{cd} = rate of heat transfer by conduction; Q_{cd} = rate of heat transfer by conduction and by turbulent motion, Btu/hr.

S = cross-sectional area perpendicular to the direction of flow, ft^2.

t = time.

T = temperature, °F.

U = overall coefficient for heat transfer; Btu/hr ft^2 °F. Also used to denote the holdup of a system in mass or molal units.

v = specific volume.

w = mass flow rate, lb/hr.

W = work, Btu/hr.

$\mathcal{W}$ = work, Btu/(hr ft).

z = length, ft.

Greek letters

Δt = increment of time; $\Delta t = t_{n+1} - t_n$, hr.

Δz = increment of length; $\Delta z = z_{j+1} - z_j$, ft.

ρ = mass density; lb/ft^3.

Subscripts

c = total number of components.

cd = conduction.

cv = convection.

i = inlet value; also component numbers ($1 \leq i \leq c$).

I = interface.

j = integer for numbering in the length domain.

n = integer for numbering primarily in the time domain.

Superscripts

L = liquid phase.

$\mathcal{L}$ = liquid phase at an interface.

V = vapor phase.

$\mathcal{U}$ = vapor phase at an interface.

REFERENCES

1. Amundson, N. R., *Mathematical Methods in Chemical Engineering; Matrices and Their Application*. Englewood Cliffs, N. J., Prentice-Hall, Inc., 1966.

2. Bird, B. B., W. E. Stewart, and E. N. Lightfoot, *Transport Phenomena*. New York: John Wiley & Sons, Inc., 1960.

3. Brown, G. G., and Associates, *Unit Operations*. New York: John Wiley & Sons, Inc., 1955.

4. Burdett, J. W. and C. D. Holland, "Dynamics of a Multiple-Effect Evaporator System," *A. I. Ch. E. Journal*, **17**, (1971), 1080.

5. Emmert, R. E., and R. L. Pigford, "A Study of Gas Absorption in Falling Liquid Films", *Chem. Eng. Prog.*, **50**, (1954), 87.

6. Holland, C. D., *Unsteady State Processes with Applications in Multicomponent Distillation*. Englewood Cliffs, N. J.: Prentice-Hall, Inc., 1966.

7. McAdams, W. H., *Heat Transmission*, 2nd ed. New York: McGraw-Hill Book Company, Inc., 1942.

8. McDaniel, Ronald and C. D. Holland "Modeling of Packed Absorbers at Unsteady State Operation-IV," *Chem. Eng. Sci.*, **25**, (1970), 1283.

9. Tung, L. N., and H. G. Drickamer, "Diffusion Through an Interface-Binary System", *Journal Chem. Phys.*, **20**, (1952), 6.

Formulation of the Equations Required to Describe Process Models

7

In this chapter a wide variety of relatively simple processes is considered. This is done for the purpose of demonstrating techniques for formulating the equations required to describe the proposed models of the respective processes.

For most of the simple models considered in this chapter, analytical solutions may be obtained for the differential or partial differential equations that are required to describe the model. Thus, the corresponding integral or integral-difference equations are first reduced to the differential or partial differential equations. This is done by first applying the mean value theorems of differential and integral calculus. Then the limit is taken as the space and/or time increments are allowed to go to zero. (Note that if the equations are to be solved numerically, this transformation step to the differential or partial differential equations may be omitted and the numerical methods applied directly to the integral or integral-difference equations.)

A process may range in size from that of a single thermometer (see Problem 7-10) to an entire plant. Accordingly, the number of equations required to describe a process may range from one to any number. The model for the large system will be seen to consist of the collection of the models for the individual parts. The component parts of the large systems are modeled, however, by use of the same techniques demonstrated in this chapter. However, the equations for the component parts can seldom be solved independently of the other equations because of the interdependence of the parts of the system.

Part 1. Use of Material Balances in the
Analysis of Rate Processes

Problems involving steady-state operation as well as unsteady-state operation are presented. The first problem, Illustrative Example 7-1, involves the rate process of mixing, and it is supposed that the mixing is perfect in all directions or that the rates of mixing are infinite in all directions.

ILLUSTRATIVE EXAMPLE 7-1. A PERFECT MIXER AT UNSTEADY-STATE OPERATION

At time $t = 0$, the mixer is at steady-state operation, that is, the mole fraction x_i° of each component in the mixer and in the liquid leaving the mixer is the same as that of the feed ($x_i^\circ = X_i^\circ$). At time $t = 0+$, an upset in the composition of the feed occurs while the rates F and B remain fixed at $F = B$. The upset consists of a step change in the composition of the feed from X_i° to X_i. (a) Set up the component-material balance for any component i over the time period from t_n to t_{n+1} for $t > 0+$. (b) Obtain the differential equation that describes the variation of the mole fraction x_i in the mixer at any time t after the upset in the composition of the feed. (c) Obtain x_i at a function of t.

Solution:

(a) A component-material balance over the time period from t_n to t_{n+1} is represented by

$$\int_{t_n}^{t_{n+1}} [\underbrace{FX_i}_{\text{input/time}} - \underbrace{Bx_i}_{\text{output/time}}] \, dt = \underbrace{Ux_i|_{t_{n+1}} - Ux_i|_{t_n}}_{\text{accumulation}} \qquad (t > 0+) \qquad (7\text{-}1)$$

(b) Equation (7-1) may be reduced to its corresponding differential equation in the same manner used to obtain the differential equation corresponding to the total material balance [see Equations (6-4) through (6-11)], that is, (1) apply the *mean value theorem of integral calculus* (B-3) to the left-hand side and the *mean value theorem of differential calculus* (B-2) to the right-hand side of Equation (7-1), (2) divide both sides of the result so obtained by Δt, and (3) take the limit as Δt goes to zero. The result is

$$(FX_i - Bx_i)|_{t_n} = \frac{d(Ux_i)}{dt}\bigg|_{t_n}$$

Since t_n was arbitrarily selected in the time domain of interest, it follows that

$$FX_i - Bx_i = \frac{d(Ux_i)}{dt} \qquad (t > 0+) \qquad (7\text{-}2)$$

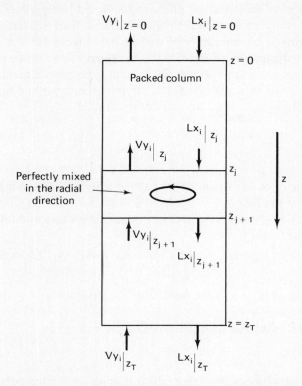

Figure 7-1. Mass transfer in a packed column—perfectly mixed in the radial direction and no mixing in the direction of z.

Since $F = B$, it follows from Equation (6-11) that

$$\frac{dU}{dt} = 0$$

Thus, Equation (7-2) reduces to

$$X_i - x_i = \frac{U}{F} \frac{dx_i}{dt} \qquad (t > 0+) \tag{7-3}$$

(c) Equation (7-3) is readily solved by separation of the variables to give

$$\int_{x_i^\circ}^{x_i} \frac{dx_i}{X_i - x_i} = \frac{F}{U} \int_0^t dt$$

Integration of this expression followed by rearrangement yields

$$\frac{X_i - x_i}{X_i - x_i^\circ} = \exp\left(-\frac{Ft}{U}\right) \tag{7-4}$$

The next problem is a classical one that involves the transfer of mass across boundaries of interfaces. In the packed column shown in Figure 7-1.

mass is transferred from the gas phase to the liquid phase. Since the rate of transfer is directly proportional to the interfacial area separating the two phases, packing is commonly employed in the column for the purpose of promoting the formation of interfacial area. Liquid flows down through the column and gas passes up through the column. The general form of the rate expression is given by Equation (6-55). Suppose the vapor and liquid phases form ideal solutions, that is, for any component i,

$$\bar{f}_i^V = f_i^V y_i \tag{7-5}$$

$$\bar{f}_i^L = f_i^L x_i \tag{7-6}$$

where

$f_i^V = $ fugacity of pure component i at the temperature and total pressure of the gas phase;

$f_i^L = $ fugacity of pure component i at the temperature and total pressure of the liquid phase.

Then, the rate expression given by Equation (6-55) may be stated in terms of the mole fractions and the fugacities of the pure components as follows:

$$N_i = K_G a_i f_i^V S(Y_i - y_i) \tag{7-7}$$

where

$a = $ ft² of interfacial area per ft³ of empty column;

$$Y_i = \left(\frac{f_i^L}{f_i^V}\right) x_i.$$

If the temperature of the vapor phase is equal to the temperature of the liquid phase, then the ratio of f_i^L / f_i^V is called K_i, the ideal solution K value. (For the relationship between K_i, vapor pressure, and total pressure, see Chapter 3.) In the following example it is supposed that the packed column is at steady-state operation. By steady-state operation it is meant that each variable of the process is independent of time. For this type of operation, the accumulation is zero and the material balance expression given by Equation (6-6) becomes

$$\begin{Bmatrix} \text{input of mass} \\ \text{to the element} \\ \text{of volume per} \\ \text{unit time} \end{Bmatrix} - \begin{Bmatrix} \text{output of mass} \\ \text{from the element} \\ \text{of volume per} \\ \text{unit time} \end{Bmatrix} = 0$$

or simply,

$$\text{Input} - \text{output} = 0$$

ILLUSTRATIVE EXAMPLE 7-2. PACKED ABSORBER AT STEADY-STATE OPERATION

Develop the component- and total-material balances for the element of volume from z_j to z_{j+1} $(0 < z_j < z_{j+1} < z_T)$ shown in Figure 7-1. Suppose that the liquid as well as the vapor is perfectly mixed in the horizontal direction (the direction normal to the flow of vapor and liquid).

Solution: A total-material balance for the element of volume at steady state is given by

$$\underbrace{V|_{z_{j+1}} + L|_{z_j}}_{\text{input/time}} - \underbrace{V|_{z_j} - L|_{z_{j+1}}}_{\text{output/time}} = 0 \tag{7-9}$$

By use of the *mean value theorem of differential calculus* and the limiting process wherein Δz is allowed to go to zero, the following differential equation is readily obtained from Equation (7-9) upon recognition that z_j was arbitrarily selected

$$\frac{dV}{dz} - \frac{dL}{dz} = 0 \qquad (0 < z < z_T) \tag{7-10}$$

For the sake of definiteness, suppose that component i under consideration is transferred from the liquid to the vapor phase as implied by Equation (7-7). A component-material balance on component i in the vapor phase for the element of volume from z_j to z_{j+1} yields

$$Vy_i|_{z_{j+1}} - Vy_i|_{z_j} + \int_{z_j}^{z_{j+1}} N_i \, dz = 0 \tag{7-11}$$

Application of the *mean value theorem of differential calculus* to the first two terms and the *mean value theorem of integral calculus* to the last term on the left-hand side of Equation (7-11) followed by the limiting process wherein Δz is allowed to go to zero, the following differential equation is obtained:

$$\frac{d(Vy_i)}{dz} + N_i = 0 \qquad (0 < z < z_T) \tag{7-12}$$

upon recognition of the fact that z_j was arbitrarily selected. Similarly, a component-material balance on the liquid phase over the volume from z_j to z_{j+1} yields

$$Lx_i|_{z_j} - Lx_i|_{z_{j+1}} - \int_{z_j}^{z_{j+1}} N_i \, dz = 0 \tag{7-13}$$

whose corresponding differential equation is

$$\frac{d(Lx_i)}{dz} + N_i = 0 \qquad (0 < z < z_T) \tag{7-14}$$

or

$$\frac{d(Lx_i)}{dz} + K_G a_i f_i^v S(Y_i - y_i) = 0 \qquad (0 < z < z_T) \tag{7-15}$$

where N_i has been replaced by its equivalent as given by Equation (7-7).

It should be observed that the direction of transfer used in the definition of N_i is arbitrary in the sense that it may be opposite that of the actual system. Once a direction is selected, however, it must be adhered to throughout the process of for-mulating the differential equations. For example, suppose that instead of Equation (7-7), the rate of mass transfer N_i is defined in terms of the transfer from the vapor to the liquid phase,

$$N_i = K_G a_i f_i S(y_i - Y_i)$$

In this case the respective component-material balances for the vapor and liquid phases are obtained by reversing the signs of the integrals that appear in Equations

(7-11) and (7-13). Thus, the corresponding differential equations are given by

$$\frac{d(Vy_i)}{dz} - N_i = 0$$

$$\frac{d(Lx_i)}{dz} - N_i = 0$$

But, when the above definition of N_i is substituted into these expressions, they are seen to be identical with the corresponding equations obtained from Equations (7-12) and (7-14) as demonstrated by Equation (7-15).

ILLUSTRATIVE EXAMPLE 7-3. THE MODIFIED MURPHREE PLATE EFFICIENCY

The differential equation corresponding to a component-material balance on the vapor passing through an element of the vapor-liquid mixture on plate j of a distillation column is given by Equation (7-12). Show that if $K_G a_i$, V, f_i^V, and Y_i do not vary as the vapor passes up through the liquid, then Equation (7-12) may be integrated to give

$$E_{ji}^M = 1 - \exp\left(-\frac{K_G a_i f_i^V S z_T}{V}\right) \tag{7-16}$$

The modified Murphree plate efficiency (4, 8) is defined by

$$E_{ji}^M = \frac{y_{ji} - y_{j+1,i}}{Y_{ji} - y_{j+1,i}} \tag{7-17}$$

Solution: For the conditions stated, this problem may be solved by separation of the variables in Equation (7-12) followed by integration. The integration is left to the student (see Problem 7-2).

The next problem is a classical one that involves the rate process of molecular diffusion. Consider the cylinder (see Figure 7-2) that is divided into two compartments by the partition at $z = 0$. Compartment B is filled with a pure gas called component B and compartment A is filled with a pure gas

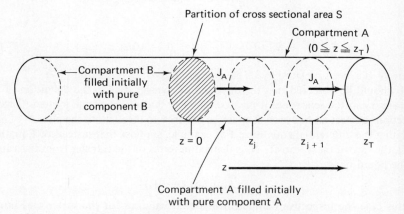

Figure 7-2. Equal countermolecular diffusion of gases A and B.

called component A. The gases in the compartments are at the same temperature and pressure. At time $t = 0+$, the partition is removed without causing any disturbance of the gases in compartments A and B.

ILLUSTRATIVE EXAMPLE 7-4. EQUAL COUNTER MOLECULAR DIFFUSION OF GASES A AND B

It is desired to formulate a quantitative description of the diffusion process (described above and in Figure 7-2) for all time t after the partition has been removed. Assume that equal counter molecular diffusion of gases A and B occurs. Thus, no bulk flow of A and B is induced as a consequence of the molecular diffusion, and the total flow rate of A at any z is J_A as given by Equation (6-45). (a) Make a material balance on component A in the volume $S(z_{j+1} - z_j)$ over the time period from t_n to t_{n+1} ($t_{n+1} > t_n > 0+$, $0 < z_j < z_{j+1} < z_T$). (b) Obtain the corresponding partial differential equations for the case in which D_V and S are independent of time and position.

Solution:

(a) The material balance for component A follows:

$$\int_{t_n}^{t_{n+1}} [J_A|_{z_j,t} - J_A|_{z_{j+1},t}] \, dt = \int_{z_j}^{z_{j+1}} C_A S \, dz|_{t_{n+1}} - \int_{z_j}^{z_{j+1}} C_A S \, dz|_{t_n} \qquad (7\text{-}18)$$

(b) This expression is reduced to the corresponding partial differential equation by application of the mean value theorems of differential and integral calculus (B-2, B-3) followed by a limiting process wherein Δz and Δt are allowed to go to zero. It is recommended that this reduction process be initiated by applying the appropriate theorem to the innermost term of the expression under consideration. Consider first the left-hand side of Equation (7-18). Application of the *mean value theorem of differential calculus* (B-2) to the integrand of this expression follows:

$$-\int_{t_n}^{t_{n+1}} [J_A|_{z_{j+1},t} - J_A|_{z_j,t}] \, dt = -\Delta z \int_{t_n}^{t_{n+1}} \frac{\partial J_A}{\partial z}\Big|_{z(t),t} dt \qquad (7\text{-}19)$$

where

$$z(t) = z_j + \alpha(t) \cdot \Delta z, \qquad 0 < \alpha(t) < 1.$$

The quantities t_n, t_{n+1}, z_j, and z_{j+1} are to be regarded as fixed values within the time and space domains ($0+ < t_n < t_{n+1}$, $0 < z_j < z_{j+1} < z_T$). Application of the *mean value theorem of integral calculus* (B-3) to the right-hand side of Equation (7-19) yields

$$-\Delta z \int_{t_n}^{t_{n+1}} \frac{\partial J_A}{\partial z}\Big|_{z(t),t} dt = -\Delta z \, \Delta t \frac{\partial J_A}{\partial z}\Big|_{z(t_p),t_p} \qquad (7\text{-}20)$$

where

$$t_p = t_n + \gamma_1 \Delta t, \quad 0 \leq \gamma_1 \leq 1.$$

(c) Continuity of $C_A S$ over the space and time domains permits the difference of integrals on the right-hand side of Equation (7-18) to be written as the integral of the difference. Application of the *mean value theorem of dif-*

ferential calculus (B-2) to the right-hand side of Equation (7-18) is represented by

$$\int_{z_j}^{z_{j+1}} [C_A S|_{t_{n+1}, z} - C_A S|_{t_n, z}]\, dz = \Delta t \int_{z_j}^{z_{j+1}} \frac{\partial(C_A S)}{\partial t}\bigg|_{t(z), z}\, dz \qquad (7\text{-}21)$$

where

$$t(z) = t_n + \beta(z)\cdot\Delta t, \qquad 0 < \beta(z) < 1.$$

Application of the *mean value theorem of integral calculus* (B-3) to the right-hand side of Equation (7-21) follows:

$$\Delta t \int_{z_j}^{z_{j+1}} \frac{\partial(C_A S)}{\partial t}\bigg|_{t(z), z}\, dz = \Delta z\, \Delta t\, \frac{\partial(C_A S)}{\partial t}\bigg|_{t(z_p), z_p} \qquad (7\text{-}22)$$

where

$$z_p = z_j + \gamma_2\, \Delta z, \qquad 0 \leqq \gamma_2 \leqq 1.$$

Thus, the following alternate but equivalent form of Equation (7-18) follows from Equations (7-19) through (7-22):

$$-\frac{\partial J_A}{\partial z}\bigg|_{z(t_p), t_p} = \frac{\partial(C_A S)}{\partial t}\bigg|_{t(z_p), z_p} \qquad (7\text{-}23)$$

If Δz and Δt are allowed to approach zero in any order whatsoever, it is readily shown that Equation (7-23) reduces to

$$-\frac{\partial J_A}{\partial z}\bigg|_{z_j, t_n} = \frac{\partial(C_A S)}{\partial t}\bigg|_{z_j, t_n} \qquad (7\text{-}24)$$

Since z_j and t_n were arbitrarily selected within the space and time domains of interest, it follows that Equation (7-24) holds for all z and t in these domains, that is,

$$-\frac{\partial J_A}{\partial z} = \frac{\partial(C_A S)}{\partial t} \qquad (0 < z < z_T, t > 0+) \qquad (7\text{-}25)$$

Since S is a constant, Equation (7-25) reduces to

$$-\frac{\partial J_A}{\partial z} = \frac{S\partial C_A}{\partial t} \qquad (0 < z < z_T, t < 0+) \qquad (7\text{-}26)$$

In view of the expression for J_A given by Equation (6-45), it follows that

$$D_V \frac{\partial^2 C_A}{\partial z^2} = \frac{\partial C_A}{\partial t} \qquad (0 < z < z_T, t > 0+) \qquad (7\text{-}27)$$

since the cross-sectional area S is equal to the area A that is normal to the direction of diffusion. This result is commonly called *Ficks second law of diffusion*.

This second order, linear partial differential with constant coefficients [Equation (7-27)] may be solved analytically for any one of several different sets of boundary conditions (3). Such problems are commonly referred to as boundary value problems. Alternately, this equation or the integral-difference form of the expression given by Equation (7-18) may be solved by use of numerical methods (1, 5, 6). When mass is also transferred by the turbulent motion of a flowing stream, the rate expression given by Equation (6-46) is needed.

When a chemical reaction occurs in a system, the expression for the reaction rate appears in the material balances. Consider first the general batch reactor shown in Figure 7-3. Such a reactor is fixed in space, and the total feed is introduced at the outset and withdrawal is not made until the reaction has reached the degree of completion desired. The reactor volume V_r is the space filled by the gas or liquid reacting mixture. The volume may vary during the course of a liquid phase reaction because of either shrinkage or expansion of the liquid phase with reaction. In the case of a gas phase reaction, the volume V_r may be either held fixed or varied in any prescribed manner. For example, the volume may be varied as required to maintain the pressure constant. The variation of the temperature T of the reaction mixture with time is also permitted.

ILLUSTRATIVE EXAMPLE 7-5.   MATERIAL BALANCE FOR A BATCH REACTOR

Let the net rate of disappearance of A by reaction be denoted by r_A and defined as follows:

$$r_A = \frac{\text{moles of } A \text{ disappearing by reaction}}{(\text{unit of reactor volume}) (\text{unit time})} \tag{7-28}$$

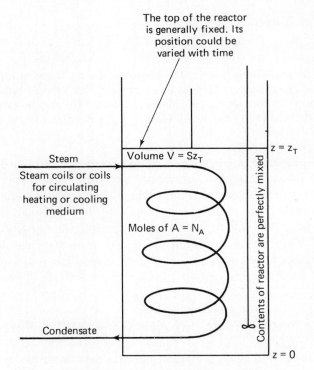

Figure 7-3. Sketch of a batch reactor.

(a) Make a material balance on reactant A over the time period from t_n to t_{n+1}. (b) Obtain the corresponding differential equation.

Solution:

(a) Let M_A denote the molecular weight of A. Then a mass balance for A over the time period from time t_n to t_{n+1} is given by

$$\int_{t_n}^{t_{n+1}} [\underbrace{0}_{\substack{\text{input/time} \\ \text{of } A}} - \underbrace{M_A r_A V_r}_{\substack{\text{output/time} \\ \text{of } A \text{ by} \\ \text{reaction}}}] \, dt = \underbrace{M_A N_A|_{t_{n+1}} - M_A N_A|_{t_n}}_{\text{accumulation}} \tag{7-29}$$

where N_A is the total moles of A in the volume V_r at any time t. Since the molecular weight of A is independent of time, Equation (7-29) reduces to

$$\int_{t_n}^{t_{n+1}} [-r_A V_r] \, dt = N_A|_{t_{n+1}} - N_A|_{t_n} \tag{7-30}$$

(b) Application of the *mean value theorem of integral calculus* to the left-hand side and the *mean value theorem of differential calculus* to the right-hand side of Equation (7-30) followed by the limiting process wherein Δt goes to zero yields

$$r_A = -\frac{1}{V_r}\frac{dN_A}{dt} \tag{7-31}$$

If V_r is independent of t, then

$$r_A = -\frac{dC_A}{dt} \tag{7-32}$$

[In general, the student is advised not to attempt to make a "concentration balance" (which would be of the same form as Equation (7-29) if it existed) because concentration is not conserved unless the volume is constant.] When the rate expression is known, Equation (7-31) may be solved for N_A as a function of time [see Problem 7-3(a)].

Consider next the case in which a reaction is carried out in a flow reactor (see Figure 7-4) at steady-state operation. Suppose that no mixing occurs in the horizontal direction and that the mixing is perfect in the radial direction. Let n_A denote the flow rate in moles per unit time of component A and v_T the total volumetric flow rate past any plane (normal to the direction of flow) in the reactor. Then the concentration of A at any plane may be stated in terms of the flow rates as follows:

$$C_A = \frac{n_A}{v_T} \tag{7-33}$$

ILLUSTRATIVE EXAMPLE 7-6. MATERIAL BALANCE FOR A FLOW REACTOR AT STEADY-STATE OPERATION

(a) For the case in which no mixing occurs in the horizontal direction of the flow reactor shown in Figure 7-4, make a component-material balance (for component A) over the element of volume from V_{rj} to $V_{r,j+1}$, where V_{rj} and $V_{r,j+1}$

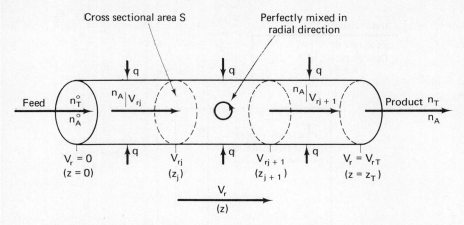

Figure 7-4. Sketch of a tubular reactor that is perfectly mixed in the radial direction and is not mixed in the horizontal direction.

are picked such that $0 < V_r < V_{rT}$. (b) Obtain the differential equation corresponding to the material balance obtained in part (a).

Solution:

(a) The component-material balance follows:

$$\underbrace{n_A|_{V_{rj}}}_{\substack{\text{input/time of } A \\ \text{to the element of} \\ \text{volume by flow}}} - \underbrace{\left[n_A|_{V_{r,j+1}} + \int_{V_{rj}}^{V_{r,j+1}} r_A \, dV_r \right]}_{\substack{\text{output/time of } A \text{ from} \\ \text{the element of volume} \\ \text{by flow and by reaction}}} = 0 \qquad (7\text{-}34)$$

Application of the mean value theorems followed by the limiting process wherein ΔV_r goes to zero yields

$$r_A = -\frac{dn_A}{dV_r} \qquad (0 < V_r < V_{rT}) \qquad (7\text{-}35)$$

When the expression for the rate of reaction is substituted for r_A, Equation (7-35) may be integrated in some instances to give n_A as a function of the reactor volume V_r, [see Problem 7-3(b)].

Illustrative Example 7-6 represents the limiting case of no horizontal mixing, Problem 7-4(b) the intermediate case of some horizontal mixing, and Illustrative Example 7-7 the limiting case of perfect mixing in both the radial and horizontal directions. A reactor which is perfectly mixed in both the radial and horizontal directions is called a *perfectly mixed* reactor and has the representation shown in Figure 7-5. Note that for a perfectly mixed flow reactor, the concentration of A in the output stream is the same as that at any point in the reactor.

Illustrative example 7-7. Perfectly mixed flow reactor at steady-state operation

Make a material balance for component *A* for the reactor shown in Figure 7-5.

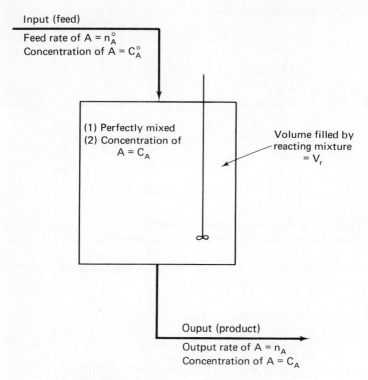

Figure 7-5. A perfectly mixed flow reactor.

Solution:

$$\underbrace{n_A^\circ}_{\substack{\text{input/time of } A \\ \text{to the reactor}}} - [\underbrace{n_A + r_A V_r}_{\substack{\text{output/time of } A \text{ by} \\ \text{flow and by reaction}}}] = 0 \qquad (7\text{-}36)$$

Thus, the material balance for component A is

$$r_A = \frac{n_A^\circ - n_A}{V_r} \qquad (7\text{-}37)$$

Part 2. Use of Energy Balances in the
Analysis of Rate Processes

In the analysis of the industrial processes considered in subsequent chapters, both mass and energy balances are employed. However, in the analysis of certain pieces of process equipment such as heat exchangers, energy balances alone are needed. Consider first the classical problem of a double-pipe exchanger at steady-state operation shown in Figure 7-6.

For a flow system at steady-state operation, the general energy balance given by Equation (6-12) reduces to

$$\left\{\begin{array}{l}\text{input of energy} \\ \text{to the system} \\ \text{per unit time}\end{array}\right\} - \left\{\begin{array}{l}\text{output of energy} \\ \text{from the system} \\ \text{per unit time}\end{array}\right\} = 0 \qquad (7\text{-}38)$$

For the arbitrarily selected direction of heat transfer from the stream in the annular space to the inner stream as shown in Figure 7-6, the energy balance on the inner stream is given by

$$(wE + wPv)|_{z_j} - (wE + wPv)|_{z_{j+1}} + \int_{z_j}^{z_{j+1}} q \, dz = 0,$$
$$(0 < z_j < z_{j+1} < z_T) \qquad (7\text{-}39)$$

where

$$q = Ua(\mathfrak{I} - T)$$

and where the kinetic and potential energy terms have been regarded as negligible. In view of Equation (6-14), it follows that

$$wH|_{z_j} - wH|_{z_{j+1}} + \int_{z_j}^{z_{j+1}} q \, dz = 0 \qquad (7\text{-}40)$$

and again $0 < z_j < z_{j+1} < z_T$. This result is recognized as a form of the well-known relationship $\Delta H = Q$, sometimes called the first law of thermodynamics for flow systems where no shaftwork is involved. Actually, Equation (7-40) could also have been obtained from Equation (6-26).

ILLUSTRATIVE EXAMPLE 7-8. THE DOUBLE-PIPE HEAT EXCHANGER AT STEADY-STATE OPERATION

(a) Make an enthalpy balance on each of the streams entering the respective elements of volume bounded by z_j and z_{j+1} $(0 < z < z_{j+1} < z_T)$ in Figure 7-6.

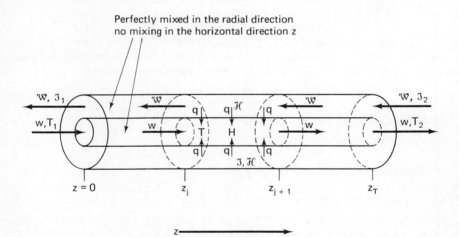

Figure 7-6. Sketch of a double-pipe heat exchanger.

(b) From the results obtained in part (a), obtain the corresponding differential equations. (c) On the basis of the assumptions that the overall heat transfer coefficient U and the specific heats of each steam are independent of temperature, show that $Q = UA\,\Delta T_{lm}$.

Solution:

(a) For the stream flowing through the inner tube, the enthalpy balance is given by Equation (7-40). For the stream flowing through the annular space, the enthalpy balance for the element of volume from z_j to z_{j+1} is given by

$$\mathcal{W}\mathcal{H}\big|_{z_{j+1}} - \mathcal{W}\mathcal{H}\big|_{z_j} - \int_{z_j}^{z_{j+1}} q\,dz = 0 \tag{7-41}$$

(b) By use of the mean value theorems, the differential equations corresponding to Equations (7-40) and (7-41) are readily shown to be

$$-w\frac{dH}{dz} + q = 0 \qquad (0 < z < z_T) \tag{7-42}$$

and

$$\mathcal{W}\frac{d\mathcal{H}}{dz} - q = 0 \qquad (0 < z < z_T) \tag{7-43}$$

(c) Since the composition of the fluid is constant throughout the length of the exchanger, the enthalpy H may be regarded as a function of temperature T and pressure P alone, that is, when no phase change occurs,

$$dH = \left(\frac{\partial H}{\partial T}\right)_P dT + \left(\frac{\partial H}{\partial P}\right)_T dP$$

Then, the derivative of H with respect to z is given by

$$\frac{dH}{dz} = C_p\frac{dT}{dz} + \left(\frac{\partial H}{\partial P}\right)_T \frac{dP}{dz} \tag{7-44}$$

where

$$C_p = \left(\frac{\partial H}{\partial T}\right)_P$$

In heat transfer processes, particularly for condensed phases, the term $(\partial H/\partial P)_T \cdot dP/dz$ is generally regarded as negligible relative to the other terms and, consequently, Equation (7-44) may be represented with good accuracy by

$$\frac{dH}{dz} = C_p\frac{dT}{dz} \tag{7-45}$$

Similarly,

$$\frac{d\mathcal{H}}{dz} = \frac{\partial\mathcal{H}}{\partial\mathcal{T}}\frac{d\mathcal{T}}{dz} = \mathcal{C}_p\frac{d\mathcal{T}}{dz} \tag{7-46}$$

Consequently, Equations (7-42) and (7-43) become

$$-wC_p\frac{dT}{dz} + q = 0 \tag{7-47}$$

$$\mathcal{W}\mathcal{C}_p\frac{d\mathcal{T}}{dz} - q = 0 \tag{7-48}$$

Thus,

$$wC_p \frac{dT}{dz} = \mathcal{W}\mathcal{C}_p \frac{d\mathfrak{I}}{dz}$$

Let

$$\alpha = \frac{\mathcal{W}\mathcal{C}_p}{wC_p}$$

$$\Delta T = \mathfrak{I} - T$$

Then

$$\frac{dT}{dz} = \alpha \frac{d\mathfrak{I}}{dz} = \alpha \left[\frac{d(\mathfrak{I} - T)}{dz} + \frac{dT}{dz} \right]$$

or

$$\frac{dT}{dz} = \left(\frac{\alpha}{1 - \alpha} \right) \frac{d\Delta T}{dz} \tag{7-49}$$

Equations (7-47), (7-49), and the definition of q [see Equation (7-39)] are readily combined to give

$$-\left(\frac{wC_p \alpha}{1 - \alpha} \right) \frac{d(\Delta T)}{dz} + Ua\,\Delta T = 0 \tag{7-50}$$

which upon integration yields

$$\left(\frac{wC_p \alpha}{1 - \alpha} \right) \log_e \frac{\Delta T_2}{\Delta T_1} = Uaz_T = UA \tag{7-51}$$

The total heat transferred per unit time Q is given by an overall enthalpy balance on either stream

$$Q = w(H_2 - H_1) = \mathcal{W}(\mathfrak{IC}_2 - \mathfrak{IC}_1)$$

$$= wC_p(T_2 - T_1) = \mathcal{W}\mathcal{C}_p(\mathfrak{I}_2 - \mathfrak{I}_1)$$

Thus,

$$\alpha = \frac{\mathcal{W}\mathcal{C}_p}{wC_p} = \frac{(T_2 - T_1)}{(\mathfrak{I}_2 - \mathfrak{I}_1)}$$

and

$$\frac{wC_p \alpha}{1 - \alpha} = \frac{wC_p(T_2 - T_1)}{(\mathfrak{I}_2 - \mathfrak{I}_1)) - (T_2 - T_1)} = \frac{Q}{\Delta T_2 - \Delta T_1}$$

In view of this result, Equation (7-51) may be rearranged to the well-known form

$$Q = UA\,\Delta T_{lm} \tag{7-52}$$

where

$$\Delta T_{lm} = \frac{\Delta T_2 - \Delta T_1}{\log_e \frac{\Delta T_2}{\Delta T_1}}$$

Other problems involving heat exchangers are listed at the end of the chapter.

Because of the similarity of the rate expressions for molecular diffusion and heat conduction, the partial differential equations that describe these processes at unsteady-state operation are also similar in form.

ILLUSTRATIVE EXAMPLE 7-9. UNSTEADY-STATE HEAT CONDUCTION THROUGH A METAL ROD

A sketch of a stationary metal rod of uniform composition is shown in Figure 7-7. The cylindrical surface of the bar is perfectly insulated. At time $t = 0$, the temperature is uniform throughout. At time $t = 0+$, the left end of the rod is brought immediately to the temperature T_W, and the right end of the rod is insulated. It may be supposed that no chemical reaction within the bar occurs, that is, the composition of the rod is constant with respect to time and position. (a) Make an energy balance on the volume bounded by z_j and z_{j+1} over the time period from t_n to t_{n+1}, where these are picked such that $0 < z_j < z_{j+1} < z_T$ and $0 < 0+ < t_n < t_{n+1}$. (b) From the result of part (a), obtain the corresponding partial differential equation.

Solution:
(a) Since the flow of heat alone is involved, the rod should be considered as a batch system. The energy balance is given by

$$\int_{t_n}^{t_{n+1}} [Q_{cd}|_{z_j,t} - Q_{cd}|_{z_{j+1},t}] \, dt = \int_{z_j}^{z_{j+1}} \rho SE \, dz|_{t_{n+1}} - \int_{z_j}^{z_{j+1}} \rho SE \, dz|_{t_n}$$

$$= \int_{z_j}^{z_{j+1}} [(\rho SE)|_{t_{n+1},z} - (\rho SE)|_{t_n,z}] \, dz \qquad (7\text{-}53)$$

(b) Application of first the *mean value theorem of differential calculus* to the integrands and then the *mean value theorem of integral calculus* to the integrals followed by a limiting process as Δz and Δt go to zero yields

$$-\frac{\partial Q_{cd}}{\partial z} = \frac{\partial(\rho SE)}{\partial t} = S \frac{\partial(\rho E)}{\partial t} \qquad (0 < z < z_T, \qquad t > 0+) \qquad (7\text{-}54)$$

Since the composition of the bar remains fixed with respect to time and position, the internal energy may be expressed as a function of the volume

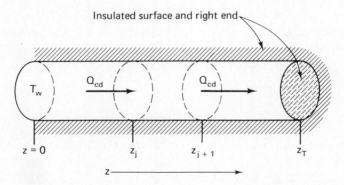

Insulated surface and right end

T_w Q_{cd} Q_{cd}

$z = 0$ z_j z_{j+1} z_T

$z \longrightarrow$

Figure 7-7. Unsteady-state heat conduction in a cylindrical bar.

V and temperature T alone, that is,

$$dE = \left(\frac{\partial E}{\partial T}\right)_V dT + \left(\frac{\partial E}{\partial V}\right)_T dV \tag{7-55}$$

Thus,

$$\frac{\partial E}{\partial t} = C_V \frac{\partial T}{\partial t} + \left(\frac{\partial E}{\partial V}\right)_T \frac{\partial V}{\partial t} \tag{7-56}$$

where

$$C_V = \left(\frac{\partial E}{\partial T}\right)_V.$$

In heat transfer processes, the second term on the right-hand side of Equation (7-56) is commonly regarded as negligible relative to the first one and, consequently, this expression can be represented with good accuracy by

$$\frac{\partial E}{\partial t} = C_V \frac{\partial T}{\partial t} \tag{7-57}$$

Now, if it is assumed that k and ρ are constant, Equation (7-54) may be reduced to the following result by use of Equation (6-33) and (7-57):

$$kA \frac{\partial^2 T}{\partial z^2} = S\rho C_V \frac{\partial T}{\partial t}$$

Since $A = S$, it follows that

$$\frac{\partial^2 T}{\partial z^2} = \left(\frac{\rho C_V}{k}\right) \frac{\partial T}{\partial t} \qquad (0 < z < z_T, \qquad t > 0+) \tag{7-58}$$

The partial differential equation for heat conduction at unsteady-state operation in three directions is derived in a manner analogous to that shown in the previous illustrative example.

ILLUSTRATIVE EXAMPLE 7-10. UNSTEADY-STATE HEAT CONDUCTION IN THREE-DIMENSIONAL SPACE

(a) Set up the energy balance for the element of volume shown in Figure 7-8 over the time period from t_n to t_{n+1}. The element of volume is composed of a material whose thermal conductivity, density, and composition are independent of time and position. (b) Obtain the partial differential equation for any point (x, y, z, t).

Solution:
(a) Let

$$q_x = -k \frac{\partial T}{\partial x}, \qquad q_{xj} = -k \frac{\partial T}{\partial x}\bigg|_{x=x_j}$$

$$q_y = -k \frac{\partial T}{\partial y}$$

$$q_z = -k \frac{\partial T}{\partial z}$$

where q_x, q_y, q_z are the rates of heat conduction (Btu/hr ft^2) in the positive directions

x, y, and z, respectively (see Figure 7-8). The energy balance is seen to be

$$\underbrace{\int_{t_n}^{t_{n+1}} \left\{ \int_{z_j}^{z_{j+1}} \int_{y_j}^{y_{j+1}} (q_{xj} - q_{x,\,j+1})\, dy\, dz \right.}_{\text{input over 1—output over 2}} + \underbrace{\int_{z_j}^{z_{j+1}} \int_{x_j}^{x_{j+1}} (q_{yj} - q_{y,\,j+1})\, dx\, dz}_{\text{input over 3—output over 4}}$$

$$+ \left. \underbrace{\int_{y_j}^{y_{j+1}} \int_{x_j}^{x_{j+1}} (q_{zj} - q_{z,\,j+1})\, dx\, dy \right\}\, dt}_{\text{input over 5—output over 6}} \tag{7-59}$$

$$= \underbrace{\int_{z_j}^{z_{j+1}} \int_{y_j}^{y_{j+1}} \int_{x_j}^{x_{j+1}} [(\rho E)|_{t_{n+1}} - (\rho E)|_{t_n}]\, dx\, dy\, dz}_{\text{accumulation}}$$

(b) Consider the first multiple integral on the left-hand side of Equation (7-59). Application of the *mean value theorem of differential calculus* yields

$$\int_{t_n}^{t_{n+1}} \int_{z_j}^{z_{j+1}} \int_{y_j}^{y_{j+1}} (q_{xj} - q_{x,\,j+1})\, dy\, dz\, dt = -\Delta x \int_{t_n}^{t_{n+1}} \int_{z_j}^{z_{j+1}} \int_{y_j}^{y_{j+1}} \left.\frac{\partial q_x}{\partial x}\right|_{\textcircled{A}}\, dy\, dz\, dt \tag{7-60}$$

where

$$\textcircled{A} = \{[x_j + \alpha(y, z, t)\cdot\Delta x], y, x, t\}, \qquad 0 < \alpha(y, z, t) < 1$$

Application of the *mean value theorem of integral calculus* three times in succession to the right-hand side of Equation (7-60) yields

$$-\Delta x \int_{t_n}^{t_{n+1}} \int_{z_j}^{z_{j+1}} \int_{y_j}^{y_{j+1}} \left.\frac{\partial q_x}{\partial x}\right|_{\textcircled{A}}\, dy\, dz\, dt = -\Delta x\, \Delta y\, \Delta z\, \Delta t \left.\frac{\partial q_x}{\partial x}\right|_{\textcircled{A}}$$

where $\textcircled{A}$ is now given by

$$\textcircled{A} = [x_p, y_p, z_p, t_p];$$

$$x_p = x_j + \alpha(y_p, z_p, t_p)\cdot\Delta x;$$

$$y_p = y_j + \beta(z_p, t_p)\cdot\Delta y, \qquad 0 \leq \beta \leq 1;$$

$$z_p = z_j + \gamma(t_p)\cdot\Delta z, \qquad 0 \leq \gamma \leq 1;$$

$$t_p = t_n + \delta\, \Delta t, \qquad 0 \leq \delta \leq 1.$$

It an analogous manner, the remaining multiple integrals in Equation (7-59) are reduced to give

$$-\Delta x\, \Delta y\, \Delta z\, \Delta t \left[\left.\frac{\partial q_x}{\partial x}\right|_{\textcircled{A}} + \left.\frac{\partial q_y}{\partial y}\right|_{\textcircled{B}} + \left.\frac{\partial q_z}{\partial z}\right|_{\textcircled{C}} \right] = \Delta x\, \Delta y\, \Delta z\, \Delta t \left.\frac{\partial(\rho E)}{\partial t}\right|_{\textcircled{D}} \tag{7-61}$$

The points of evaluation $\textcircled{B}$, $\textcircled{C}$, and $\textcircled{D}$ are of the same form as that shown for $\textcircled{A}$. Since it is given that the density ρ, the thermal conductivity k, and the composition of the element remain fixed with respect to time and position, then Equation (7-61) may be further reduced in the usual way to give

$$\frac{\partial^2 T}{\partial x^2} + \frac{\partial^2 T}{\partial y^2} + \frac{\partial^2 T}{\partial z^2} = \left(\frac{\rho C_V}{k}\right) \frac{\partial T}{\partial t} \tag{7-62}$$

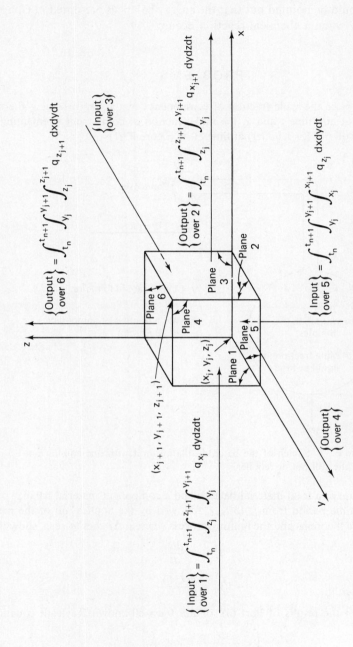

Figure 7-8. Transient heat transfer by conduction in three directions.

235

Although models involving chemical reactors are not considered in this book, it should be pointed out that the energy balances presented in Chapter 6 also apply when a chemical reaction occurs.

PROBLEMS

7-1. Let y_i be the mole fraction of component i in the vapor leaving the still pot at any time t and x_i the mole fraction of component i remaining in the still pot (or reboiler) at time t (see Figure P7-1).

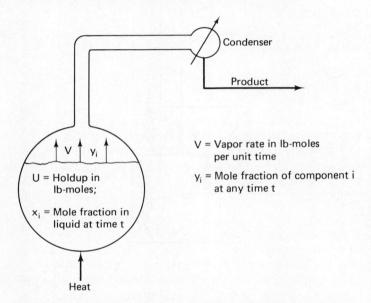

Condenser

Product

V = Vapor rate in lb-moles per unit time

U = Holdup in lb-moles;

x_i = Mole fraction in liquid at time t

y_i = Mole fraction of component i at any time t

Heat

Figure P7-1. Sketch of the batch distillation unit with one equilibrium stage, the still pot or reboiler.

(a). By use of a total-material balance and a component-material balance over the time period from t_n to t_{n+1}, followed by the application of the mean value theorems and the limiting process wherein Δt goes to zero, show that

$$V = -\frac{dU}{dt}$$

$$-Vy_i = \frac{d(Ux_i)}{dt}$$

(b). From the results of Part (a), obtain the well-known Rayleigh equation.

$$\int_{x_i^\circ}^{x_i} \frac{dx_i}{y_i - x_i} = \log_e \frac{U}{U^\circ}$$

where x_i° and U° are the values of these variables at time $t = 0$.

(c). From the results of Part (a), show that

$$u_i = u_i^\circ \left(\frac{u_b}{u_b^\circ}\right)^{\alpha_i}$$

where $\alpha_i = K_i/K_b$, and α_i is assumed to be constant throughout the distillation process and where

K_i = Henry law constant, $y_i = K_i x_i$; K_b denotes the value for the base component;

$u_i = U x_i$;

$u_b = U x_b$.

7-2. Verify the result given by Equation (7-16).

7-3(a). Suppose that it has been determined experimentally that the rate of reaction for the reaction

$$A \longrightarrow R$$

is given by

$$r_A = k_c C_A \qquad (A)$$

where k_c is a constant. Further suppose that it is desired to carry this reaction out in an isothermally operated batch reactor and that it is desired to find an expression for N_A versus time. Show that the elimination of r_A from Equations (A) and (7-31) followed by integration over the time period from $t = 0$ to any t gives

$$N_A = N_A^\circ e^{-k_c t} \qquad (B)$$

where N_A° is the moles of A in the reactor at time $t = 0$.

7-3(b). If the same reaction described in Part (a) is carried out isothermally at steady-state operation in a flow reactor (perfectly mixed in the radial direction but no mixing in the horizontal direction), show that when the total volumetric flow rate v_T remains constant throughout the reactor,

$$n_A = n_A^\circ e^{-k_c \theta}$$

where

$\theta = V_r/v_T$, the true residence time.

7-4(a). If the net rate of reaction for component A is defined in terms of appearance rather than disappearance,

$$r_A = \frac{\text{moles of } A \text{ appearing by reaction}}{(\text{unit of reactor volume})(\text{unit time})} \qquad (A)$$

show that for this definition of r_A, the following expressions are obtained:

$$r_A = \frac{1}{V_r}\frac{dN_A}{dt} \quad \text{and} \quad r_A = \frac{dn_A}{dV_r} \qquad (B)$$

rather than Equations (7-31) and (7-35), respectively.

7-4(b). Consider a flow reactor at steady-state operation in which the mixing is assumed to be perfect in the radial direction Figure P7-4. If it is supposed that the mixing along the axis (in the z-direction) may be described by

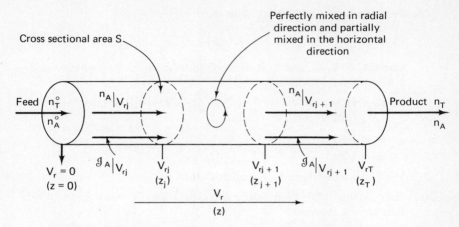

Figure P7-4. Sketch of a tubular reactor that is perfectly mixed in the radial direction and partially mixed in the horizontal direction.

Equation (6-46), show that

$$r_A = -\frac{dn_A}{dV_r} + \mathfrak{D} S^2 \frac{d^2 C_A}{dV_r^2} \qquad (0 < V_r < V_{rT}) \qquad \text{(C)}$$

where r_A is defined in terms of the disappearance of A. (*Hint:* Follow the same general procedure used in Illustrative Example 7-6.)

7-4(c). Suppose that the flow reactor shown in Figure 7-4 is at unsteady-state operation at time $t = 0$. Make a component-material balance (for component A) on the element of volume from V_{rj} to $V_{r,j+1}$ ($0 < V_{rj} < V_{r,j+1} < V_{rT}$) over the time period from t_n to t_{n+1} ($t > 0$) for the case in which no mixing occurs in the horizontal direction. Then show that the following partial differential is obtained upon applying the mean value theorems and the limiting process (wherein Δz and Δt are allowed to go to zero) to the component-material balance

$$r_A = -\frac{\partial n_A}{\partial V_r} - \frac{\partial C_A}{\partial t} \qquad (0 < V_r < V_{rT}, \qquad t > 0) \qquad \text{(D)}$$

where r_A is defined in terms of the disappearance of A.

7-5(a). Produce the set of integral-difference equations required to describe the component-material balances for the system of perfect mixers at unsteady operation (see Figure P7-5).

7-5(b). From the results of Part (a), obtain the corresponding set of differential-difference equations. The symbols shown in Figure P7-5 have the following meanings

f, q = volumetric flow rates, ft³/hr;

$C_m(t)$ = concentration of the component under consideration that leaves the mth mixer at time t, lb/ft³; $C_i(t)$ = inlet concentration.

V = holdup of each mixer, ft³.

This problem has been considered in detail by Holland (5).

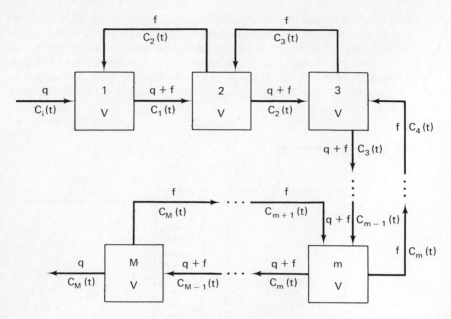

Figure P7-5. Sketch of perfect mixers with feedback.

7-6. Colburn (2) suggested that the variation of U throughout a double-pipe heat exchanger be accounted for by use of a linear variation of U with ΔT (the temperature difference between the two streams). Repeat Illustrative Example 7-8 for this case, that is, for

$$U = m\,\Delta T + b$$

where m and b are constants, and show that

$$Q = A\frac{U_1\,\Delta T_2 - U_2\,\Delta T_1}{\log_e \dfrac{U_1\,\Delta T_2}{U_2\,\Delta T_1}}$$

U_1, U_2 = values of U at $z = 0$ and $z = z_T$, respectively;
$\Delta T_2 = (\mathfrak{I}_2 - T_2)$ and $\Delta T_1 = (\mathfrak{I}_1 - T_1)$ (see Figure 7-6).

7-7. In some applications of heat transfer, it appears that U may be approximated with good accuracy by a logarithmic function of ΔT. Repeat Illustrative Example 7-8 on the basis that

$$U = c + g\log_e \Delta T$$

where c and g are constants, and show that

$$Q = U_{lm} A\,\Delta T_{lm}$$

where

$$U_{lm} = \frac{U_2 - U_1}{\log_e \dfrac{U_2}{U_1}}$$

7-8. This problem and the next one are based on a problem posed by Marshall and Pigford (7). A liquid stream is fed to the perfect mixer (shown in Figure P7-8) where it is simultaneously mixed and heated by means of steam coils. At time $t = 0$, the process is at steady state with no heating. The temperature of the feed is T_i for all time t. At time $t = 0+$, the steam at its saturation temperature T_s is turned on. The flow rate in and out, w, is held fixed for all time t.

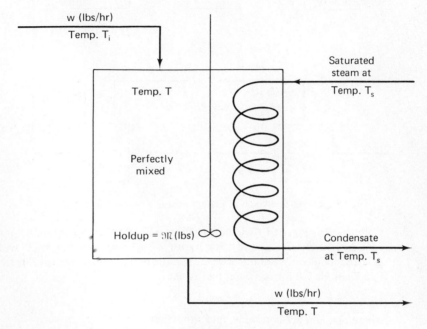

Figure P7-8. A perfect mixer with steam coils.

7-8(a). Show that if PE and KE effects as well as W of the mixer are neglected, $d(P_s v_s)/dt = 0$, and C_p is constant, then the differential equation corresponding to Equation (6-30) is

$$wC_p(T_i - T) + UA(T_s - T) = \mathfrak{M}C_p \frac{dT}{dt}$$

where A is the surface area of the heating coils.

7-8(b). If C_p and U are taken to be independent of temperature, show that at any time t,

$$\phi = \left\{ \frac{\dfrac{UA}{wC_p}}{1 + \dfrac{UA}{wC_p}} \right\} \left\{ 1 - \exp\left[-\left(1 + \frac{UA}{wC_p} \right) \frac{wt}{\mathfrak{M}} \right] \right\}$$

where

$$\phi = \frac{T - T_i}{T_s - T_i}$$

7-9. Consider again Problem 7-8 with the following revisions. Initially, at time $t = 0$, there is no liquid in the tank. At time $t = 0+$, the steam at its saturation temperature T_s is turned on, and feed at temperature T_i begins entering the mixer at the rate w. No liquid is withdrawn from the tank until it is filled. The total capacity of the tank is $\mathfrak{M}$ lb of liquid, and the total heating surface of the steam coils is A sq ft. The amount of heating surface per unit height as well as the capacity per unit height are constant as implied in Figure P7-8.

7-9(a). Show that the differential equation corresponding to the enthalpy balance is given by

$$wH_i + U\mathfrak{C}(T_s - T) = \frac{d(MH)}{dt} \qquad \left(0+ < t < \frac{\mathfrak{M}}{w}\right) \tag{A}$$

where

$\mathfrak{C} = $ surface area covered by liquid at any time $t < \mathfrak{M}/w$;

$M = $ mass of liquid in the mixer at any time $t < \mathfrak{M}/w$.

7-9(b). If C_p is taken to be independent of temperature show that Equation (A) may be stated in the form

$$\frac{d\phi}{df} + \left(\frac{UA}{wC_p} + \frac{1}{f}\right)\phi = \frac{UA}{wC_p} \qquad \left(0+ < t < \frac{\mathfrak{M}}{w}\right) \tag{B}$$

where

$$f = \frac{\mathfrak{C}}{A} = \frac{M}{\mathfrak{M}} = \frac{wt}{\mathfrak{M}}$$

$$\phi = \frac{T - T_i}{T_s - T_i}$$

7-9(c). Show that the solution that satisfies Equation (B) and the initial condition is

$$\phi = 1 - \frac{\left[1 - \exp\left(-\dfrac{UAf}{wC_p}\right)\right]}{\dfrac{UAf}{wC_p}} \tag{C}$$

Problems 7-8 and 7-9 are similar to ones proposed by Marshall and Pigford (7).

7-10. A thermometer with an initial reading of T_i is immersed in a bath at the constant temperature T_0 ($T_0 > T_i$) at time $t = 0+$ (see Figure P7-10). Assume that the temperature of the fluid in the bulb is independent of position at any time t. (1) Make an energy balance on the fluid in the bulb, and show that if the small variation in the mass of fluid in the bulb is neglected, the following result is obtained:

$$UA(T_0 - T) = \mathfrak{M}C_v \frac{dT}{dt} \tag{A}$$

Suppose that the ratio $\mathfrak{M}C_v/UA$ is a constant, and let it be denoted by L. Show that Equation (A) may be integrated to give

$$\frac{T - T_i}{T_0 - T_i} = 1 - \exp\left(-\frac{t}{L}\right)$$

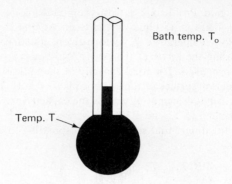

Bath temp. T_o

Temp. T

Figure P7-10. Sketch of the thermometer.

7-11. At time $t = 0$, the right circular cone composed of a pure metal (shown in Figure P7-11) is at a uniform temperature T_i. At time $t = 0+$, the base (at $z = 0$) is brought to some temperature $T_W \neq T_i$, and the remainder of the cone is immersed in a gas in turbulent flow at a uniform temperature of T_0 ($T_0 \neq T_W$, $T_0 \neq T_i$). (a) On the basis of the supposition that the diameter of the base of the cone is small relative to its length (that is, the temperature gradient in the radial direction is negligible relative to the gradient in the axial direction of z), show that

$$-\frac{\partial Q_{cd}}{\partial z} + q_{cv} = \rho S C_V \frac{\partial T}{\partial t} \qquad (0+ < t, 0 < z < z_T)$$

where T is the temperature of the cone at any time t. (b) By use of the geometrical relationships for a right-circular cone, show that the result of part (a) reduces to

$$k(z_T - z)\frac{\partial^2 T}{\partial z^2} - 2k\frac{\partial T}{\partial z} + h(2 \csc \alpha)(T_0 - T) = \rho(z_T - z)C_V\frac{\partial T}{\partial t}$$

where h is the convection coefficient.

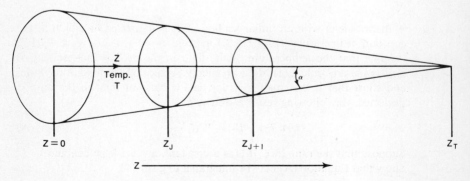

z

Temp. T

α

$z = 0$ z_J z_{J+1} z_T

$z \longrightarrow$

Figure P7-11. Heat transfer by conduction in a right circular cone.

7-12. The base of the fin shown in Figure P7-12 is maintained at a constant
temperature T_W. The fin is surrounded by air at the constant temperature
T_0. Heat is transferred from the top and bottom of the fin to the air by
convection. The surface area of the sides is negligible relative to that of
the top and bottom. Also, since $b \ll z_T$ (see Figure P7-12), the temperature gradient is negligible in the planes perpendicular to z relative to the
gradient in the direction z. Show that at steady state

$$\frac{d^2\phi}{dz^2} + \frac{1}{z}\frac{d\phi}{dz} - \frac{B}{z}\phi = 0$$

where

$$\phi = T - T_0$$

$$B = \left(\frac{2hx_T}{bk}\right)$$

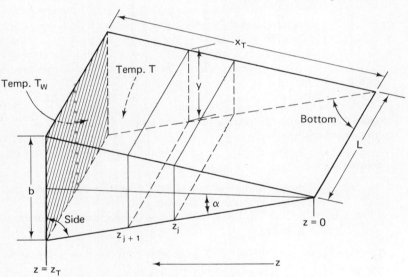

Surroundings at Temp. T_o

Figure P7-12. Sketch of the wedge-shaped fin.

7-13. Fill in all the intermediate steps involved in the derivation of Equation
(7-62).

7-14. If the liquid on each plate of a distillation column as well as the vapor
above each plate is taken to be perfectly mixed, set up the following integral-difference equations for unsteady-state operation: (a) component-
material balance, (b) total-material balance, and (c) energy balance. A
sketch of a typical plate j is shown in Figure P7-14. The symbols that
appear in this figure which are not defined in Chapter 3 are defined as

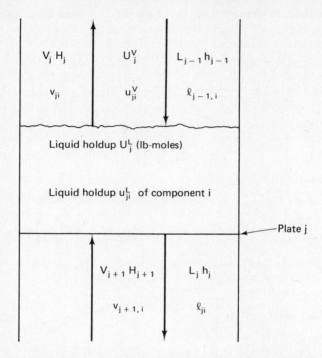

Figure P7-14. Sketch of streams entering and leaving a typical plate of a distillation column.

follows:

u_{ji}^L, u_{ji}^V = molal holdup of component i in the liquid and vapor phases, respectively, on plate j;

U_j^L, U_i^V = holdup of liquid and vapor, respectively, on plate j in lb moles.

[This problem and Problem 7-15 have been considered in detail in Reference (5).]

7-15. From the results obtained in Problem 7-14, obtain the corresponding differential-difference equations.

7-16(a). Repeat Problem 7-4(b) for a reactor at unsteady-state operation and show that the partial differential equation representing this operation is given by

$$-\frac{\partial n_A}{\partial V_r} + \mathfrak{D}S^2 \frac{\partial^2 C_A}{\partial V_r^2} - r_A = \frac{\partial C_A}{\partial t}$$

7-16(b). For the case of unsteady operation, show that the material balance for the perfectly mixed reactor at constant V_r in Figure 7-5 is given by

$$n_A^o - n_A - r_A V_r = V_r \frac{dC_A}{dt}$$

7-17. The diffusion of a gas A through a stagnant film of B followed by the absorption of A but not of B by a nonvolatile liquid gives rise to bulk flow, and the total rate of transfer of A at any z is equal to the rate of transfer of A by diffusion plus the rate of transfer of A by bulk flow. Show that the total rate of transfer of A through a stagnant film of B is given by *Stefan's law*

$$J_{AT} = -D_V A \left(\frac{C_T}{C_B}\right) \frac{dC_A}{dz}$$

where

$C_T = C_A + C_B$, total concentration or density, moles/ft³;

J_{AT} = total rate of transfer of A by diffusion and by bulk flow in the positive direction of z, lb moles/hr.

Hints:

1. Pick any z, say a z next to the interface. Then, the total rate of flow of A or the rate of absorption of A is given by

$$J_{AT} = J_A + J'_A$$

and since B is not absorbed,

$$J_{BT} = J_B + J'_B = 0$$

where J'_A and J'_B are the bulk flow rates of A and B.

2. Suppose that the total concentration $(C_A + C_B)$ is constant both with respect to time and position.

3. Note that the bulk flow rates are related by

$$J'_A = \frac{C_A}{C_B} J'_B$$

7-18. Show that the equation of continuity for any point (x, y, z) in a flowing fluid is given by

$$\frac{\partial(u_x \rho)}{\partial x} + \frac{\partial(u_y \rho)}{\partial y} + \frac{\partial(u_z \rho)}{\partial z} = -\frac{\partial \rho}{\partial t}$$

where

u_x, u_y, u_z = components of velocity in the x, y, and z directions, respectively;

ρ = density of the fluid.

Hints:

1. Choose an arbitrary point (x_j, y_j, z_j) in the flowing fluid, an arbitrary set of increments, $\Delta x, \Delta y$, and Δz, and construct the element of volume shown in Figure P7-18.

2. Make a material balance on this element of volume over the time period from t_n to time $t_n + \Delta t$.

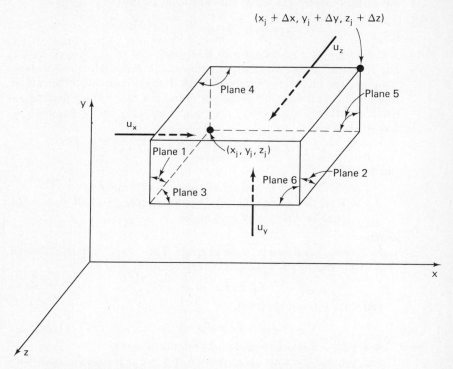

Figure P7-18. Element of volume used to make a material balance over the time period Δt.

NOTATION*

B = product rate; moles/hr.

C_p = heat capacity at constant pressure; $C_p = (\partial H/\partial T)_P$.

C_V = heat capacity at constant volume; $C_V = (\partial E/\partial T)_V$.

E_{ji}^M = modified Murphree plate efficiency; defined by Equation (7-17).

F = feed rate, moles/hr.

L = total flow rate of a liquid stream, moles/hr.

M_A = molecular weight of compound A, lb/(lb mole).

n_A, n_B = flow rates of components A and B, respectively; n_T = total flow rate of the entire stream, moles/hr.

N_A, N_B = total moles of A and B, respectively; N_T = total moles of a mixture.

v_T = total volumetric flow rate, ft³/hr.

*See also Chapter 6 notations.

V = total flow rate of a vapor stream, moles/hr.

V_r = volume of a reactor, ft^3.

x = a coordinate axis in three-dimensional space.

x_i = mole fraction of component i in the liquid phase.

X_i = mole fraction of component i in the feed.

y = a coordinate axis in three-dimensional space.

y_i = mole fraction of component i in the vapor phase.

Y_i = a function of fugacities and mole fractions; defined after Equation (7-7).

z = a coordinate axis in three-dimensional space; also used to denote length in one-dimensional space.

Superscripts

$\circ$ = initial value or value at the beginning of the time period under consideration.

M = modified Murphree efficiency.

REFERENCES

1. Carahan, B., H. A. Luther, and J. O. Wilkes, *Applied Numerical Methods* (New York: John Wiley & Sons, Inc., 1969).

2. Colburn, A. P., "High Temperature Difference and Heat Transfer Coefficient in Liquid Heat Exchangers," *Ind. Eng. Chem.*, **25**, (1933), 873.

3. Crank, J., *The Mathematics of Diffusion*. London E. C. 4: Cambridge University Press, 1955.

4. Holland, C. D., *Multicomponent Distillation*. Englewood Cliffs, N. J.: Prentice-Hall, Inc., 1963.

5. ———, *Unsteady State Processes with Applications in Multicomponent Distillation*. Englewood Cliffs, N. J.: Prentice-Hall, Inc., 1966.

6. Lapidus, L., *Digital Computation for Chemical Engineers*. New York: McGraw-Hill Book Company, Inc., 1962.

7. Marshall, W. R., Jr., and R. L. Pigford, *The Application of Differential Equations to Chemical Engineering Problems*. Newark, Delaware: University of Delaware, 1947.

8. Murphree, E. V., "Rectifying Column Calculations," *Ind. Eng. Chem.*, **17**, (1925), 747.

Mass Transfer Relationships and Their Application to the Classical Models for Absorption Columns

8

The topic of rates of mass transfer that was introduced in Chapters 6 and 7 is considered in greater detail in this chapter. Rate expressions utilizing concentrations and partial pressures as driving forces are introduced as well as other concepts such as the number of mass transfer units and the height of a transfer unit. After these fundamental units have been presented, they are used to formulate a classical model that is used in the design of packed columns for the absorption of a soluble component from a gas stream (composed primarily of insoluble components plus a small amount of the soluble component) by contacting it countercurrently with a liquid stream. To provide good contact between the gas and liquid phases, the columns are filled with packing. Pictures of some of the various kinds of packing used in these columns are shown in Figure 8-1.

A brief review of the major developments of mass transfer concepts and their applications to packed columns follows. In 1923, Whitman (11) proposed the *two-film* concept. Two films are visualized through which a gas must pass when it is transferred from the vapor (or liquid) phase to the liquid

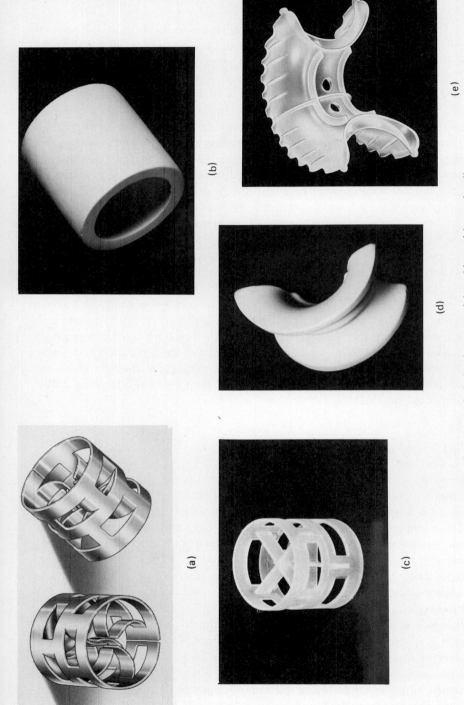

Figure 8-1. Some typical kinds of commercial packings: (a) metal pall rings, (b) plastic pall ring, (c) Raschig ring, (d) ceramic intalox saddle, (e) plastic intalox saddle. (*Courtesy of Norton Company, Chemical Process Products Division*).

(or vapor) phase. Each film provides a resistance to mass transfer. The total resistance to mass transfer from one phase to another is taken equal to the sum of the two-film resistances (3, 11). The reciprocal of the total resistance is called the overall mass transfer coefficient. A graphical procedure based on the use of mass transfer coefficients was developed by Walker, Lewis, and McAdams (10) and used for the solution of problems that could be characterized by the three components: an inert gas, and inert liquid, and an *active* component. In 1927, this procedure was applied by Lewis (4) to solve problems involving the separation of hydrocarbons in natural gasoline plants. Constant total flow rates and isothermal operation were assumed. In 1935, a procedure for obtaining the necessary height of packing required to achieve a specified separation was presented by Chilton and Colburn (1). This procedure involved the use of the concept of the *height of a transfer unit*, or HTU. In 1922, the concept of the *height equivalent to a theoretical plate*, or HETP, was introduced by Peters (6). The HETP is that height of packing required to effect the same separation as that which could be achieved by an equilibrium stage (or a perfect plate).

Recently, Rubac *et al.* (8, 9) introduced the concept of the *mass transfer section* to account for the behavior of packed distillation columns and packed absorbers. When this concept is employed, the resulting equations required to describe packed columns are identical in form to those required to describe columns with plates. This approach is used in subsequent chapters which are concerned with the modeling of columns in the process of absorbing multicomponent mixtures.

Part 1 : Mass Transfer Relationships

At steady state, the rate of mass transfer from the vapor to the liquid phase consists of a sequence of rate processes that occur in series as implied by Figure 8-2. This process may be described in a pictorial manner as follows. First, component i must be transferred from a position corresponding to the bulk conditions of the gas phase to the interface at the gas side. The fugacity potential or driving force for this transfer is $\bar{f}_i^V - \bar{f}_i^\mathcal{V}$. The superscript V refers to the bulk conditions of the vapor phase, the superscript $\mathcal{V}$ to the vapor side of the interface, the superscript $\mathcal{L}$ to the liquid side of the interface, and the superscript L to the bulk conditions of the liquid phase. Next, component i must be transferred across the interface, and the driving force for this transfer is $\bar{f}_i^\mathcal{V} - \bar{f}_i^\mathcal{L}$. Next, component i must be transferred from the liquid side of the interface to a position corresponding to the bulk conditions of the liquid phase, and the driving force for this transfer is $\bar{f}_i^\mathcal{L} - \bar{f}_i^L$. The rate N_i may be defined either in terms of the transfer of component i from the gas to the liquid phase or from the liquid to the gas phase. If N_i is defined as the moles

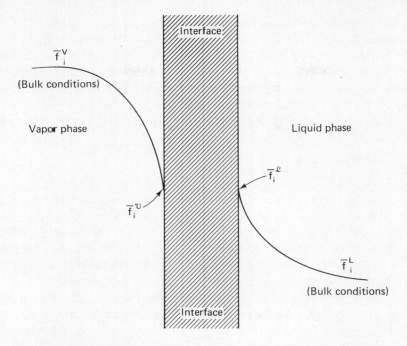

Figure 8-2. Sketch of the fugacity profile involved in the transfer of a component from the vapor phase to the liquid phase.

of component i transferred from the liquid phase to the vapor phase per unit time per unit height of packing, then N_i is given by

$$N_i = k_L a_i S(\bar{f}_i^L - \bar{f}_i^\ell) = k_I a_i S(\bar{f}_i^\ell - \bar{f}_i^v) = k_G a_i S(\bar{f}_i^v - \bar{f}_i^V) \qquad (8\text{-}1)$$

Consider now the system shown in Figure 8-3 in which the distance z is measured from the top of the column down, the liquid flows in the positive direction of z, and the vapor flows in the counter direction to the liquid (or in the negative direction of z). The limit of the component-material balance as the distance over which the balance is made is allowed to go to zero gives the following differential equation for component i in the gas phase:

$$\frac{d(Vy_i)}{dz} + N_i = 0 \qquad (8\text{-}2)$$

where N_i is defined by Equation (8-1). For component i in the liquid phase, the corresponding expression is given by

$$\frac{d(Lx_i)}{dz} + N_i = 0 \qquad (8\text{-}3)$$

Equations (8-2) and (8-3) assume that perfect mixing occurs in the radial direction but that no mixing occurs in the direction z. The derivation of these equations is outlined in Illustrative Example 7-2.

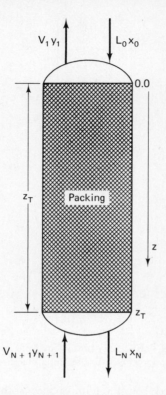

Figure 8-3. Sketch of a packed absorption column.

Again, as in Chapter 6, it is supposed in all subsequent developments that k_I is so large that a dynamic equilibrium may be assumed to exist at the interface, that is,

$$\bar{f}_i^v \cong \bar{f}_i^\ell \tag{8-4}$$

Note that setting $\bar{f}_i^v \cong \bar{f}_i^\ell$ does not imply that the rate of mass transfer across the interface is equal to zero. On the contrary, the rate of transfer across the interface is equal to the same value of N_i as the one for each of the other steps, that is, in the limit as k_I approaches infinity, the difference $\bar{f}_i^v - \bar{f}_i^\ell$ approaches zero and the corresponding indeterminate form $k_I a_i S(\bar{f}_i^v - \bar{f}_i^\ell)$ approaches the finite number N_i, the rate of mass transfer across the vapor and liquid phases. The use of only two rates or two resistances ($1/k_L a_i S$ and $1/k_G a_i S$) is known as the two-film theory (3, 10) of mass transfer.

Instead of using fugacities as the driving forces in the rate expressions, concentrations and partial pressures are commonly used in the older literature. Again, dynamic equilibrium is assumed to exist at the interface and the rate of transfer across the interface is omitted in this and all subsequent developments. Then the rate of transfer of component i from the liquid to vapor phase is given by

$$N_i = k'_L a_i S(C^L_i - C^{\mathcal{L}}_i) = k'_G a_i S(p^{\upsilon}_i - p^V_i) \tag{8-5}$$

where C_i has the units of moles per unit volume.

Relationships between the mass transfer coefficients

At a given z in the column, let the following assumptions be made:

$$
\begin{aligned}
&1. \ P^V = P^{\upsilon} = P^{\mathcal{L}} = P^L = P \\
&2. \ T^V = T^{\upsilon} = T^{\mathcal{L}} = T^L = T \\
&3. \ \frac{\gamma^{\upsilon}_i}{\gamma^V_i} = 1, \quad \frac{\gamma^{\mathcal{L}}_i}{\gamma^L_i} = 1 \\
&4. \ \rho^L = \rho^{\mathcal{L}}
\end{aligned}
\tag{8-6}
$$

where ρ^L is equal to the molar density of the liquid. Since the fugacities of pure components depend on temperature and pressure alone, it follows from conditions 1 and 2 that $f^{\upsilon}_i = f^V_i$ and $f^{\mathcal{L}}_i = f^L_i$ at any z. Thus, on the basis of the assumptions given by Equation (8-6), it is possible to restate Equation (8-5) as follows:

$$N_i = \frac{k'_L a_i \rho^L S}{\gamma^L_i f^L_i}(\bar{f}^L_i - \bar{f}^{\mathcal{L}}_i) = \frac{k'_G a_i PS}{\gamma^V_i f^V_i}(\bar{f}^{\upsilon}_i - \bar{f}^V_i) \tag{8-7}$$

Comparison of Equations (8-1) and (8-7) gives the relationships shown in Item I of Table 8-1 for $k_L a_i$, $k'_L a_i$, $k_G a_i$, and $k'_G a_i$.

TABLE 8-1
SUMMARY OF THE MASS TRANSFER RELATIONSHIPS

I. *Relationships based on the assumptions given by Equation (8-6)*

  1. *Mass Transfer Coefficients*

$$k_L a_i = \left(\frac{\rho^L}{\gamma^L_i f^L_i}\right)k'_L a_i; \quad k_G a_i = \left(\frac{P}{\gamma^V_i f^V_i}\right)k'_G a_i; \quad K_G a_i = \left(\frac{P}{\gamma^V_i f^V_i}\right)K'_G a_i$$

$$\frac{1}{K_G a_i} = \frac{1}{k_G a_i} + \frac{1}{k_L a_i}; \quad \frac{1}{K'_G a_i} = \frac{1}{k'_G a_i} + \left(\frac{\gamma^L_i K_i P}{\gamma^V_i \rho^L}\right)\frac{1}{k'_L a_i}$$

$$\frac{1}{K'_L a_i} = \frac{1}{k'_L a_i} + \left(\frac{\gamma^V_i \rho^L}{\gamma^L_i P K_i}\right)\frac{1}{k'_G a_i}$$

  2. *Number of Transfer Units*

$$n_{Li} = k_L a_i \gamma^L_i f^L_i \left(\frac{Sz_T}{L}\right) = -\int_{①}^{②}\frac{dx^L_i}{x^L_i - x^{\mathcal{L}}_i}, \text{ where } ① = \text{value of } x^L_i \text{ at } z = 0;$$
$$② = \text{value of } x^L_i \text{ at } z = z_T.$$

$$n_{Gi} = k_G a_i \gamma^V_i f^V_i \left(\frac{Sz_T}{V}\right) = -\int_{①}^{②}\frac{dy^V_i}{y^{\upsilon}_i - y^V_i}, \text{ where } ① = \text{value of } y^V_i \text{ at } z = 0;$$
$$② = \text{value of } y^V_i \text{ at } z = z_T.$$

$$n_{0Gi} = K_G a_i \gamma^V_i f^V_i \left(\frac{Sz_T}{V}\right) = -\int_{①}^{②}\frac{dy^V_i}{Y_i - y^V_i}, \text{ where } Y_i = \frac{\gamma^L_i K_i x^L_i}{\gamma^V_i}.$$

$$n'_{Li} = k'_L a_i \rho^L \left(\frac{Sz_T}{L}\right) = -\int_{①}^{②}\frac{dx^L_i}{x^L_i - x^{\mathcal{L}}_i}$$

TABLE 8-1 (Cont.)

$$n'_{Gi} = k'_G a_i P\left(\frac{Sz_T}{V}\right) = -\int_{①}^{②} \frac{dy_i^V}{y_i^0 - y_i^V}$$

$$n'_{0Gi} = K'_G a_i P\left(\frac{Sz_T}{V}\right) = -\int_{①}^{②} \frac{dy_i}{Y_i - y_i^V}$$

$$n'_{0Li} = K'_L a_i \rho^L\left(\frac{Sz_T}{L}\right) = -\int_{①}^{②} \frac{dx_i^L}{x_i^L - X_i}, \text{ where } X_i = \frac{\gamma_i^V y_i^V}{\gamma_i^L K_i}$$

II. *Relationships based on the use of the tangent line $y_i^* = m_i x_i + b_i$ and associated assumptions*
  1. *Mass Transfer Coefficients*

$$\frac{1}{K'_G a_i} = \frac{1}{k'_G a_i} + \left(\frac{m_i P}{\rho^L}\right)\frac{1}{k'_L a_i}$$

  2. *Number of Transfer Units*
  n'_{Gi} = same form as Item I.
  n'_{Li} = same form as Item I.

$$n'_{0Gi} = K'_G a_i P\left(\frac{Sz_T}{V}\right) = -\int_{①}^{②} \frac{dy_i^V}{y_i^* - y_i^V}, \text{ where } y_i^* \text{ is a hypothetical mole fraction defined by Equation (8-12).}$$

$$n'_{0Li} = K'_L a_i \rho^L\left(\frac{Sz_T}{L}\right) = -\int_{①}^{②} \frac{dx_i^L}{x_i^L - x_i^*}, \text{ where } x_i^* \text{ is a hypothetical mole fraction defined below Equation (8-15).}$$

III. *Relationships based on Henry's law, $p_i = H_i C_i$ and associated assumptions*
  1. *Mass Transfer Coefficients*

$$\frac{1}{K'_G a_i} = \frac{1}{k'_G a_i} + \frac{H_i}{k'_L a_i}$$

$$\frac{1}{K'_L a_i} = \frac{1}{k'_L a_i} + \frac{1}{H_i k'_G a_i}$$

  2. *Number of Transfer Units*
  n'_{Gi} = same form as Item I.
  n'_{Gi} = same form as Item I.

$$n'_{0Gi} = K'_G a_i P\left(\frac{Sz_T}{V}\right) = -\int_{①}^{②} \frac{dy_i^V}{y_i^* - y_i^V}, \text{ where } y_i^* \text{ is defined below Equation (8-17).}$$

$$n'_{0Li} = K'_L a_i \rho^L\left(\frac{Sz_T}{L}\right) = -\int_{①}^{②} \frac{dx_i^L}{x_i^L - x_i^*}, \text{ where } x_i^* \text{ is defined below Equation (8-17).}$$

When the rates of mass transfer across the vapor and liquid films as given by Equation (8-1) are divided by $k_G a_i$ and $k_L a_i$, respectively, and added, one obtains

$$N_i\left[\frac{1}{k_G a_i} + \frac{1}{k_L a_i}\right] = S[(\bar{f}_i^L - \bar{f}_i^\varrho) + (\bar{f}_i^\varrho - \bar{f}_i^V)] \tag{8-8}$$

This equation may be restated in the following form:

$$N_i = K_G a_i S(\bar{f}_i^L - \bar{f}_i^V) = K_G a_i S f_i^V(Y_i - y_i^V) \tag{8-9}$$

where $K_G a_i$ is defined in Item I of Table 8-1, $Y_i = (\gamma_i^L K_i x_i^L)/\gamma_i^V$, and $K_i = f_i^L/f_i^V$ (the ideal solution K value).

In a manner analogous to that used to develop Equation (8-9) from Equation (8-1), the following expression for N_i may be developed:

$$N_i = K'_G a_i S(p_i^{\neq} - p_i^V) \tag{8-10}$$

where $K'_G a_i$ is defined in Item I of Table 8-1, and

$$p_i^{\neq} = \frac{\gamma_i^L P K_i x_i}{\gamma_i^V}$$

$$K_i = \frac{f_i^L}{f_i^V}$$

Similarly, N_i may be stated in terms of concentrations as follows:

$$N_i = K'_L a_i S(C_i^L - C_i^{\neq}) \tag{8-11}$$

where $K'_L a_i$ is defined in Item I of Table 8-1 and

$$C_i^{\neq} = \frac{\gamma_i^V \rho^L y_i^V}{\gamma_i^L K_i}$$

[Note that $\sum_{i=1}^{c} p_i^{\neq}$ is not necessarily equal to P and that $\sum_{i=1}^{c} C_i^{\neq}$ is not necessarily equal to ρ^L because the liquid and vapor were not assumed to be at the bubble point and dew point temperature, respectively.]

By comparison of the above equations, relationships other than those mentioned above are readily obtained.

Use of the tangent line to the equilibrium curve in the definitions of $K'_G a_i$ and $K'_L a_i$

An alternate method that has been used to relate the mole fractions in the vapor and liquid phases is to assume a linear relationship between the mole fraction of component i in a given phase and the mole fraction that it would have in the other phase if the two phases were in equilibrium. For the case of a binary mixture, the linear relationship is represented by the tangent line to the equilibrium curve,

$$y_i^* = mx_i + b_i \tag{8-12}$$

where the quantity y_i^* is defined as the mole fraction that component i would have in the vapor phase if the vapor were in equilibrium with a liquid having the mole fraction x_i. From this definition of y_i^*, it follows that $y_i^* = K_i x_i$ and $\sum_{i=1}^{c} y_i^* = 1 = \sum_{i=1}^{c} K_i x_i$, which implies that the liquid is assumed to be at its bubble point temperature. Let it be assumed that the values of m_i and b_i apply with good accuracy for both x_i^L and x_i^{ℓ} (that is, it is supposed that the equilibrium curve is straight over this range of values of x_i) and that $\rho^L = \rho^{\ell}$. Then Equation (8-5) may be restated in the following form:

$$N_i = \frac{k'_L a_i \rho^L S}{m_i}[(m_i x_i^L + b_i) - (m_i x_i^{\ell} + b_i)] = k'_G a_i PS(y_i^{\upsilon} - y_i^V) \tag{8-13}$$

In a manner analogous to the development of Equation (8-9) from (8-1), the following expression for N_i is obtained from Equation (8-13)

$$N_i = K'_G a_i S(p_i^* - p_i^V) = K'_G a_i PS(y_i^* - y_i^V) \tag{8-14}$$

where $K'_G a_i$ is defined in Item II of Table 8-1. In a similar manner, Equation (8-5) may be used to obtain

$$N_i = K'_L a_i S(C_i^L - C_i^*) = K'_L a_i \rho^L S(x_i^L - x_i^*) \tag{8-15}$$

where $K'_L a_i$ is defined in Item II of Table 8-1 and x_i^* is the mole fraction that component i would have in the liquid phase if the liquid phase were in equilibrium with a vapor phase having the mole fraction y_i^V. From this definition of x_i^*, it follows that $x_i^* = y_i^V/K_i$ and $\sum_{i=1}^{c} y_i^V/K_i = 1$ which implies that the vapor is assumed to be at its dew point temperature.

Use of Henry's law, $p_i = H_i C_i$, in the definitions of $K'_G a_i$ and $K'_L a_i$

If it is assumed that the same value of H_i may be used for both C_i^L and C_i^g, then expressions of the same form as Equation (8-14) and (8-15) may be obtained by commencing with Equation (8-5) and carrying out developments analogous to those outlined above with the exception of the use of $p_i = H_i C_i$ instead of $y_i^* = m_i x_i + b_i$. The corresponding expressions obtained for $K'_G a_i$ and $K'_G a_i$ are listed in Item III of Table 8-1 and the corresponding rate expressions are given by

$$N_i = K'_G a_i S(p_i^* - p_i^V) = K'_G a_i PS(y_i^* - y_i^V) \tag{8-16}$$

$$N_i = K'_L a_i S(C_i^L - C_i^*) = K'_L a_i \rho^L S(x_i^L - x_i^*) \tag{8-17}$$

where

$$y_i^* = \frac{H_i C_i^L}{P};$$

$$x_i^* = \frac{C_i^L}{H_i \rho^L}.$$

Definitions of the number of mass transfer units

To demonstrate the development of the defining equations for the number of transfer units listed in Table 8-1, the steps involved in the case of n_{Gi} are enumerated. Elimination of N_i from Equations (8-1) and (8-2) followed by rearrangement yields

$$\frac{d(Vy_i)}{dz} = k_G a_i S(\bar{f}_i^V - \bar{f}_i^v) = k_G a_i \gamma_i^V f_i^V S(y_i^V - y_i^v) \tag{8-18}$$

If it is now assumed that V, $k_G a_i$, and $\gamma_i^V f_i^V$ are constant over the interval from $z = 0$ to $z = z_T$, then the variables in Equation (8-18) may be separated and integrated over the interval from $z = 0$ to $z = z_T$ to give the expression for

n_{Gi} listed in Item I of Table 8-1. The remaining defining equations for the n_i's given in this table are developed in a similar manner. The concept of the number of transfer units was proposed by Chilton and Colburn (1).

Another unit proposed by Chilton and Colburn (1) is called the *height of a transfer unit*, and for the case of the gas phase, it is defined as follows:

$$(HTU)_{Gi} = \frac{z_T}{n_{Gi}} \tag{8-19}$$

Such a unit may be defined for each n_i listed in Table 8-1.

By use of the relationship shown in Table 8-1, may others may be deduced. For example, the number of overall mass transfer units may be stated in terms of the number of transfer units for each phase. By use of the relationships given in Item I of Table 8-1, it can be shown that

$$\frac{1}{n_{0Gi}} = \frac{1}{n_{Gi}} + \left(\frac{\gamma_i^L K_i V}{\gamma_i^V L}\right)\frac{1}{n_{Li}} = \frac{1}{n_{Li}'} + \left(\frac{\gamma_i^L K_i V}{\gamma_i^V L}\right)\frac{1}{n_{Li}'} = \frac{1}{n_{0Gi}'} \tag{8-20}$$

and from this development it is evident that $n_{Li} = n_{Li}'$, $n_{Gi} = n_{Gi}'$, and $n_{0Gi} = n_{0Gi}'$. If the tangent line is used, the relationship given in Item II of Table 8-1 may be used to show that

$$\frac{1}{n_{0Gi}'} = \frac{1}{n_{Gi}'} + \left(\frac{m_i V}{L}\right)\frac{1}{n_{Li}'} \tag{8-21}$$

The relationships given by Equations (8-20) and (8-21) are utilized in a subsequent chapter in the prediction of plate efficiencies.

Although the sets of assumptions upon which the relationships presented in Table 8-1 are based differ somewhat, the resulting expressions give approximately the same results for the types of problems considered in this chapter.

Also, as shown below, the relationships presented in Table 8-1 may be used in the design of columns in the process of removing a single component from one of the two streams.

Part 2: Classical Models for Absorption Columns

For the case in which a solute such as NH_3 or SO_2 is to be removed from a carrier gas such as air by contacting the gas stream with a solvent, say water, the above equations may be applied as demonstrated below.

Since most such absorption processes are carried out at pressures near atmospheric, the deviation of the gas phase from a perfect gas may be neglected. For this case, Equation (8-14) reduces to

$$\frac{-d(Vy)}{dz} = K_G' aSP(y^* - y) \tag{8-22}$$

where countercurrent flow as shown in Figure 8-3 is assumed. The subscript i has been dropped in Equation (8-22) in the interest of simplicity and the symbols y and y^* represent mole fractions of the solute. Since it is assumed that only one component, the solute, is transferred from the vapor to the liquid phase, the derivative of the product appearing on the left-hand side of Equation (8-22) may be restated in terms of dy/dz as follows. The total flow rate of the solute plus carrier gas V may be stated in terms of the flow rate V' of the carrier gas as follows:

$$V = \frac{V'}{1 - y} \tag{8-23}$$

When V in Equation (8-22) is replaced by its equivalent as given by Equation (8-23), the left-hand side of this equation may be restated as follows:

$$\frac{d(Vy)}{dz} = \frac{V'd\left(\frac{y}{1-y}\right)}{dz} = \frac{V'}{(1-y)^2}\frac{dy}{dz} = \frac{V}{1-y}\frac{dy}{dz} \tag{8-24}$$

Equations (8-22) and (8-24) may be combined and the resulting variables separated to give

$$\int_0^{z_T} \left(\frac{K'_G a PS}{V}\right)dz = \int_{y_1}^{y_{N+1}} \frac{dy}{(1-y)(1-y^*)} \tag{8-25}$$

where y_{N+1} is the mole fraction of the solute in the inlet gas stream, y_1 is the mole fraction of the solute in the gas leaving the column, and z is measured down from the top of the column as shown in Figure 8-3. Use of an appropriate mean value of the integrand on the left-hand side of Equation (8-25) followed by the replacement of y^* by its equivalent $y^* = mx + b$, yields

$$n'_{0G} = K'_G a P\left(\frac{Sz_T}{V}\right) = \int_{y_1}^{y_{N+1}} \frac{dy}{(1-y)[y - (mx + b)]} \tag{8-26}$$

Obviously, the integral on the right-side of Equation (8-26) may be evaluated by graphical integration. If, as is usually the case, both the vapor and liquid phases are very dilute in the solute, then the equilibrium pairs y^* and x will be located near the origin. Consequently, the tangent line $y^* = mx + b$ will generally intersect the y-axis near the origin, and the tangent line may be approximated with good accuracy by $y = mx$. Also, for streams dilute in the solute, the term $(1 - y)$ is very nearly equal to unity. If the process is also carried out isothermally, then the integrand on the left-hand side of Equation (8-25) is very nearly constant for all z. When all of these assumptions are imposed on Equation (8-26), one obtains

$$n'_{0G} = K'_G a P\left(\frac{Sz_T}{V}\right) = \int_{y_1}^{y_{N+1}} \frac{dy}{y - mx} \tag{8-27}$$

The assumption that $1 - y = 1$ amounts to assuming that V and L do not vary throughout the column. Thus, for this case the component-material balance enclosing the top of the column and the packing down to any z is given by

$$x = \frac{V}{L}y + x_0 - \frac{V}{L}y_1 \tag{8-28}$$

After x in Equation (8-27) has been replaced by its equivalent as given by Equation (8-28) and the indicated integration carried out, the following result is obtained upon rearrangment:

$$n'_{0G} = K'_G aP\left(\frac{Sz_T}{V}\right)\left(\frac{1}{1 - \frac{mV}{L}}\right)\log_e \frac{y_{N+1} - mx_N}{y_1 - mx_0} \tag{8-29}$$

[To obtain this final form, use is made of the component-material balance that encloses the entire column, $x_N = (V/L)y_{N+1} + x_0 - (V/L)y_1$.]

   To demonstrate the use of Equation (8-29), the following example is solved.

ILLUSTRATIVE EXAMPLE 8-1

   Air containing 2 mole % NH_3 is to be scrubbed with water in a column packed with 2-in. quartz. The water and gas streams are to flow countercurrently and their respective rates are to be 13.3 and 8.2 lb-moles/(hr ft²). The streams enter at 68°F, and the column may be assumed to operate isothermally at this temperature and at an absolute pressure of 1 atm. At this temperature and at these concentrations, the following approximate form of the equilibrium relationship may be used:

$$y_{NH_3} = 0.79\, x_{NH_3}$$

Also, for this system b may be taken equal to zero, and thus,

$$y_{NH_3} = mx_{NH_3} = K_{NH_3} x_{NH_3}$$

If $K'_G a = 9.6$ lb-moles/(hr ft³ atm) for the quartz packing, find the height of packing required to absorb 98% of the NH_3 in the inlet gas stream.

Solution:
1. Calculation of y_1 and x_N:

$$\frac{\text{lb-moles of } NH_3 \text{ absorbed}}{(\text{hr}) (\text{ft}^2)} = (0.98)(Vy_{N+1})$$

$$\frac{\text{lb-moles of } NH_3 \text{ in outlet gas}}{(\text{hr}) (\text{ft}^2)} = Vy_{N+1} - 0.98\, Vy_{N+1}$$

Then

$$y_1 = \frac{Vy_{N+1}[1 - 0.98]}{V} = (0.02)(0.02) = 0.0004$$

and

$$x_N = \frac{\text{lb-moles of } NH_3 \text{ absorbed}}{\text{lb-mole of liquid}} = \frac{(0.98)(Vy_{N+1})}{L}$$

$$= (0.98)(0.02)\left(\frac{8.2}{13.3}\right) = 0.0121$$

2. Calculation of the height of packing:

Since $y_{N+1} = 0.02$, $V/L = 8.2/13.3 = 0.617$, $m = 0.79$, $y_1 = 0.0004$, $x_N = 0.0121$, $V/S = 8.2$, $K'_G a = 9.6$, and $P = 1$ atm, the height of packing may be computed by use of Equation (8-29) as follows:

$$z_T = \left(\frac{1}{K'_G a}\right)\left(\frac{\dfrac{V}{S}}{1 - \dfrac{mV}{L}}\right) \log_e \frac{y_{N+1} - mx_N}{y_1 - mx_1}$$

$$z_T = \frac{(.82)}{(9.6)[1 - (0.79)(0.617)]} \log_e \frac{0.02 - (0.079)(0.0121)}{0.0004 - (0.79)(0.0)}$$

$$z_T = 5.44 \text{ ft}$$

Experimental values for the mass transfer rate constants and height transfer units

Collections of the experimentally determined values of the rate constants or the height transfer units have been presented by various authors [see, for example, references (2, 5, 7)]. Various methods for presenting the data are used. Plots of the height transfer units for the vapor (or liquid) versus mass flow rate of the vapor per square foot of cross-sectional area (or the mass flow rate of the liquid per square foot of cross-sectional area) are common. Also, plots of the height transfer units for the vapor phase versus mV/L are frequently used. Attempts to correlate the rate constants as functions of the Reynolds' number and the Schmidt number have met with some success.

PROBLEMS

8-1. An air stream containing 4.0 mole % n-heptane is to be scrubbed by contacting it countercurrently with a nonvolatile oil in a tower packed with 1-in. Raschig rings. The air stream is to enter the column at a rate of 8.3 lb-moles per hr per sq ft of cross-sectional area and at 160°F. The nonvolatile oil enters the column at a rate of 6.7 [lb-moles/(hr ft²)] and at 160°F. The column is to be operated isothermally at 160°F and at an absolute pressure of 1 atm. The vapor pressure of n-heptane at 160°F may be taken equal to 6 psia, and $K'_G a$ = 5.0 lb-moles/(hr ft³ atm). Assume $b = 0$, that is, $y^* = mx = Kx$.
   (a) Find the height of packing required to remove 99.5% of the n-heptane from the entering air stream.
   (b) If the feed rate is doubled and V/S and L/S held fixed, how will the column dimensions be changed?

8-2. Air containing 3 mole % benzene at 68°F is to be scrubbed with kerosene in a column which is packed with 1-in. Raschig rings. The kerosene enters the column at 68°F, and the column is to operate isothermally at 68°F and 1 atm and $K'_G a = 15$ lb-moles/(hr ft^3 atm). At 68° F, the vapor pressure of benzene may be taken equal to 95 mm, and it may be assumed that $y^* = mx = Kx$. It is desired to remove 98% of the benzene from the entering gas stream.

(a) For an air flow rate $V/S = 1.7$ lb-moles/(hr ft^2) and a liquid rate $L/S = 16.7$ lb-moles/(hr ft^2), find the height of packing required to effect the specified separation when countercurrent flow is used.

(b) Repeat Part (a) for the case of parallel flow.

8-3. Verify the results given by Equations (8-20) and (8-21).

NOTATION

$a_i =$ interfacial area per unit volume of empty tower.

$C_i =$ concentration, lb-moles/unit volume.

$C_i^{\neq} =$ concentration; defined below Equation (8-11), lb-moles/unit volume.

$f_i^V, f_i^L =$ fugacities of pure component i in the vapor and liquid states, respectively; evaluated at the total pressure and at the temperature of the vapor and liquid phases, respectively, atm.

$\bar{f}_i^V, \bar{f}_i^L =$ fugacities of component i in the vapor and liquid streams, respectively; evaluated at the total pressure, temperature, and compositions of these respective phases, atm.

$k_I a_i =$ product of the mass transfer coefficient for the interface and the interfacial area unit volume of empty tower.

$k_G a_i, k'_G a_i =$ product of the mass transfer coefficients for the gas phase and the interfacial area per unit volume of empty tower; defined by Equations (8-1) and (8-5), respectively.

$k_L a_i, k'_L a_i =$ product of the mass transfer coefficients for the liquid phase and the interfacial area per unit volume of empty tower; defined by Equations (8-1) and (8-5), respectively.

$K_G a_i, K'_G a_i =$ products of the overall mass transfer coefficients and the interfacial area; defined in Table 8-1.

$K_i =$ the ideal solution K value, $K_i = f_i^L/f_i^V$, for component i where the fugacities are evaluated at the same total pressure and temperature.

$L =$ total flow rate of liquid at any z, lb-moles/hr.

$n_G, n'_G =$ number of gas phase transfer units; defined in Table 8-1.

$n_L, n'_L =$ number of liquid phase transfer units; defined in Table 8-1.

$n_{0G}, n'_{0G} =$ number of overall gas phase transfer units; defined in Table 8-1.

$n'_{0L} =$ number of overall liquid phase transfer units; defined in Table 8-1.

$N_i =$ rate of mass transfer of component i from the liquid to the vapor phase; lb-moles/(unit time) (unit height of packing).

$p_i =$ partial pressure of component i, atm.

$p_i^{\neq} =$ a quantity having the units of pressure (atm); defined below Equation (8-10), atm.

P = total pressure for a given plate, atm.

S = cross-sectional area of the column, ft^2.

V = total flow rate of the vapor at any z, lb-moles/hr.

x_0 = mole fraction of the solute in the solvent entering the column.

x_N = mole fraction of the solute in the solvent stream that leaves the column.

x_i = mole fraction of component i in the liquid at any z.

y_1 = mole fraction of the solute in the gas stream leaving the column.

y_{N+1} = mole fraction of the solute in the gas stream entering the column.

y_i = mole fraction of component i in the vapor at any z.

y_i^* = mole fraction that component i would have if the vapor at any z were in equilibrium with the liquid at the same z.

Y_i = a product defined in Table 8-1.

z = depth of liquid on a given plate; measured from the surface ($z = 0$) to the floor of the plate ($z = z_T$), ft.

Greek letters

γ_i^V, γ_i^L = activity coefficients for component i in the vapor and liquid phases, respectively; evaluated at the temperature, pressure, and composition of these respective streams at any given z.

$\rho^L, \rho^{\mathcal{L}}$ = total molar density of the liquid; evaluated at the bulk conditions and at the conditions at the interface, respectively.

Subscripts

i = component number.

Superscripts

L = to be evaluated at the bulk conditions of the liquid phase.

$\mathcal{L}$ = to be evaluated at the conditions of the liquid at the interface.

V = to be evaluated at the bulk conditions of the vapor phase.

$\mathcal{U}$ = to be evaluated at the conditions of the vapor at the interface.

REFERENCES

1. Chilton, T. H., and A. P. Coluburn, "Distillation and Absorption in Packed Columns A Convenient Design and Correlation Method," *Ind. Eng. Chem.*, **27**, (1935), 255.

2. Colburn, A. P., "The Simplified Calculation of Diffusional Process. General

Consideration of Two-Film Resistances," *Trans. Am. Inst. Chem. Engrs.*, **35**, (1939) 211.

3. Lewis, W. K., and W. G. Whitman, "Principles of Gas Absorption," *Ind. Eng. Chem.*, **16**, (1924), 1215.

4. ———, "Oil Absorption of Natural Gasoline," *Trans. Am. Inst. Chem. Engrs.*, **20**, (1927), 1.

5. Perry, J. H. (ed.), *Chemical Engineer's Handbook*, 3rd ed., 4th ed. New York, N. Y.: McGraw-Hill Book Co., Inc., 1950, 1963.

6. Peters, W. A., Jr., "The Efficiency and Capacity of Fractionating Columns," *Ind. Eng. Chem.*, **14**, (1922), 476.

7. Molstad, M. C., J. F. Mckinney, and R. G. Abbey, "Performance of Drip-Point Grid Tower Packings—III. Gas-Film Mass Transfer Coefficients; Additional Liquid-Film Mass Transfer Coefficients," *Trans. Am. Inst. Chem. Engrs.*, **39**, (1943), 605.

8. Rubac, R. E., "Determination of Vaporization Efficiencies for Packed Columns at Steady State Operation," Ph.D. Dissertation, Texas A & M University, College Station, Texas, 1968.

9. ———, R. McDaniel, and C. D. Holland, "Packed Distillation Columns and Absorbers," *A.I.Ch.E. Journal*, **15**, (1969), 568.

10. Walker, W. H., W. K. Lewis, and W. H. McAdams, *Principles of Chemical Engineering*, 2nd ed. New York: McGraw-Hill, 1927.

11. Whitman, W. G., "The Two-Film Theory of Gas Absorption," *Chem. & Met. Eng.*, **29**, (1923), 146.

Fundamentals and Procedures for Modeling Packed Distillation Columns

9

For columns having relatively small diameters of 3 or 4 ft or less, the use of packing material rather than plates is quite common. Some of the various kinds of packing that are commercially available are shown in Figure 8-1.

In Part I of this chapter, the fundamental relationships describing packed columns are developed. In Part II, a procedure for modeling packed distillation columns is presented. In Part III, problems involved in the determination of the overall mass transfer coefficients and Murphree-type efficiencies from experimental compositions obtained from a packed distillation column in the process of separating multicomponent mixtures are shown to exist. On the other hand, the vaporization plate efficiencies are shown to be nonzero, finite and positive numbers, provided certain rather general conditions exist.

Part 1 : Fundamental Relationships

Although considerable effort has been directed toward the solution of problems involving distillation columns and absorbers with plates (1, 4, 12, 13,

15, 16, 23, 25, 26, 27), little attention has been given to the improvement of calculational procedures for packed columns since the proposal of some of the original concepts of mass transfer (7, 10, 19, 20, 22, 24, 33, 34). Although the classical models presented in Chapter 8 may be used for the case in which the solute is dilute in the gas and liquid streams, these models are not applicable to the general case of a multicomponent mixture.

A sketch of a typical packed column is shown in Figure 9-1. The notation employed in the description of packed columns (see Figures 9-2 and 9-3) corresponds to that used in the treatment of distillation columns with plates (15).

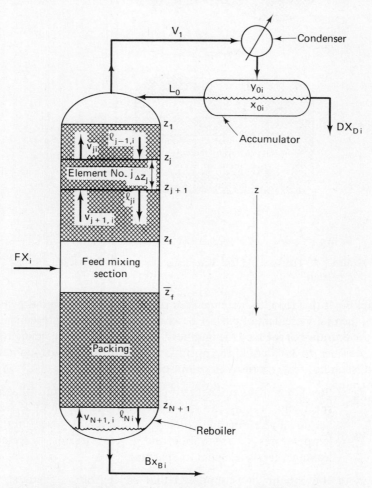

Figure 9-1. Diagram of a packed distillation column (Taken from R. E. Rubac, R. McDaniel, and C. D. Holland, "Packed Distillation Columns and Absorbers at Steady State Operation," *A.I.Ch.E. Journal, 15,* (1969), 568.)

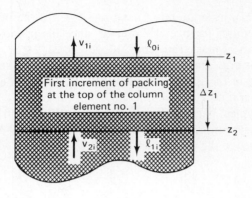

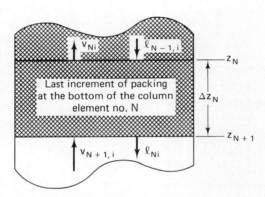

Figure 9-2. The first and last increments of packing of a packed distillation column.

The packing is divided into increments or sections and numbered down from the top; the first increment of packing is assigned the number 1, the increment below the entrance of the feed the number f, and the bottom element of packing the number N. To identify the positions of the flow rates throughout the column, each flow rate carries a position subscript j, that is,

v_{ji}, l_{ji} = flow rates of component i in the vapor and liquid phases, respectively, leaving the jth element of packing, and V_j and L_j are the corresponding total flow rates, moles per unit time.

T_j^V, T_j^V = temperatures of the vapor and liquid streams, respectively, leaving the jth element of packing.

Throughout the column, it is supposed that concentration gradients exist only in the direction z (see Figure 9-1).

In the column used to make the field tests, the section where the feed was introduced ($z_f \leqq z \leqq \bar{z}_f$) contained a liquid-distributor plate but no

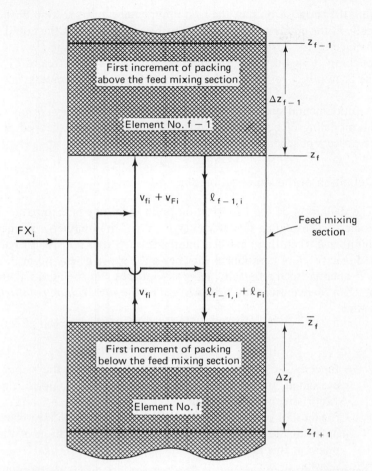

Figure 9-3. Model 2 of the feed mixing section and the increments of packing immediately above and below it (Taken from R. E. Rubac *et al.*, "Packed Distillation Columns and Absorbers at Steady State Operation," 569.)

packing, as indicated in Figures 9-1 and 9-3. This section of the column is called the *feed mixing section*. The packed section ($z_1 \leq z \leq z_f$) above and the packed section ($\bar{z}_f \leq z \leq z_{N+1}$) below the feed mixing section are referred to as the packed-rectifying and packed-stripping sections, respectively.

Two possible schemes may be used for the modeling of the feed mixing section, namely, Models 1 and 2 of Figure 3-13. In this chapter, Model 2 (see Figure 9-3) was used for the feed mixing section. This model is based on the same assumptions as Model 2 shown in Figure 3-13 for columns with plates.

In Model 2, the behavior of the feed mixing section is approximated by making the supposition that the feed upon entering this section flashes adiabatically at the column pressure. It is further supposed that the liquid stream, L_F, formed by the flash mixes perfectly with the liquid stream L_{f-1} and enters the packing at $\bar{z}_f$. Likewise, it is supposed that the vapor stream, V_F, formed by the flash mixes perfectly with the vapor stream V_f and enters the packing at z_f.

A fundamental quantity called the *vaporization plate efficiency* was used in the modeling of both packed distillation columns and packed absorbers in the process of separating multicomponent mixtures.

Definition of the vaporization plate efficiency

The definition of the vaporization point efficiency is axiomatic. That is, the logic used to define this efficiency is that if the number a is unequal to the number b, then there exists a number c such that $a = cb$. In particular, if the fugacity $\bar{f}_i^V$ of component i in the vapor phase at any point in the column is unequal to the fugacity $\bar{f}_i^L$ of component i in the liquid at the same point, then there exists a multiplier E_i, called the *vaporization point efficiency*, such that

$$\bar{f}_i^V = E_i \bar{f}_i^L \qquad (9\text{-}1)$$

where

$\bar{f}_i^V =$ fugacity of component i at a given point in the vapor phase, evaluated at the temperature, pressure, and composition of the vapor stream at this point;

$\bar{f}_i^L =$ fugacity of component i at a given point in the liquid phase, evaluated at the temperature, pressure, and composition of the liquid phase at this point.

The vaporization point efficiency E_i is used in a subsequent chapter in the modeling of a plate of a distillation column.

The *vaporization plate efficiency* is defined in a manner analogous to the vaporization point efficiency. In particular, if the fugacity $\bar{f}_{ji}^V$ of component i in the vapor phase leaving any element Δz_j (or plate j of a column with plates) is unequal to the fugacity $\bar{f}_{ji}^L$ of component i in the liquid leaving the element j, then there exists a multiplier E_{ji}, called the *vaporization plate efficiency*, such that

$$\bar{f}_{ji}^V = E_{ji} \bar{f}_{ji}^L \qquad (9\text{-}2)$$

where

$\bar{f}_{ji}^V =$ fugacity of component i in the total vapor stream leaving element Δz_j, evaluated at the temperature, pressure, and composition of this stream;

$\bar{f}_{ji}^L =$ fugacity of component i in the liquid stream leaving element

Δz_j, evaluated at the temperature, pressure, and composition of this stream.

When the mass transfer that occurs in an element Δz_j is described by Equation (9-2) for each component, it is called a *mass transfer section*. Since Equation (9-2) does not necessarily represent an equilibrium relationship, it is called the *mass transfer relationship*. If, however, a state of equilibrium does exist between the two streams leaving element Δz_j, it follows from the thermodynamics of mixtures (9, 15) that $E_{ji} = 1$ for all i, since $\bar{f}^V_{ji} = \bar{f}^L_{ji}$ for all i for the given element Δz_j. If the vapor phase forms an ideal solution, but the temperatures of the vapor and liquid phases differ, then Equation (9-2) reduces to

$$y_{ji} = E_{ji}\gamma_{ji}\left(\frac{f^L_{ji}}{f^V_{ji}}\right)x_{ji} \qquad (9\text{-}3)$$

where f^L_{ji} and f^V_{ji} denote the fugacities of pure component i evaluated at the respective temperatures and pressures at which the liquid and vapor streams leave the element Δz_j. The activity coefficient γ_{ji} of component i is evaluated at the conditions of the liquid stream as it leaves the element Δz_j. When both the vapor and liquid phases form ideal solutions, the ratio f^L_i/f^V_i is commonly denoted by K_i (9, 15). Note that in this definition, the fugacities are evaluated at the same temperature and pressure. Thus, if the temperatures and pressures of the vapor and liquid streams leaving the given element Δz_j are equal, then Equation (9-3) becomes

$$y_{ji} = E_{ji}\gamma_{ji}K_{ji}x_{ji} \qquad (9\text{-}4)$$

The vaporization plate efficiency was perhaps first employed by either McAdams in the analysis of the batch-steam distillation of a volatile component from a dissolved nonvolatile component present in large amounts as described in Reference (6) or by Carey (5). In 1949, Edmister (11) proposed a similar efficiency and called it an *absorption efficiency*. Later the vaporization plate efficiency was applied to the batch-steam distillation of multicomponent mixtures (17) and subsequently to the separation of multicomponent mixtures in conventional distillation columns with plates (15, 31, 32).

The concept of a mass transfer section may be applied to packed columns in the following manner. Let the packing of the column be divided into sections or elements $\Delta z_1, \Delta z_2, \ldots, \Delta z_N$, where the elements are numbered down from the top of the column as shown in Figures 9-1, 9-2, and 9-3. Let y_{ji} in Equation (9-4) be the mole fraction of component i in the vapor leaving element Δz_j and x_{ji} the mole fraction of component i in the liquid leaving element Δz_j. Similarly, γ_{ji} is evaluated at the conditions of the liquid leaving Δz_j, and K_{ji} is evaluated at the temperature and pressure of the vapor and liquid streams leaving element Δz_j. It should be noted that if E_{ji} is equal to unity for any one component for a given element of packing, then the particular Δz_j might be regarded as the *partial* HETP for that component at that particular location in the column. The HETP (height equivalent to a theoretical plate) was introduced by Peters (6) and defined as that height of packing required to effect the same separation of a binary mixture as that which could be achieved by use of a perfect plate.

One of the most important characteristics of the E_{ji}'s is that they are always nonzero, finite, and positive numbers, provided the components to which they apply are detectable and have equilibrium constants which are nonzero, finite, and positive. More precisely, the vaporization plate efficiency as defined by Equation (9-4) takes on bounded positive values provided

$$0 < x_{ji} \leq 1$$
$$0 < y_{ji} \leq 1 \tag{9-5}$$
$$0 < \gamma_{ji} K_{ji} < \infty$$

[or by Equation (9-3) the more general condition is: $0 < \gamma_{ji}(f_{ji}^V/f_{ji}^L) < \infty$]. Also observe that Equation (9-3) constitutes a functional definition of E_{ji} in that for any given model proposed for the interaction of the vapor and liquid phases within a mass transfer section, an expression for E_{ji} in terms of the operating variables exists as demonstrated in Chapter 11.

Simultaneous mass and heat transfer in packed distillation columns

Let the rate of mass transfer of component i from the liquid to the vapor phase at any z in the packing be denoted by N_i [moles of component i transferred/(unit time) (unit height of packing)]. Without specifying the precise form of N_i other than to imply that it depends on the compositions of the two phases and the interfacial area separating the two phases, the equations required to describe simultaneous mass and heat transfer in a packed distillation column may be formulated.

Since the interfacial area a between the vapor and liquid phases is approximately equal to zero outside of the packing, no appreciable mass transfer occurs. This condition is described with good accuracy by

$$N_i = 0 \qquad (z < z_1, z_f < z < \bar{z}_f, z > z_{N+1}) \tag{9-6}$$

where the meanings of the symbols are portrayed in Figures 9-2 and 9-3.

In the following developments, it is supposed that N_i is continuous, finite, and positive throughout the interior and on the boundaries of the closed intervals $z_1 \leq z \leq z_f, \bar{z}_f \leq z \leq z_{N+1}$, but that it may have discontinuities as the boundaries $z_1, z_f, \bar{z}_f$, and z_{N+1} are approached from the exterior of the closed intervals.

MATERIAL AND ENERGY BALANCES FOR THE PACKED SECTIONS. When the boundaries z_j and z_{j+1} of each element of packing are also selected for each component-material balance in the closed intervals $z_1 \leqq z \leqq z_f$ and $\bar{z}_f \leqq z \leqq z_{N+1}$, the following expressions are obtained:

$$v_{j+1,i} - v_{ji} + \int_{z_j}^{z_{j+1}} N_i \, dz = 0$$

$$l_{j-1,i} - l_{ji} - \int_{z_j}^{z_{j+1}} N_i \, dz = 0 \qquad (9\text{-}7)$$

$$v_{j+1,i} + l_{j-1,i} - v_{ji} - l_{ji} = 0$$

where N_i is defined in terms of the transfer of component i from the liquid to the vapor phase [see for example Equations (8-1) and (8-9)]. When Model 2 of Figure 9-3 is used to describe the behavior of the feed-distributor plate (or the feed mixing section), the component-material balances for element Δz_{f-1} ($z_{f-1} \leqq z \leqq z_f$) are obtained by setting $j = f - 1$ and replacing v_{fi} wherever it appears in Equation 9-7 by $\bar{v}_{fi}$ (defined below). Similarly, the balances for element Δz_f ($\bar{z}_f \leqq z \leqq z_{f+1}$) are obtained by setting $j = f$ in Equation (9-7) and replacing $l_{f-1,i}$ wherever it appears by $\bar{l}_{f-1,i}$, where,

$$\bar{v}_{fi} = v_{fi} + v_{Fi}; \qquad \bar{l}_{f-1,i} = l_{f-1,i} + l_{Fi}; \qquad FX_i = v_{Fi} + l_{Fi} \qquad (9\text{-}8)$$

As in the analysis of multicomponent distillation problems (15) and described in Chapter 3, v_{Fi} and l_{Fi} represent the flow rates of component i in the vapor and liquid streams V_F and L_F, respectively, formed by flashing the feed F, and thus $V_F + L_F = F$. Also, the streams V_F and L_F are taken to be in equilibrium,

$$y_{Fi} = K_{Fi} x_{Fi}$$

The flow rates V_F and L_F as well as the sets of compositions $\{y_{Fi}\}$ and $\{x_{Fi}\}$ may be found as described previously in the analysis of flash distillation problems in Chapter 3. Note that of the three equations given by Equation (9-7) only two are independent for each component and any given element of length Δz. Model 2 for the feed plate may be used to describe a feed of any thermal condition by placing the following interpretations on v_{Fi} and l_{Fi}. Namely, for a boiling point liquid and subcooled feeds (at the column pressure), $v_{Fi} = FX_i$ and $l_{Fi} = 0$; for dew point vapor and superheated feeds (at the column pressure), $l_{Fi} = FX_i$ and $v_{Fi} = 0$.

The total-material balances corresponding to the component-material balances given by Equation (9-7) are as follows:

$$V_{j+1} - V_j + \int_{z_j}^{z_{j+1}} \sum_{i=1}^{c} N_i \, dz = 0$$

$$L_{j-1} - L_j - \int_{z}^{z_{j+1}} \sum_{i=1}^{c} N_i \, dz = 0 \qquad (9\text{-}9)$$

$$V_{j+1} + L_{j-1} - V_j - L_j = 0$$

For $j = f - 1$ (element Δz_{f-1}), V_{j+1} in Equation (9-9) should be replaced by

$\bar{V}_f$ and for $j = f$ (element Δz_f), L_{f-1} should be replaced by $\bar{L}_{f-1}$,

$$\bar{V}_f = V_f + V_F; \qquad \bar{L}_{f-1} = L_{f-1} + L_F$$

Again, it is evident that only two of the three total-material balance expressions are independent.

For each element of packing in the intervals $z_1 \leqq z \leqq z_f$ and $\bar{z}_f \leqq z \leqq z_{N+1}$, the enthalpy balances are developed in the following manner. Suppose that thermal equilibrium exists at the vapor-liquid interface at steady-state operation, that is, the limiting temperature T^υ of the vapor phase is equal to the limiting temperature $T^\mathscr{L}$ of the liquid phase as the interface is approached from each phase,

$$T^\upsilon = T^\mathscr{L} = T^I$$

The energy balance enclosing the vapor phase over the element Δz_j and over to the *vapor side* of the vapor film (see Enclosure ① of Figure 9-4) is given by

$$V_{j+1}H^V_{j+1} - V_j H^V_j + \int_{z_j}^{z_{j+1}} \sum_{i=1}^{c} N_i H^V_i \, dz - \int_{z_j}^{z_{j+1}} q_G \, dz = 0 \qquad (9\text{-}10)$$

where

$H^V_j = \sum_{i=1}^{c} H^V_{ji} y_{ji}$, for an ideal solution;

H^V_{ji} = enthalpy of pure component i in the vapor state at the bulk temperature T^V_j and pressure P, Btu/lb-mole;

$q_G = h_G a(T^V - T^I)$, Btu/(hr) (ft of packing).

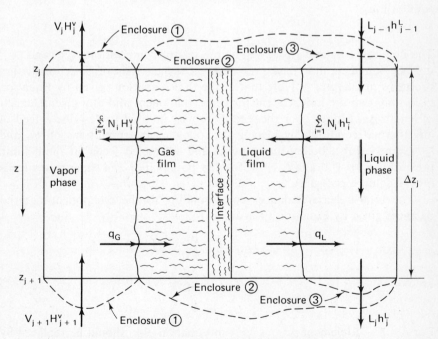

Figure 9-4. Mass and energy balance enclosures for the element Δz_j.

Similarly, when the energy-balance enclosure for the liquid phase includes the liquid phase, the liquid film, the interface, and the vapor film over Δz_j (see Enclosure ② of Figure 9-4), the following expression is obtained:

$$L_{j-1}h_{j-1}^L - L_j h_j^L - \int_{z_j}^{z_{j+1}} \sum_{i=1}^c N_i H_i^V \, dz + \int_{z_j}^{z_{j+1}} q_G \, dz = 0 \qquad (9\text{-}11)$$

where

$h_j^L = \sum_{i=1}^c h_{ji}^L x_{ji}$, for an ideal solution;

$h_{ji}^L =$ enthalpy of pure component i at the bulk temperature T_j^L and pressure P, Btu/lb-mole.

The energy balance enclosing both streams over the element Δz_j is given by

$$V_{j+1}H_{j+1}^V + L_{j-1}h_{j-1}^L - V_j H_j^V - L_j h_j^L = 0 \qquad (9\text{-}12)$$

Again, only two of the expressions given by Equations (9-10), (9-11), and (9-12) are independent.

Further insight into the process of simultaneous mass and heat transfer is achieved by making an energy balance on the liquid phase according to Enclosure ③ of Figure 9-4. This enclosure encompasses the liquid phase over Δz_j and intersects the boundary of the *liquid film*. The energy balance so obtained is given by

$$L_{j-1}h_{j-1}^L - L_j h_j^L - \int_{z_j}^{z_{j+1}} \sum_{i=1}^c N_i h_i^L \, dz + \int_{z_j}^{z+1} q_L \, dz = 0 \qquad (9\text{-}13)$$

where

$$q_L = h_L a(T^I - T^L), \text{ Btu/(hr) (ft of packing)}.$$

The rates of heat transfer q_G and q_L are related by use of an energy balance over the two films. This enclosure includes both films over the length Δz_j and encloses the outer boundaries of the two films. The result so obtained is

$$\int_{z_j}^{z_{j+1}} \left[-\sum_{i=1}^c N_i(H_i^V - h_i^L) + (q_G - q_L) \right] dz = 0$$

Since this result holds for each choice of z_j and z_{j+1} in the rectifying and stripping sections ($z_1 \leq z \leq z_f, \bar{z}_f \leq z \leq z_{N+1}$), it follows that the integrand is zero for each z in the packed sections, or

$$-\sum_{i=1}^c N_i(H_i^V - h_i^L) + (q_G - q_L) = 0 \qquad (9\text{-}14)$$

Note that this result may also be obtained by subtracting the members of Equation (9-13) from the corresponding members of Equation (9-11). Equation (9-14) emphasizes the fact that for processes in which simultaneous mass and heat transfer are involved, the heat transfer rates across the resistances are not equal as they are for the case of heat transfer alone across resistances in series.

THE HEAT TRANSFER SECTION. If the temperature of the vapor leaving each element of packing Δz_j is equal to the temperature of the liquid leaving the element, then the element Δz_j is said to be a *perfect heat transfer* section. For the general case in which the temperatures of the streams leaving a given element Δz_j are not necessarily equal, the heat transfer section is defined by *heat transfer relationship*

$$T_j^V = e_j T_j^L \tag{9-15}$$

where

$e_j =$ the deviation of an actual heat transfer section from a perfect heat transfer section.

Heat transfer efficiencies analogous to Murphree plate efficiencies have been defined and employed by Ravicz (26) and Holland (15).

There follows in Part 2 a modeling procedure for packed columns that makes use of the concepts of mass transfer sections and perfect heat transfer sections.

Part 2: Modeling of Packed Distillation Columns

When rates of mass transfer are used directly in the modeling of packed columns, two of the three expressions given by Equation (9-7) are used, and at least one of these will contain the rate of mass transfer N_i. Since the rate constants (mass transfer coefficients) that appear in N_i are not generally well known and because of certain problems that may arise in the use of the rate expressions in modeling (discussed below), it was elected not to use the rate expressions directly in the modeling of packed columns. That is, instead of using either one of the first two expressions with the third expression of Equation (9-7) as the independent set, the mass transfer relationship (Equation (9-4)) and the third expression of Equation (9-7) were selected as the independent set. The details of the formulation of this model follows.

The first step of the modeling procedure consisted of the transformation of the continuous mass transfer process into an equivalent stage process through the use of the mass transfer relationship. The transformation of a continuous mass transfer process into an equivalent stage process is effected as follows. Since it is generally difficult to differentiate experimentally between vapor and liquid temperatures, it is supposed that each element of packing Δz_j is a perfect heat transfer section, $T_j^V = T_j^L = T_j$. Possible errors resulting from this supposition are accounted for by absorbing them in the vaporization efficiency.

The mass transfer relationship [Equation (9-4)] plays the same role in the modeling techniques for packed columns that the equilibrium relationship, $y_i = K_i x_i$, plays in the calculational procedure presented in Chapter 3 for a conventional distillation column with perfect plates. For example, the temperature function that corresponds to the bubble point function for a perfect plate is obtained by summing both sides of Equation (9-4) over all components i, rearranging, and stating in functional notation,

$$f_j = \sum_{i=1}^{c} E_{ji}\gamma_{ji}K_{ji}x_{ji} - 1 \qquad (9\text{-}16)$$

Then for the given sets $\{x_{ji}\}$, $\{\gamma_{ji}\}$, $\{E_{ji}\}$, and the specified pressure P, the temperature of mass transfer section j is that $T_j > 0$ which makes $f_j = 0$.

The mass transfer relationship [Equation (9-4)] makes it possible to relate the flow rate v_{ji} of each component in the vapor leaving element Δz_j to its flow rate l_{ji} in the liquid leaving Δz_j. Beginning with Equation (9-4), it is readily shown that

$$v_{ji} = S_{ji}l_{ji} \quad \text{and} \quad l_{ji} = A_{ji}v_{ji} \qquad (9\text{-}17)$$

where the absorption factor A_{ji} and stripping factor S_{ji} are defined as follows:

$$A_{ji} = \frac{L_j}{E_{ji}\gamma_{ji}K_{ji}V_j}: \quad S_{ji} = \frac{1}{A_{ji}} \qquad (9\text{-}18)$$

Then, the component-material balance enclosing any element Δz_j [given by the last expression of Equation (9-7)] may be rearranged to the following form by use of Equation (9-17):

$$A_{j-1,i}v_{j-1,i} - (1 + A_{ji})v_{ji} + v_{j+1,i} = 0 \qquad (9\text{-}19)$$

which is precisely the same expression as that obtained for plate j for a column with plates. In fact, the component-material balances for a packed column become identical with those for a column with plates, where for each element of packing Δz_j there exists a plate j. The condenser and reboiler are treated in precisely the same manner used in the analysis of distillation columns with plates. Similarly, the enthalpy balances presented in Chapter 3 for a distillation column with plates are applicable for a packed distillation column. Thus, the calculational procedures for distillation columns with plates become applicable to packed distillation columns.

Calculational procedure for the determination of vaporization plate efficiencies from the results of field tests

Generally, all of the values of the variables needed to compute a unique set of E_{ji}'s by use of Equation (9-3) are seldom available from the results of field tests. This lack of data led to the proposal of a simple product model [Equation (9-22)]. By use of this model, first-order approximations of the vaporization plate efficiencies may be determined by use of the data commonly available from the results of field tests. The calculational procedures used to compute the vaporization plate efficiencies by the product model for columns with plates (8, 15, 16, 32) may be applied to packed columns without making any major modifications as outlined below and demonstrated in Chapters 10 and 12.

Although composition profiles are seldom determined in making field tests, the compositions of the terminal streams (feed, distillate, bottoms) are usually available from the results of field tests or they may be obtained without making major modifications to existing equipment. The temperature profile obtained by inserting thermocouples in the packing are taken to be those of the vapor and liquid phases, that is, perfect heat transfer sections ($T_j^V = T_j^L = T_j$) are assumed.

When perfect heat transfer sections are assumed, a convention is needed for the assignment of temperatures. A convention such as the following one is recommended. If the temperature of the terminal stream leaving a terminal element of packing is measured, then the temperature of the terminal element should be taken equal to the temperature of the terminal stream leaving it; otherwise, the temperature of the terminal element may be assigned in the same manner as recommended below for the interior elements of packing. For example, if the temperatures T_1 and T_N of the terminal streams V_1 and L_N are measured, then the first element Δz_1 is assigned the temperature T_1 and the Nth element Δz_N is assigned the temperature T_N. This particular choice of temperatures for the terminal elements is seen to place the overall energy balances of the model and the existing column in a one-to-one correspondence, provided, of course, that the compositions and total flow rates are identical. For the interior elements such as Δz_2, it is recommended that each element be assigned that temperature given by the temperature profile at the midpoint of the element.

The calculational procedure that follows for the determination of the E_{ji}'s for a given set of Δz_j's is based on the knowledge of the additional specifications of all of the T_j's and all of the $(b_i/d_i)_{\exp}$'s. The term *additional specifications* is used to identify those variables which are known in addition to the usual operating conditions and definition of the column.

In addition to the T_j's and $(b_i/d_i)_{\exp}$'s, it is, of course, supposed that all of the variables included in the complete definition of the column and the operating conditions are known. These specifications are enumerated in Chapter 3 under the heading "Separation of Multicomponent Mixtures by Use of Conventional Distillation Columns with Multiple Stages." The subscript "exp" is used to emphasize that these are the experimental or specified values of the product distributions $\{b_i/d_i\}$. [In previous work (8, 15, 16, 32) these experimental values of the product distributions were called the corrected values and given the subscript "co." However, the subscript "exp" is perhaps more descriptive, and this notation is used throughout the following analysis.]

Since FX_i is known for each component and D is known, the set $\{X_{Di}\}$ could have been specified instead of the product distributions. The equivalence of the two sets is shown as follows. An overall component-material balance yields

$$FX_i = (d_i)_{\text{exp}} + (b_i)_{\text{exp}} \tag{9-20}$$

Thus,

$$(d_i)_{\text{exp}} = \frac{FX_i}{1 + \left(\dfrac{b_i}{d_i}\right)_{\text{exp}}} \tag{9-21}$$

from which the set $\{X_{Di}\}$ readily follows from the definition of a mole fraction.

For a distillation column with a partial condenser, the set $\{E_{ji}\}$ contains at most $c(N + 2)$ elements. For a distillation column with a total condenser, the set $\{E_{ji}\}$ contains $c(N + 1)$ elements. The c product distributions and the total distillate rate D constitute a total of only c degrees of freedom instead of $(c + 1)$. For, given c product distributions, the total distillate rate may be computed by use of overall component-material balances, and thus D becomes a dependent variable. The column used to make the field tests was equipped with a total condenser; hence, the set $\{T_j\}$ contained $N + 1$ elements. Therefore, a total of $(c + N + 1)$ degrees of freedom is available from the results of the field tests.

To calculate the set $\{E_{ji}\}$ of $c(N + 1)$ elements from the $(c + N + 1)$ degrees of freedom, the following simple product model was proposed (15):

$$E_{ji} = \bar{E}_i \beta_j \tag{9-22}$$

where $\bar{E}_i$ is the component efficiency factor and β_j is the plate or element factor. Since there are c elements in the set $\{\bar{E}_i\}$ and $N + 1$ elements in the set $\{\beta_j\}$, there are $c + N + 1$ variables to be determined from the $(c + N + 1)$ independent specifications. The introduction of an additional restriction called the normalization procedure, which is presented below, has the effect of reducing the number of independent β_j's from $(N + 1)$ to N. This model for the vaporization efficiency is obviously only a first-order approximation of E_{ji}. Other models which are of a predictive nature are presented in Chapter 11.

The equations to be satisfied by the solution set of E_{ji}'s consist of the component-material balances, the total-material balances, the enthalpy balances, and the temperature functions. The following calculational procedure was used for the determination of the solution sets $\{\bar{E}_i\}$ and $\{\beta_j\}$. In this procedure, the specification of the b_i/d_i's is restated in terms of θ_i as follows:

$$\theta_i = \frac{\left(\dfrac{b_i}{d_i}\right)_{\text{exp}}}{\left(\dfrac{b_i}{d_i}\right)_{\text{ca}}} \qquad (1 = i \leq c) \tag{9-23}$$

where the subscript "ca" refers to calculated values of the given variables, obtained on the basis of an assumed set of E_{ji}'s. Thus, the solution set $\{E_{ji}\}$ must give $\theta_i = 1 (1 \leq i \leq c)$.

Let the $\bar{E}_i$'s be regarded as the independent variables, and the β_j's, V_j's, and L_j's as dependent variables. To initiate the calculational procedure, a complete set of variables $\{\bar{E}_i, \beta_j, L_j, V_j\}$ is assumed. Then Equation (3-64) of Chapter 3 is solved for the $(v_{ji})_{ca}$'s from which the corresponding values of $(b_i)_{ca}$ and $(d_i)_{ca}$ are readily obtained. Then θ_i is calculated by use of Equation (9-23). An improved value of $\bar{E}_i$ for each component i is obtained by use of Newton's method. The functional form $\log_e \theta_i$ is used instead of the functional form θ_i because the logarithmic form led to a more rapid rate of convergence for all problems considered by Taylor et al. (32) than did the simple functional form θ_i.

$$ 0 = \log_e \theta_i + \left(\frac{\partial \log_e \theta_i}{\partial \bar{E}_i} \right) \Delta \bar{E}_i \qquad (1 \leq i \leq c) \qquad (9\text{-}24) $$

where

$$ \Delta \bar{E}_i = \bar{E}_{i,n+1} - \bar{E}_{i,n} $$

The values of $\bar{E}_i$ used to make the nth trial carry the subscript n, and the improved set to be determined the subscript $n + 1$. The values of the functions ($\log_e \theta_i$) and partial derivatives in the set of equations denoted by Equation (9-24) were evaluated at the last set of values of the variables $\{\bar{E}_{i,n},$ $\beta_{j,n}, V_{j,n}, L_{j,n}\}$. Each derivative was evaluated numerically as described by Holland (16) (see Problem 9-9).

After the $\bar{E}_{i,n+1}$'s had been determined by use of Equation (9-24), they were used to calculate a new set of compositions. This set of compositions was found by solving the component-material balances on the basis of the following set of variables: $\{E_{i,n+1}, \beta_{j,n}, V_{j,n}, L_{j,n}\}$. As suggested previously (15, 16), the component-flow rates so obtained were used to compute the compositions as follows:

$$ (x_{ji})_{ca} = \frac{\left(\dfrac{l_{ji}}{d_i} \right)_{ca} (d_i)_{ca}}{\sum\limits_{i=1}^{c} \left(\dfrac{l_{ji}}{d_i} \right)_{ca} (d_i)_{ca}} = \frac{(l_{ji})_{ca}}{\sum\limits_{i=1}^{c} (l_{ji})_{ca}} \qquad (9\text{-}25) $$

Let the x_{ji}'s so obtained be identified by the subscript "$n + 1$." On the basis of these $x_{ji,n+1}$'s and the $\bar{E}_{i,n+1}$'s, the $\beta_{j,n+1}$'s were found by setting each $f_j = 0$ [see Equation (9-16)] and solving to obtain

$$ \beta_{j,n+1} = \frac{1}{\sum\limits_{i=1}^{c} \bar{E}_{i,n+1} K_{ji} x_{ji,n+1}} \qquad (9\text{-}26) $$

where the activity coefficient has been omitted in the interest of simplicity. Then on the basis of the sets $\{\bar{E}_{i,n+1}\}$, $\{\beta_{j,n+1}\}$, and $\{x_{ji,n+1}\}$, the total flow rates for the next trial were computed by use of Equations (3-77) through (3-81) of Chapter 3.

The normalization procedure that reduces the number of independent

β_j's from $(N + 1)$ to N was proposed by Taylor *et al.* (32), and it consisted of the requirement that the product of the elements of the set $\{\beta_j\}$ be equal to unity, that is,

$$\prod_{j=1}^{N+1} \beta_j = 1 \qquad (9\text{-}27)$$

In this normalization procedure, the normalized value β_j is defined by

$$\beta_j = \frac{(\beta_j)_{\text{ca}}}{\sqrt[N+1]{\prod_{j=1}^{N+1} (\beta_j)_{\text{ca}}}} \qquad (9\text{-}28)$$

The subscript "ca" is used to identify the most recent set of calculated values of β_j and $\bar{E}_i$. Note that the normalization factor or denominator on the right-hand side of Equation (9-28) is the geometric mean of the calculated values of the β_j's.

To keep the set $\{E_{ji}\}$ unchanged by the normalization procedure, that is, to retain the relationship

$$E_{ji} = (\bar{E}_i)_{\text{ca}}(\beta_j)_{\text{ca}} = \bar{E}_i\beta_j \qquad (9\text{-}29)$$

it is evident from Equation (9-28) that the normalized value of $\bar{E}_i$ must be computed by use of the following formula:

$$\bar{E}_i = (\bar{E}_i)_{\text{ca}} \sqrt[N+1]{\prod_{j=1}^{N+1} (\beta_j)_{\text{ca}}} \qquad (9\text{-}30)$$

This particular normalization procedure has the desirable characteristic that at total reflux (or total recycle), the product distribution b_i/d_i for any component i depends only on its $\bar{E}_i$ and the product of its K_{ji}'s over all sections j. The calculational procedure described was repeated until convergence had been achieved (the change in the $\bar{E}_i$'s and β_j's between successive trials was less than some small preassigned number). The sets $\{\bar{E}_i\}$ and $\{\beta_j\}$ so obtained satisfy all of the additional specifications as well as the material balances, energy balances, and mass transfer relationships.

Modifications of the above procedure which are useful in the modeling of existing columns follow. A summary of the steps of the calculational procedure follows the presentation of Modification 1.

MODIFICATIONS OF THE CALCULATIONAL PROCEDURE FOR THE DETERMINATION OF THE VAPORIZATION PLATE EFFICIENCIES. The following modifications were formulated for the purpose of making it possible to model existing columns on the basis of information which is either presently available or can be readily obtained (21). This information consists of the values of the operating variables (called the usual specifications) and the additional specifications, the product distributions and the temperatures of the terminal streams. This set of values of the variables may be used to place the model and the existing column in a one-to-one correspondence with respect to an overall-component-material balance and an overall-energy balance.

Modification 1: Additional specifications consist of all or part of the set $\{b_i/d_i\}$ and the set $\{T_1, T_{N+1}\}$; the E_i's, β_1, and β_{N+1} are to be determined

Consider first the case in which all of the b_i/d_i's ($1 \leq i \leq c$) are known. To obtain a one-to-one correspondence between the observed and predicted product distributions, a set of $\bar{E}_i$'s is picked by use of the following modified version of the procedure described above. The observed and predicted values of T_1 and T_{N+1} are placed in agreement by computing β_1 and β_{N+1} by use of Equation (9-26). For all intermediate sections, one may select the β_j's ($2 \leq j \leq N$) arbitrarily and compute the corresponding temperatures by use of the K_b method (see Chapter 3). Let β_j ($2 \leq j \leq N$) be selected as the geometric mean of β_1 and β_{N+1}, that is,

$$(\beta_j)_{\mathrm{ca}} = [(\beta_1)_{\mathrm{ca}}(\beta_{N+1})_{\mathrm{ca}}]^{1/2} \qquad (2 \leq j \leq N) \tag{9-31}$$

Also, let the calculated values of $\bar{E}_i$ and β_j be normalized as indicated by use of Equations (9-25) and (9-27). The choice of β_j ($2 \leq j \leq N$) as given by Equation (9-31) plus the normalization procedure lead to the following sets of normalized values for the β's and $\bar{E}_i$'s.

$$\beta_1 = \left(\frac{\beta_1}{\beta_{N+1}}\right)^{1/2}_{\mathrm{ca}}; \qquad \beta_{N+1} = \left(\frac{\beta_{N+1}}{\beta_1}\right)^{1/2}_{\mathrm{ca}} \tag{9-32}$$

$$\beta_j = 1 \qquad (2 \leq j \leq N)$$

and

$$\bar{E}_i = (E_i)_{\mathrm{ca}}[(\beta_1)_{\mathrm{ca}}(\beta_{N+1})_{\mathrm{ca}}]^{1/2} \tag{9-33}$$

For all mass transfer sections for which the temperatures are regarded as unknown (namely, $T_2 \leq T_j \leq T_N$), the temperatures are computed each trial by use of the K_b method. This calculation replaces the determination of β_j as implied by Equation (9-26). The calculational procedure as modified is repeated until convergence has been achieved. It is of interest to note that for this case, the geometric mean of β_1 and β_{N+1} approaches unity in the limit as convergence is approached. A summary of the steps of the calculational procedure (as amended by Modification 1) for the determination of the vaporization efficiencies for a conventional column follows:

1. Assume a set of component efficiencies $\{\bar{E}_i; 1 \leq i \leq c\}$, a set of vapor rates $\{V_j; 2 \leq j \leq N + 1\}$, a set of temperatures $\{T_j; 2 \leq j \leq N\}$ and β_1. Compute $\beta_{N+1} = 1/\beta_1$. Compute the liquid rates $\{L_j\}$ by use of the assumed vapor rates and the total-material balances. For $2 \leq j \leq N$, take $\beta_j = 1$.
2. Solve the component-material balances [Equation (3-64) of Chapter 3)] for each component i to obtain $\{(v_{ji})_{\mathrm{ca}}\}$ and $\{(b_i/d_i)_{\mathrm{ca}}\}$.
3. Compute $\{\theta_i\}$ by use of Equation (9-23).
4. Compute an improved set of component efficiencies $\{(\bar{E}_i)_{\mathrm{ca}}\}$ by use

of Newton's method [see Equation (9-24)]. [The partial derivative appearing in Equation (9-24) may be evaluated numerically by use of the procedure described in Problem 9-7.]

5. On the basis of the improved set of component efficiencies $\{(\bar{E}_i)_{ca}\}$ found in Step 4 and the values of the other variables assumed in Step 1, repeat Step 2 to obtain a new set of component flow rates $\{(v_{ji})_{ca}\}$. Compute the corresponding set of liquid rates for each component i by use of the relationship $l_{ji} = A_{ji}v_{ji}$.

6. Use the results of Step 5 to compute a new set of mole fractions by use of Equation (9-25).

7. Compute $(\beta_1)_{ca}$ and $(\beta_{N+1})_{ca}$ by use of Equation (9-26). Compute the unknown temperatures T_j; $2 \leq j \leq N$ by use of the K_b method [see Equations (3-73) and (3-74) of Chapter 3].

8. Use the results of Steps 4 through 7 to compute a new set of total flow rates by use of Equations (3-77) through (3-81) of Chapter 3.

9. Normalize $(\beta_1)_{ca}$ and $(\beta_{N+1})_{ca}$ found in Step 7 and the $(E_i)_{ca}$ found in Step 4 as indicated by Equations (9-32) and (9-33).

10. If each element of the set $\{\theta_i\}$, $\{\bar{E}_i\}$, $\{\beta_1, \beta_{N+1}\}$, $\{T_j\}$, and $\{V_j\}$, satisfies the preassigned tolerances, convergence has been achieved; otherwise, repeat Steps 2 through 9 on the basis of the most recent sets of values of the variables.

In the determination of the efficiencies for the packed absorber described in Chapter 10, the determination of the unknown temperatures by use of the enthalpy balances and the Newton-Raphson method gave a more rapid rate of convergence than that obtained when the K_b method was used to determine these temperatures as described in Step 7 above.

OTHER MODIFICATIONS. When part of the b_i/d_i's are unknown, appropriate variations of the calculational procedure for the determination of the E_{ji}'s have been proposed (16, 28, 29, 32). If it is impossible to determine certain of the b_i/d_i's because of the impossiblity of detecting and measuring with any degree of accuracy the amounts of these components in one of the terminal streams D or B, the corresponding $(\bar{E}_i)_{ca}$'s were set equal to unity. This modification was used in the analysis of the results of the field tests made on the packed distillation column (2). If, however, the components whose b_i/d_i's are unknown are not separated, a single efficiency $\bar{E}_i$ may be computed on the basis of the known contribution of this group of components to either the distillate rate D or the bottoms rate B as described previously (29, 32).

Another modification consists of finding a set of E_{ji}'s corresponding to a known set of b_i/d_i's. For this case, the calculational procedure is the same as the one described for Modification 1 with the exception that a temperature is computed for each section j on the basis that $\beta_j = 1$ for all j. This procedure was used in the analysis of the field tests made on a liquid-liquid extractor (see Chapter 12).

A preliminary analysis of the characteristics of vaporization efficiencies, overall mass transfer coefficients, and Murphree-type efficiencies follows. A further development of this subject is presented in Chapter 11.

Part 3: Characteristics of Vaporization Plate Efficiencies, Murphree-type Efficiencies, and Overall Mass Transfer Coefficients

In this section, some of the problems involved in the determination of the overall mass transfer coefficients from experimentally determined compositions are demonstrated. Similar problems arise in the determination of Murphree-type efficiencies from experimentally determined compositions. The determination of the vaporization plate efficiencies is shown to be free of these problems.

Problems involved in the determination of overall mass transfer coefficients from experimental compositions

In the analysis that follows, it is supposed that experimental data in the form of compositions and flow rates are available, and it is desired to use these data to determine the mass transfer coefficients. That is, the overall mass transfer coefficients are regarded as the dependent variables and the compositions and total flow rates as the independent variables; in Chapter 12 in the prediction of vaporization efficiencies, the choice of independent and dependent variables is in effect reversed. The conclusions reached for each of these two choices of independent and dependent variables need not agree, and, in fact, they do not.

The following examples demonstrate that a complete set of nonzero, finite, and positive mass transfer coefficients may not exist for a packed column in the process of separating a multicomponent mixture. In the examination that follows, inquiry is made into the existence of the integral of the overall mass transfer coefficient over the interval $z_j \leqq z \leqq z_{j+1}$ for various types of operating conditions. To carry out this analysis, the second expression of Equation (9-7) is first transformed to its corresponding differential equation in the usual way to give

$$\frac{d(Lx_i)}{dz} + (K_G aSf^v)_i(Y_i - y_i) = 0 \qquad (9\text{-}34)$$

where N_i has been replaced by the expression

$$N_i = (K_G aSf^V)_i(\gamma_i K_i x_i - y_i) \tag{9-35}$$

where

$$Y_i = \gamma_i K_i x_i$$

In the interest of simplicity, ideal solutions ($\gamma_i = 1$) are used in the analysis. Also, since the validity of the examples that follow is not affected by regarding the total flow rate as a constant, this assumption is made in the interest of simplicity. Separation of the variables in Equation (9-34) followed by integration over the interval from z_j to z_{j+1} yields

$$\int_{z_j}^{z_{j+1}} (K_G aSf^V)_i \, dz = -L \int_{x_{j-1,i}}^{x_{ji}} \frac{dx_i}{Y_i - y_i} \tag{9-36}$$

The first example is based on the operating condition of total reflux of the type $F = D = B = 0$ [or total recycle (15)]. At this condition a total-material balance enclosing the condenser and accumulator at the top of the column and the packing down to any z is given by $V = L$, and the corresponding component-material balance is given by $Vy_i = Lx_i$. Thus, at any z at total reflux $y_i = x_i$. Since the K values for a multicomponent mixture at its bubble point range from positive numbers less than unity to finite and positive numbers greater than unity, it is not unreasonable to suspect that for some component i, $K_i = 1$ at some z, say z_k. In fact, if $K_i < 1$ at the top of the column and $K_i > 1$ at the bottom of the column, then it is evident that there exists some z at which $K_i = 1$. One or more components of this type are common for columns in the process of separating multicomponent mixtures.

For $K_i = 1$ at total reflux, the integrand $1/(K_i x_i - y_i)$ of the integral on the right-hand side of Equation (9-36) has an infinite discontinuity. However, the existence of the integral depends on the manner in which $(K_i x_i - y_i)$ approaches zero as demonstrated by the following illustrative example.

ILLUSTRATIVE EXAMPLE 9-1

Given:

1. $y_i = x_i$ $(z_j \leqq z \leqq z_{j+1})$
2. $0 < K_i < 1$ $(z_j \leqq z < z_k)$
   $K_i = 1$ $(z = z_k)$
   $1 < K_i < \infty$ $(z_k < z \leqq z_{j+1})$
3. K_i and x_i are continuous, nonzero, finite, and positive in the closed interval $z_j \leqq z \leqq z_{j+1}$, and $x_{j-1,i} < x_{ki} < x_{ji}$, where x_{ki} denotes the value of x_i at $z = z_k$.
4. Since x_i varies with z and K_i varies with z, K_i may be regarded as a function

of x_i. Suppose

$$\frac{1}{x_i(1 - K_i)} \geq \frac{N_1}{(x_{ki} - x_i)^{v_1}} \qquad (x_{j-1,i} \leq x_i < x_{ki})$$

$$\frac{1}{x_i(K_i - 1)} \geq \frac{N_2}{(x_i - x_{ki})^{v_2}} \qquad (x_{ki} < x_i \leq x_{ji})$$

where v_1 and v_2 are positive numbers equal to or greater than unity and N_1 and N_2 are fixed numbers independent of x_i.

To Prove:

The integral of the overall mass transfer coefficient over the closed interval does not exist for component i.

Proof:

By use of condition (1), the integral on the right-hand side of Equation (9-36) may be stated as follows:

$$\int_{x_{j-1,i}}^{x_{ji}} \frac{dx_i}{Y_i - y_i} = \int_{x_{j-1,i}}^{x_{ji}} \frac{dx_i}{x_i(K_i - 1)} \qquad (9\text{-}37)$$

When condition (2) is imposed, the integrand is seen to have an infinite discontinuity in the interior of the interval of integration. Such an integral is said to exist provided each of the following limits exists (see Definition B-3):

$$\int_{x_{j-1,i}}^{x_{ji}} \frac{dx_i}{x_i(K_i - 1)} = -\lim_{\epsilon \to 0} \int_{x_{j-1,i}}^{x_{ki}-\epsilon} \frac{dx_i}{x_i(1 - K_i)} + \lim_{\delta \to 0} \int_{x_{ki}+\delta}^{x_{ji}} \frac{dx_i}{x_i(K_i - 1)} \qquad (9\text{-}38)$$

Consider the first integral on the right-hand side of Equation (9-38). From conditions (3) and (4), it follows that

$$\lim_{\epsilon \to 0} \int_{x_{j-1,i}}^{x_{ki}-\epsilon} \frac{dx_i}{x_i(1 - K_i)} \geq \lim_{\epsilon \to 0} \int_{x_{j-1,i}}^{x_{ki}-\epsilon} \frac{N_1 \, dx_i}{(x_{ki} - x_i)^{v_1}} \qquad (9\text{-}39)$$

By Theorem B-6, the integral on the right-hand side of this expression does not exist. Thus, the integral on the left-hand side does not exist. Then, by Definition B-3, the integral on the left-hand side of Equation (9-38) does not exist. Although it can be shown that the second integral on the right-hand side of Equation (9-38) also fails to exist, it was not necessary to do so to complete the proof.

In certain instances, the physical conditions within a column may be such that the variation of the function $1/[x_i(K_i - 1)]$ is in agreement with condition (4) of the above example. In other instances, the variation of this function may be such that the integral of the overall mass transfer coefficient exists. For example, if the inequalities given by the fourth condition of Illustrative Example 9-1 are reversed, it can be shown that the overall mass transfer coefficient exists.

Thus, at total reflux, the integral of the overall mass transfer coefficient over the closed interval, which contains the discontinuity resulting form $K_i =$

1, may or may not exist. The existence of this integral depends on the precise variation of the function $1/[x_i(K_i - 1)]$ throughout the interval of integration. In general, it appears plausible to expect the variation of this function to satisfy condition (4) of Illustrative Example 9-1 in some instances and not to satisfy this condition in other instances. Other examples are given as Problems 9-8 and 9-9 which are based on the work of Bassyoni *et al.* (3).

Infinite discontinuities of the type described above do not generally occur in the separation of binary mixtures because the K value for one component is always less than unity and the K value for the other component is always greater than unity. Obvious exceptions to this statement are the azeotropes of binary mixtures. In this case, $\gamma_1 K_1$ approaches $\gamma_2 K_2$ which approaches unity as the azeotrope is approached.

From the standpoint of the precise physical conditions existing in a column, the existence or nonexistence of the integral on the right-hand side of Equation (9-36) appears equally likely. That is, because of the random motion of the liquid and vapor along any vertical of an actual column, it would be almost impossible to evaluate the right-hand side of Equation (9-36) at a point of apparent discontinuity by use of experimental results. It can be equally well argued, however, that if one supposes that the transfer of mass is represented by Equation (9-36), then the limit of the integral does exist at infinite discontinuities because perfect separations are not observed experimentally.

Problem 9-9 is formulated on the basis of the experimentally observed fact that the composition profiles for certain components may pass through either a maximum or minimum in columns in the process of separating multicomponent mixtures. Since the composition profiles of the remaining components do not generally pass through either a maximum or minimum at the same Δz at which a particular component passes through a maximum composition, the integrals of the overall mass transfer coefficients for the remaining components can be expected to be nonzero numbers of almost any magnitude. Since it is reasonable to suppose that the mechanism of transfer of all components is roughly the same, one would expect to find mass transfer coefficients of the same order of magnitude. Again, however, in the case of a binary mixture, problems of the type demonstrated by Problem 9-9 do not arise because equilibrium curves are characterized by the fact that y increases monotonically with x. It is to be recognized, however, that for columns in the process of separating binary mixtures at minimum reflux, $x_{j-1,i}$ does approach x_{ji} for those plates at and adjacent to the feed plate. Another exception is, of course, azeotropic binary mixtures.

For each of the cited examples above, it is to be observed that for all components that are detectable and have finite K values [for which Equation (9-5) is satisfied], a set of nonzero, finite, and positive vaporization efficiencies exist. Furthermore, for any actual column, there exists a unique set of E_{ji}'s for each choice of the number of packed sections.

EXISTENCE OF A UNIQUE SET OF VAPORIZATION PLATE EFFICIENCIES. Suppose that for a completely defined packed distillation column operating at a known set of steady-state conditions, the additional specifications stated below are available from the results of field tests. By a completely defined column and a known set of operating conditions is meant that the following information is available: the diameter of the column, the type of packing, the depth of packing, the type of condenser (partial or total), the column pressure, the distillate rate D, the reflux rate L_0, the feed rate F as well as the composition and thermal condition of the feed, and the location of the point of introduction of the feed. The additional specifications follow: x_{ji} $(0 \leq j \leq N + 1,$ and $1 \leq i \leq c)$; X_{Di} $(1 \leq i \leq c)$; T_j^L $(0 \leq j \leq N + 1)$; and T_j^V $(0 \leq j \leq N + 1)$.

Although this particular set of specifications is seldom known in practice, the set is of interest because it does yield a unique set of E_{ji}'s, which may be computed directly without resorting to the use of trial-and-error procedures (16).

First the L_j's are determined by use of enthalpy balances as shown in Chapter 5. (In this application of these equations, the liquid enthalpies h_{ji} and the vapor enthalpies H_{ji} are evaluated at T_j^L and T_j^V, respectively.) After the L_j's have been determined, the corresponding V_j's are found by use of total-material balances. On the basis of this set of total flow rates and the known values of x_{ji}, T_j^L, and T_j^V, the component-material balances may be solved explicitly for the E_{ji}'s in the same manner as demonstrated previously (16).

Illustrative Example 9-1 and Problems 9-8 through 9-10 lead to the conclusion that in spite of the fact that the rate expression for mass transfer [Equations (8-9) and (9-35)] and the definition for the vaporization efficiency [Equation (9-2)] involve the same fundamental thermodynamic quantities $\bar{f}_i^V$ and $\bar{f}_i^L$, a nonzero, finite, and positive set of E_{ji}'s always exists, provided the conditions of Equation (9-5) are satisfied; whereas, under these same conditions, the existence of nonzero, finite, and positive set of overall mass transfer coefficients for a packed distillation column in the process of separating a multicomponent mixture is open to serious question. Consequently, in the subsequent mathematical modeling of packed distillation columns, vaporization efficiencies are used exclusively instead of the rate expression for mass transfer.

Instead of using mass transfer rates or vaporization plate efficiencies in the modeling of a packed column, the possibility exists of regarding each element as a plate whose deviation from a perfect plate is accounted for by use of a *Murphree-type* efficiency. This method was discarded because of some of the problems encountered in the use of Murphree-plate efficiencies as demonstrated below.

Problems involved in the determination of Murphree-type efficiencies from experimental compositions

In the analysis that follows, it is supposed that experimental data in the form of compositions are available, and that it is desired to use these to determine the plate efficiencies, that is, the actual compositions are regarded as the independent variables and the efficiencies as the dependent variables.

The difficulties encountered in the use of mass transfer rate expressions are reflected in the Murphree-type efficiency. The term Murphree-type efficiency is used to denote that class of efficiencies defined in terms of a ratio of differences [see, for example, the modified Murphree efficiency defined by Equation (7-17)] and others (14, 18, 30). When a Murphree-type efficiency is applied to a multicomponent mixture, the possibility of obtaining a zero for either the numerator or the denominator exists.

Problems encountered in the use of the Murphree-type efficiencies for the description of the degree of completion of the transfer processes occurring in an element of packing or on a plate of a distillation column in the process of separating a multicomponent mixture are demonstrated by selected examples. For purposes of demonstration, the Murphree efficiency for the vapor is selected for consideration.

The Murphree plate efficiency (22) for the vapor, E_{Mji}, for component i on plate or element Δz_j is defined as follows:

$$E_{Mji} = \frac{y_{ji} - y_{j+1,i}}{y_{ji}^* - y_{j+1,i}} \tag{9-40}$$

where y_{ji} is the actual mole fraction of component i in the vapor leaving plate j (or element Δz_j), $y_{j+1,i}$ is the actual mole fraction of component i in the vapor entering plate j (or element Δz_j), and y_{ji}^* is the hypothetical mole fraction that component i would have if the vapor leaving plate j (or element Δz_j) were in equilibrium with the liquid leaving, or more precisely,

$$y_{ji}^* = K_{Mji} x_{ji} \tag{9-41}$$

Since y_{ji}^* was defined as a mole fraction, it follows that

$$\sum_{i=1}^{c} y_{ji}^* = 1 \tag{9-42}$$

or

$$1 = \sum_{i=1}^{c} K_{Mji} x_{ji} \tag{9-43}$$

Thus, K_{Mji} is evaluated at that bubble point temperature required to satisfy Equation (9-43) at the column pressure. The Murphree plate efficiency for component i in the liquid was defined in a manner similar to that shown for the vapor.

ILLUSTRATIVE EXAMPLE 9-2

Given:

1. The column is at total reflux ($y_{j+1,i} = x_{ji}$) in the process of separating a multicomponent mixture.
2. The actual vapor leaving plate j is not in equilibrium with the liquid leaving $y_{ji} \neq y_{ji}^*$.
3. For some component i, $K_{Mji} = 1$.

To Prove:
E_{Mji} is undefined.

Proof:
Equations (9-40) and (9-41) may be combined and rearranged to give

$$E_{Mji} = \frac{\dfrac{y_{ji}}{y_{j+1,i}} - 1}{\dfrac{K_{Mji}x_{ji}}{y_{j+1,i}} - 1} \tag{9-44}$$

Since it is given that the column is at total reflux, Equation (9-44) reduces to

$$E_{Mji} = \frac{\dfrac{y_{ji}}{y_{j+1,i}} - 1}{K_{Mji} - 1} \tag{9-45}$$

To complete the proof, it is first necessary to show that

$$\frac{y_{ji}}{y_{j+1,i}} \neq 1 \tag{9-46}$$

This inequality is shown to exist by commencing with Equation (9-41) and then applying conditions (1) and (3) as follows:

$$y_{ji}^* = K_{Mji}x_{ji} = K_{Mji}y_{j+1,i} = y_{j+1,i} \tag{9-47}$$

Thus,

$$\frac{y_{ji}^*}{y_{ji}} = \frac{y_{j+1,i}}{y_{ji}} \tag{9-48}$$

Since it is given by condition (2) that $y_{ji}^* \neq y_{ji}$, the inequality given by Equation (9-46) follows immediately from Equation (9-48). Thus, the numerator of Equation (9-45) is seen to be finite while its denominator is zero. Since the division of a finite number by zero is undefined, E_{Mji} is undefined.

It should be observed that the condition $K_{Mji} = 1$ is not at all severe for some component i of a multicomponent mixture composed of distinguishable components (the value of K at a given temperature and pressure differs for each component). For it follows from Equation (9-45) that for any set of x_{ji}'s, the K values at the bubble point temperature must take on a set of numbers ranging from less than unity to greater than unity. Thus, the occurrence of condition (3) is certainly a likely possibility.

Throughout the development in Illustrative Example 9-2, it was supposed, of course, that the conditions of Equation (9-5) were satisfied. Thus, for this example, the E_{ji} as given by Equation (9-4) is finite and positive.

Numerous other examples for which the Murphree efficiency is either zero or unbounded while the corresponding vaporization efficiencies are nonzero, finite, and positive may be produced [see Reference (18)].

Although the Murphree efficiency for the vapor was singled out for examination in the above analysis, other Murphree-type efficiencies would appear to be subject to similar difficulties for multicomponent mixtures. For binary mixtures, however, the difficulty demonstrated by Illustrative Example 9-2 is not generally encountered because K is always less than unity for one component and greater than unity for the other component except for the special case of an azeotrope.

PROBLEMS

9-1. If the vapor phase forms an ideal solution while the liquid phase forms a nonideal solution, show that the definition of E_{ji} given by Equation (9-2) reduces to the one given by Equation (9-4) provided $T_j^V = T_j^L$ and $P^V = P^L$.

9-2. Begin with the mass transfer relationship $y_{ji} = E_{ji}K_{ji}x_{ji}$ and the fact K_{jb} is any arbitrarily selected positive number, and eliminate the y_{ji}'s to obtain

$$K_{jb}|_{T_{j,n+1}} = \frac{1}{\sum\limits_{i=1}^{c} E_{ji}\alpha_{ji}|_{T_{jn}}x_{ji}}$$

9-3. Show that when K_{ji} is stated in terms of the y_{ji}'s, the following formula is obtained for K_{jb}, namely,

$$K_{jb}|_{T_{j,n+1}} = \sum_{i=1}^{c} \frac{y_{ji}}{(E_{ji}\alpha_{ji}|_{T_{jn}})}$$

9-4. Show that the expressions given in Problems 9-2 and 9-3 yield equal values for K_{jb}.

9-5. Let $(\beta_j)_{ca}$ for a packed distillation column be defined by Equation (9-31) and let the β_j's and $\bar{E}_i$'s be normalized as indicated by Equations (9-28) and (9-30). On this basis, produce the formulas given by Equations (9-32) and (9-33).

9-6. The purpose of this problem and the next one is to provide the student some numerical experience in the calculation of the vaporization efficiencies as proposed in Modification 1. To simplify the calculational procedure, it will be assumed that the total flow rates remain fixed at the values stated in the following table.

Comp. No.	X_i	K_i	$(b_i/d_i)_{\exp}$	Other Specifications
1	1/3	T/P	3.0	Same as stated in Problem 3-11 plus
2	1/3	2T/P	0.9	the following specifications: $T_{1,\exp}$
3	1/3	3T/P	0.383636	$= 50°F$, and $T_{3,\exp} = 58°F$. Also, the
				total flow rates remain fixed at the
				following values: $V_1 = V_2 = V_3 =$
				100, $L_0 = L_1 = 50$, and $L_2 = 150$,
				where all flow rates are in lb-moles
				per hour.

On the basis of the following set of initial values of the variables, namely, $T_2 = 55°F$, $\bar{E}_1 = \bar{E}_2 = \bar{E}_3 = 1$, and $\beta_1 = \beta_2 = \beta_3 = 1$, compute the corresponding values of θ_i. Ans. $\theta_1 = 0.79200$, $\theta_2 = 1.0956$, $\theta_3 = 1.1343$.

9-7. To evaluate the partial derivatives in Equation (9-24), the following numerical procedure is recommended in Reference (16).

$$\frac{\partial \log_e \theta_i}{\partial \bar{E}_i} = \frac{\log_e \dfrac{\theta_{i,p}}{\theta_{i,n}}}{p} \tag{A}$$

where $\theta_{i,p}$ is the value of θ_i found by repeating the component-material balances with $\bar{E}_i$ replaced by $\bar{E}_i + p$. For the first and second trials through the column, p is taken equal to 0.1, and for the third and each trial thereafter, use the value p given by

$$p = \sum_{i=1}^{c} X_i |\log_e \theta_{i,n}| \tag{B}$$

provided that $10^{-3} \leq p \leq 0.1$, where $\theta_{i,n}$ is the value of the θ_i's obtained by use of the nth set of values of the variables. If the value of p given by Equation (B) is greater than 0.1, take $p = 0.1$. A value of $p < 10^{-3}$ indicates that convergence has been obtained.

(a). On the basis of the set of vaporization efficiencies $\bar{E}_1 = \bar{E}_2 = \bar{E}_3 = 1.1$ and the intial set of β_j's, $\beta_1 = \beta_2 = \beta_3 = 1.0$, compute $\theta_{i,p}$ and the partial derivatives as given by Equation (A).

Ans. $\theta_{1,p} = 0.97721$, $\theta_{2,p} = 1.35078$; $\theta_{3,p} = 1.39544$,

$$\frac{\partial \log_e \theta_1}{\partial \bar{E}_1} = 2.1014; \qquad \frac{\partial \log_e \theta_2}{\partial \bar{E}_2} = 2.0937; \qquad \frac{\partial \log_e \theta_3}{\partial \bar{E}_3} = 2.0716$$

(b). Calculate the improved sets $\{\bar{E}_i\}$ and $\{\beta_1, \beta_3\}$ to be used for the next trial through the column. To reduce the labor involved for this particular example, omit the repetition of Step 2 called for in Step 5 of the calculational procedure.

Ans. $(\bar{E}_1)_{ca} = 1.1110$, $(\bar{E}_2)_{ca} = 0.95639$, $(\bar{E}_3)_{ca} = 0.93916$, $(\beta_1)_{ca} = 1.0036$, $(\beta_3)_{ca} = 1.050$, from which the following normalized values are obtained: $\bar{E}_1 = 1.1405$, $\bar{E}_2 = 0.9818$, $\bar{E}_3 = 0.9642$, $\beta_1 = 1.0265$, $\beta_3 = 1.0228$.

9-8. Suppose that the inequalities of the fourth condition of Illustrative Example 9-1 are reversed and v_1 and v_2 are positive constants less than unity. If the other conditions remain unchanged, show that the overall mass transfer coefficient exists.

9-9. Suppose that for some component i, maxima in the x_i and y_i profiles occur at some z_k in the open interval $z_j < z_k < z_{j+1}$, and that throughout the closed interval $z_j \leqq z \leqq z_{j+1}$, $K_i x_i > y_i$. Further suppose that by chance the initial mole fraction $x_{j-1,i}$ (at z_j) is equal to the final mole fraction x_{ji} (at z_{j+1}). Show that the integral of the overall mass transfer coefficient over the closed interval $z_j \leqq z \leqq z_{j+1}$ has the value zero.

9-10. Suppose that the results of a field tests listed under "Existence of a Unique Set of Vaporization Plate Efficiencies" are available for a completely defined column. (a) Solve the component-material balances explicitly for the E_{ji}'s. (b) Show that this set of E_{ji}'s satisfy the bubble point function for each plate.

NOTATION

a = interfacial area for mass transfer in ft² of surface per ft³ of empty column; interfacial area for heat transfer in ft² of surface per ft of packing. (Although these areas are not necessarily equal, the same symbol is used for each in the interest of simplicity.)

A_{ji} = adsorption factor; defined by Equation (9-18).

b_i = flow rate of component i in the bottoms, moles/unit time.

B = total flow rate of bottoms, moles/unit time.

d_i = flow rate of component i in the distillate. For a partial condenser, $d_i = v_{oi}$, moles/unit time.

D = total flow rate of the distillate, moles/unit time.

e_j = heat transfer efficiency for the element of packing Δz_j; defined by Equation (9-15).

E_i = vaporization point efficiency; defined by Equation (9-1).

$\bar{E}_i$ = component efficiency; defined by Equation (9-22).

E_{ji} = vaporization plate efficiency; defined by Equation (9-2).

E_{Mji} = Murphree plate efficiency; defined by Equation (9-40).

E_{ji}^M = modified Murphree plate efficiency; defined by Equation (7-17).

f_j = temperature function for the jth mass transfer section; defined by Equation (9-16).

f_{ji}^V, f_{ji}^L = fugacities of pure component i in the vapor and liquid streams, evaluated at the total pressure and temperature at which these respective streams leave the jth mass transfer section.

K_{ji}^L = equilibrium vaporization constant; evaluated at the temperature and pressure of the liquid leaving the jth mass transfer section.

K_{Mji} = value of K_{ji} when evaluated at the bubble point temperature of a liquid having the composition $\{x_{ji}\}$.

l_{ji} = flow rate at which component i in the liquid phase leaves the jth mass transfer section, moles/unit time.

l_{0i} = flow rate of component i in the liquid reflux, moles/unit time.

l_{Fi}, v_{Fi} = flow rates of component i in the liquid and vapor parts, respectively, of a partially vaporized feed. For bubble point liquid and subcooled feeds, $l_{Fi} = FX_i$ and $v_{Fi} = 0$. For dew point vapor and superheated feeds, $v_{Fi} = FX_i$ and $l_{Fi} = 0$, moles/unit time.

L_j = total flow rate at which liquid leaves the jth mass transfer section, moles/unit time.

N = total number of elements of packing.

N_i = moles of component i transferred from the liquid to the vapor phase per unit time per unit length of packing at any $z(z_1 \leqq z \leqq z_{N+1})$.

N_1, N_2 = fixed constants.

S = internal cross-sectional area of the column.

S_{ji} = $(E_{ji}\gamma_{ji}K_{ji}V_j)/L_j$, stripping factor for component i; evaluated at the conditions of the liquid leaving the jth mass transfer section.

T_j^L, T_j^V = temperature of the vapor and liquid streams leaving the jth mass transfer section. When these temperatures are taken to be equal, the subscripts are omitted.

v_{ji} = flow rate at which component i in the vapor phase leaves the jth mass transfer section moles/unit time.

V_j = total flow rate of vapor leaving the jth mass transfer section moles/unit time.

x_{ji} = mole fraction of component i in the liquid leaving the jth mass transfer section.

x_{Bi} = mole fraction of component i in the bottoms.

X_i = total mole fraction of component i in the feed, regardless of state.

X_{Di} = total mole fraction of component i in the distillate, regardless of state.

y_{ji} = mole fraction of component in the vapor leaving the jth mass transfer section.

y_{ji}^* = mole fraction of component i in the vapor as computed by use of the temperature convention given by Equation (9-41).

$Y_{ji} = \gamma_{ji}K_{ji}x_{ji}$; γ_{ji} activity coefficient for component i and element Δz_j.

z = depth of packed bed, measured down from the top of the column.

Greek letters

α_{ji} = relative volatility, $\alpha_{ji} = K_{ji}/K_{jb}$.

β_j = mass transfer factor for the jth mass transfer section.

$\gamma_{ji}^L, \gamma_{ji}^V$ = activity coefficients for component i in the liquid and vapor phases; evaluated at the respective conditions of these phases.

θ_i = multiplier for each component i; defined by Equation (9-23).

ν_1, ν_2 = positive constants equal to or greater than unity.

Subscripts

ca = calculated value.

exp = experimental value.

f = first element of packing below the entrance of the feed.

F = variables associated with a partially vaporized feed.

i = component number ($1 \leq i \leq c$).

j = integer for numbering the mass transfer sections of the column; the first increment of packing at the top of the column is assigned the number 1 and the last increment at the bottom, the number N. At the top of the column where the packing begins $z = z_1$, and at the bottom where the packing ends, $z = z_{N+1}$.

k = general counting integer.

N = total number of elements of packing.

Superscripts

I = to be evaluated at the conditions of the interface.

L = to be evaluated at the bulk conditions of the liquid phase.

$\mathcal{L}$ = to be evaluated at the limiting conditions of the interface when approached from the interior of the liquid phase.

V = to be evaluated at the bulk conditions of the vapor phase.

$\mathcal{V}$ = to be evaluated at the limiting conditions of the interface when approached from the vapor phase.

Mathematical symbols

$\prod_{j=1}^{N} x_j$ = product of the quantities x_j ($1 \leq j \leq N$).

$\sum_{i=1}^{c} x_i$ = sum over all values x_i ($1 \leq i \leq c$).

$\{x_j\}$ = set of all values x_j belonging to the particular set under consideration.

REFERENCES

1. Amundson, N. R., and A. J. Pontinen, "Multicomponent Distillation Calculations on a Large Digital Computer," *Ind. Eng. Chem.*, **50**, (1958), 730.

2. Bassyoni, A. A., "Formulation of a Model for Packed Distillation Columns on the Basis of Field Tests," Ph. D. Dissertation, Texas A&M University, College Station, Texas, 1969.

3. ——, R. McDaniel, and C. D. Holland, "Examination of the Use of Mass Transfer Rate Expressions in the Description of Packed Distillation Columns-II," *Chem. Eng. Sci.*, **25**, (1970), 437.

4. Billingsley, D. S., "On the Equations of Holland in the Solution of Problems in Multicomponent Distillation," *IBM J. Res. Develop.*, **14**, (1970), 33.

5. Carey, J. S., "Plate Efficiency of Bubble Cap Rectifying Columns," Sc. D. Thesis, Mass. Inst. Tech., 1930.

6. *Chemical Engineer's Handbook*, 3rd ed., J. H. Perry (ed). New York: McGraw-Hill Book Co., Inc., 1950, p. 582.

7. Chilton, T. H., and A. P. Colburn, "Distillation and Absorption in Packed Columns—A Convenient Design and Correlation Method," *Ind. Eng. Chem.*, **27**, (1935), 255.

8. Davis, Parke, D. L. Taylor, and C. D. Holland, "Determination of Plate Efficiencies from Operation Data," Part II, *A. I. Ch. E. J.*, **11**, (1965), 678.

9. Denbigh, Kenneth, *The Principles of Chemical Equilibrium*. New York: Cambridge University Press, 1955.

10. Eckert, J. S., "A New Look at Distillation-4 Tower Packings, Comparative Performance," *Chem. Eng. Progr.*, **59**, (1963), 76.

11. Edmister, W. C., "Hydrocarbon Adsorption and Fractionation Process Design Methods-Part 18, Plate Efficiency," *Petroleum Engineer*, **21**, 1 (1949), C-45.

12. Greenstadt, John, Yonathan Bard, and Burt Morse, "Multicomponent Distillation Calculations on the IBM 704," *Ind. Eng. Chem.*, **50**, (1958), 1644.

13. Hanson, D. N., Duffin, J. H., and G. F. Somerville, *Computation of Multistage Separation Processes*. New York: Reinhold Publishing Corp., 1962.

14. Hausen, H., "The Definition of the Degree of Exchange on Rectifying Plates for Binary and Ternary Mixtures," *Chem. Eng. Tech.*, **25**, (1953), 747.

15. Holland, C. D., *Multicomponent Distillation*. Englewood Cliffs, N. J.: Prentice-Hall, Inc., 1963.

16. ———, *Unsteady State Processes with Applications in Multicomponent Distillation*. Englewood Cliffs, N. J.: Prentice-Hall, Inc., 1966.

17. ———, and N. E. Welch, "Steam Batch Distillation Calculation," *Petroleum Refiner*, **36**, 5 (1957), 251.

18. ———, and K. S. McMahon, "Comparison of Vaporization Efficiencies with Murphree-Type Efficiencies in Distillation-I," *Chem. Eng. Sci.*, **25**, (1970), 431.

19. Lewis, W. K., "Oil Absorption of Natural Gasoline," *Trans. Am. Inst. Chem. Engrs.*, **20**, (1927), 1.

20. Lewis, W. K., and W. G. Whitman, "Principles of Gas Absorption," *Ind. Eng. Chem.*, **16**, (1924), 1215.

21. McDaniel, R., A. A. Bassyoni, and C. D. Holland, "Use of the Results of Field Tests in the Modeling of Packed Distillation Columns and Packed Absorbers-III," *Chem. Eng. Sci.*, **25**, (1970), 633.

22. Murphree, E. V., "Rectifying Column Calculations," *Ind. Eng. Chem.*, **17**, (1925), 747.

23. Nartker, T. A., J. M. Srygley, and C. D. Holland, "Solution of Problems Involving Systems of Distillation Columns," *Can. J. Chem. Engr.*, **44**, (1966), 217.

24. Peters, W. A., Jr., "The Efficiency and Capacity of Fractionating Columns," *Ind. Eng. Chem.*, **14**, (1922), 476.

25. Petryschuk, W. R., and A. I. Johnson, "Simulation of the Steady State Behavior of a Multicomponent Multifeed Reboiled-Absorber," *Can. J. of Chem. Eng.*, **43**, (1965), 209.

26. Ravicz, A. E., "Non-Ideal State Multicomponent Absorber Calculations by Automatic Digital Computer," Ph. D. Dissertation, University of Michigan, Ann Arbor, Michigan, 1958.

27. Rose, A., R. F. Sweeny, and V. N. Schrodt, "Continuous Distillation Calculations by Relaxation Method," *Ind. Eng. Chem.*, **50**, (1958), 737.

28. Rubac, R. E., "Determination of Vaporization Efficiencies for Packed Columns at Steady State Operation," Ph. D. Dissertation, Texas A&M University, College Station, Texas, 1968.

29. ———, R. McDaniel, and C. D. Holland, "Packed Distillation Columns and Absorbers at Steady State Operation," *A. I. Ch. E. Journal*, **15**, (1969), 568.

30. Standart, G., "Studies on Distillation-V, Generalized Definition of a Theoretical Plate or Stage of Contacting Equipment," *Chem. Eng. Sci.*, **20**, (1965), 611.

31. Taylor, D. L., "Use of Plate Efficiencies in the Treatment of Conventional Columns," M. S. Thesis, Texas A&M University, College Station, Texas, 1962.

32. ———, Parke Davis, and C. D. Holland, "Determination of Plate Efficiencies from Operation Data," *A. I. Ch. E. Journal*, **10**, (1964), 864.

33. Walker, W. H., W. K. Lewis, and W. H. McAdams. *Principles of Chemical Engineering*, 2nd ed. New York: McGraw-Hill, 1927.

34. Whitman, W. G., "The Two-Film Theory of Gas Absorption," *Chem. & Met. Eng.*, **29**, (1923), 146.

Use of the Results of Field Tests in the Modeling of Packed Distillation Columns and Packed Absorbers

10

The modeling procedure that was developed in Chapter 9 is demonstrated by modeling a packed distillation column and a packed absorber. These columns were located at the Zoller Gas Plant near Refugio, Texas. The modeling procedure utilized for these two columns makes use of the results of a series of field tests. This particular modeling procedure was selected because it was free of the problems involved in the use of other possible procedures (2, 6) and because its order of approximation was consistent with the data available.

In Part 1, the modeling of the packed distillation is described, and in Part 2 the modeling of the packed absorber is described.

Part 1 : Modeling of a Packed Distillation Column

The packed distillation column used in the field tests is shown in Figure 10-1. The plant shown in this photograph was used to recover propane and heavier hydrocarbons from feed stocks consisting of natural gas (1, 8, 9, 14).

Figure 10-1. The Zoller Gas Plant. [Taken from R. McDaniel, A. A. Bassyoni, and C. D. Holland, "Use of the Results of Field Tests in the Modeling of Packed Distillation Columns and Packed Absorbers III," *Chem. Eng. Sci.*, **25**, (1970), 634.]

A sketch of the packed distillation column is shown in Figure 10-2. The column consisted of two packed sections, a feed mixing section, a total condenser, and a reboiler. This column was designed by the Stearns-Roger Manufacturing Company. The upper section of the column had an inside diameter of 3 ft and the bottom section had an inside diameter of 4 ft. The packing consisted of 2-in. metallic Pall Rings. The upper section contained 3,340 lb, and the lower section contained 5,920 lb of packing.

The two packed sections were each 17 ft in length. A packing hold-down grating and a packing support grating were located at the top and bottom, respectively, of each of the packing sections. The feed mixing section consisted of the space between the packing support grating of the rectifying section (the upper section) and the hold-down grating for the packed stripping section (the lower section). As indicated in Figure 10-2, the feed mixing section contained the liquid distributor plate upon which the feed was introduced.

In addition to the temperatures of the terminal streams, the temperature profile for the packed sections was determined for each field test by means of 17 iron-constantan thermocouples (18 in. in length, which extended about 15 in. into the packing).

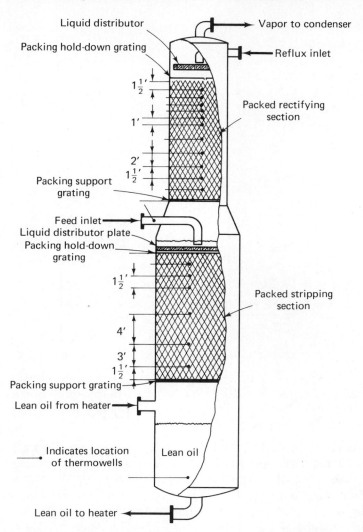

Figure 10-2. The packed distillation column at the Zoller Gas Plant. (Taken from R. McDaniel, A. A. Bassyoni, and C. D. Holland, "Use of the Results of Field Tests in the Modeling of Packed Distillation Columns and Packed Absorbers III," 634.)

Orifice meters were used to measure the flow rates of the distillate and the bottoms, and the feed rate was taken to be equal to the sum of these two rates. The reflux rate was measured by means of an orifice meter. For each run, samples were withdrawn from the feed, distillate, and bottoms lines. These samples were analyzed by means of a very sensitive gas chromatograph as described by Bassyoni (1) and McDaniel (9). The composition of the feed was computed on the basis of the analyses of the distillate and the bottoms and the respective total flow rates. The feed compositions so obtained were in good agreement with those determined by analyzing the feed sample.

A summary of the operating conditions for the runs made on the packed distillation column are presented in Table 10-1, and typical feed compositions are shown in Table 10-2. A complete listing of all experimental results obtained on the packed distillation column has been provided by Bassyoni (1).

TABLE 10-1
SUMMARY OF THE OPERATING CONDITIONS FOR THE FIELD TESTS
MADE ON THE PACKED DISTILLATION COLUMN (1, 10)

Field Test No.	Flow Rate lb-moles/hr			Temperature, °F				Pressure (psia)
	Feed	Reflux	Distil-late	Feed	Re-flux	Reboil-er	Over-head Vapor	
001	408.9	130.3363	161.6	358.3	82.0	493.0	176.0	165
002	411.3	184.5878	161.8	356.6	97.0	498.0	181.0	165
003	408.3	238.9106	161.3	360.0	107.0	499.0	180.0	165
101	405.4	182.9833	153.8	360.0	93.0	494.0	175.0	163
102	368.3	180.3270	146.0	356.0	99.0	490.0	174.0	165
103	373.3	180.1612	146.7	360.3	100.0	490.0	174.0	162
104	420.7	184.0207	155.2	364.5	109.0	490.0	175.0	165
105	405.6	183.4153	153.7	356.0	98.0	495.0	175.0	165
201	407.6	220.9822	155.8	351.8	90.0	494.0	170.0	176
202	390.8	221.6499	152.6	353.0	90.0	500.0	170.0	175
203	372.6	223.5266	151.6	342.5	90.0	494.0	170.0	175
204	425.8	227.1104	157.0	362.5	88.0	502.0	171.0	176
205	439.3	229.5392	157.4	364.5	88.0	502.0	171.0	176
301	438.1	149.0619	156.3	363.0	88.0	494.0	180.0	177
303	385.9	147.8764	147.9	355.0	89.0	500.0	179.0	175
304	370.9	147.3064	144.4	352.3	88.0	500.0	178.0	175
305	352.1	144.1691	138.0	375.8	88.0	500.0	180.0	175

TABLE 10-2
FEED RATES (lb-moles/hr)

Component	Field Test Number		
	001	101	201
C_2H_6	1.3496	0.8330	1.3222
C_3H_8	81.4189	78.4683	75.3087
$i\text{-}C_4H_{10}$	33.5991	31.1295	30.6927
$n\text{-}C_4H_{10}$	23.7421	24.4780	24.2832
$i\text{-}C_5H_{12}$	9.7414	8.1443	8.9942
$n\text{-}C_5H_{12}$	5.5896	4.7172	5.7121
$n\text{-}C_6H_{14}$	10.0958	6.9601	6.4633
$n\text{-}C_7H_{16}$	24.8791	32.7723	31.1026
$n\text{-}C_8H_{18}$	71.5403	90.5138	94.6198
$n\text{-}C_9H_{20}$	66.1016	68.6985	80.3184
$n\text{-}C_{10}H_{22}$	80.8034	58.7195	48.7687

Formulation of the model and the optimization functions for a packed distillation column

The packed rectifying and stripping sections were divided into elements as indicated in Figure 10-3. However, instead of Model 2 (see Figures 3-13 and 9-3), Rubac *et al.* (13, 14) found that the feed mixing section could be represented best by Model 1 shown in Figure 10-4. The behavior of the liquid

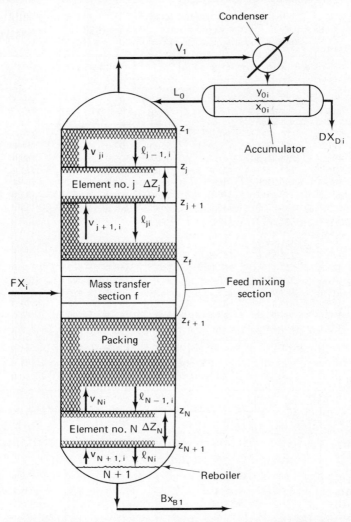

Figure 10-3. Sketch of a packed distillation column. (Taken from R. McDaniel, A. A. Bassyoni, and C. D. Holland, "Use of the Results of Field Tests in Modeling of Packed Distillation Columns and Packed Absorbers III," 638.)

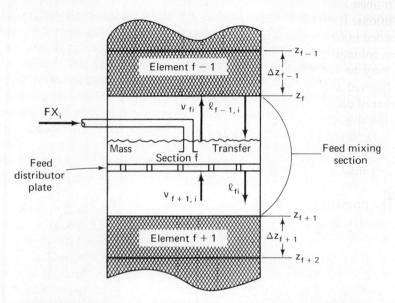

Figure 10-4. Model 1 of the feed mixing section for a submerged-feed entrance. (Taken from R. McDaniel, A. A. Bassyoni, and C. D. Holland, "Use of the Results of Field Tests in the Modeling of Packed Distillation Columns and Packed Absorbers III," 639.)

distributor plate of the feed mixing section was analyzed as follows. At relatively high flow rates where the outlet of the feed-entrance pipe (see Figure 10-4) is submerged, all of the feed tends to contact all of the liquid on the feed distributor plate. For this case, which is representative of the field tests, the feed distributor plate is perhaps best represented as a plate or mass transfer section as indicated in Figure 10-4. At relatively low flow rates, where the holdup on the feed distributor plate is low, the feed mixing section is perhaps best represented by a mixing zone without mass transfer as shown in Figure 9-3.

In the analysis of the runs on the distillation column, the reboiler and condenser duties were taken to be those computed by use of the customary energy balances as shown in Chapter 3. This procedure permitted errors in the enthalpies as well as certain other errors to be absorbed in the calculated values of the condenser and reboiler duties.

The enthalpy and K data used in all of the calculations for the packed distillation columns are presented in Tables C-1 and C-2 of Appendix C.

FORMULATION OF OBJECTIVE FUNCTIONS. Several conceivable methods for using the results of field tests to formulate suitable mathematical models for a packed distillation column exist. For example, consider the case in which the additional specifications consisting of the sets $\{T_j\}$ and $\{b_i/d_i\}$ have been determined experimentally as well as the usual specifications or operating conditions. It is evident from the description of the calculational procedure described in Chapter 9 that for each choice of $N \geqq 3$, an exact representation of the column exists. That is, for any given $N \geq 3$, a set of $\bar{E}_i$'s and a set of β_j's may be found such that a one-to-one correspondence exists between all observed and calculated values of the variables (5). Thus, in general, the number of packed sections may be picked as required to best satisfy a particular objective. For example, in the formulation of a model for the purpose of process control, the objective might well be to minimize the number of equations required to describe the column, in which case the number of packed sections may be taken as two, one above and one below the feed distributor plate.

The primary objective of the work described herein was the formulation of a model, possessing relatively few parameters, which could be used to predict accurate product distributions and terminal temperatures over wide ranges of operating conditions. In an effort to reduce the total number of parameters, the basic procedure for the determination of the E_{ji}'s was modified so that any number of β_j's could be set equal to unity (see Chapter 9). In the applications which follow, Modification 1, which involves the determination of β_1 and β_{N+1} is employed.

The problem then consists of finding the feed distributor plate location f, the number of mass transfer sections N, the sets $\{\bar{E}_i\}$ and $\{\beta_j\}$ such that the deviation of each calculated product distribution b_i/d_i for its corresponding experimental value $(b_i/d_i)_{\exp}$ is minimized. In addition, it is desired that the deviation of the calculated values of the terminal temperatures from the experimental values be minimized, although the deviations of these temperatures do not appear explicitly in the determination of the β's. For any given run, the objective function to be minimized is given by

$$0_1(f, N, \{\bar{E}_i\}, \{\beta_j\}) = \frac{1}{c} \sum_{i=1}^{c} |\log_e \theta_i| \qquad (10\text{-}1)$$

where

$$\theta_i = \frac{\left(\dfrac{b_i}{d_i}\right)_{\exp}}{\left(\dfrac{b_i}{d_i}\right)_{\text{ca}}}$$

The symbol $(b_i/d_i)_{\text{ca}}$ refers to the product distribution computed for component i by use of the calculational procedure described in Chapter 3. Note that for any given set $\{f, N\}$, there exist a set of $\bar{E}_i$'s and a set of β_j's such that $0_1 = 0$. Since there exists infinitely many sets $\{f, N\}$, it follows that the function 0_1 has infinitely many zeroes. However, this statement is not necessarily true if it is required that the same values of the arguments of 0_1 be used over all runs R, which represents the problem of interest. That is, the objective is to find the sets $\{f, N\}, \{\bar{E}_i\}, \{\beta_j\}$ such that the deviation of each calculated product distribution from its corresponding experimental value is minimized. This objective is represented by the averages of 0_1 over all runs R,

$$\bar{0}_1(f, N, \{\bar{E}_i\}, \{\beta_j\}) = \frac{1}{Rc} \sum_{r=1}^{R} \sum_{i=1}^{c} |\log_e \theta_{ir}| \qquad (10\text{-}2)$$

where the subscript r denotes the run number. The particular form of the argument ($|\log_e \theta_{ir}|$) was selected because it assigns the same weight to deviations resulting from values of $\theta_{ir} < 1$ as it does to deviations resulting from values of $\theta_{ir} > 1$. The problem of finding the sets $\{\bar{E}_i\}, \{\beta_j\}$, and $\{f, N\}$ for which $\bar{0}_1$ takes on its minimum value is recognized as a problem in optimization. Such a problem may be solved by use of multivariable-sequential search techniques (7, 16), but, such a search was considered and regarded as impractical because of the large amount of computer time that would have been required to find the global minimum (the smallest of all possible minima) of $\bar{0}_1$. Instead of searching sequentially over all variables, the function $\bar{0}_1$ was searched over all f and N at perfect plates or perfect mass transfer sections ($E_{ji} = 1$ for all i and j). Then on the basis of the solution set $\{f, N\}$ so obtained, the $\bar{E}_i$'s and β_j's, one set of $\bar{E}_i$'s and one set of β_j's, were computed by means of an appropriate averaging procedure developed below. Although the approach described may not produce the global minimum in $\bar{0}_1$, this disadvantage is more than offset by the fact that both 0_1 and $\bar{0}_1$ were unimodal (possessed a single minimum) at perfect mass transfer sections. The desire to choose an objective function that always has one and only one minimum with respect to f and N stems from the fact that commercial packings are commonly rated with respect to their "equivalent number of theoretical plates."

Actually, the combination of the objective function $\bar{0}_1$ and the decision to search it over f and N at perfect mass transfer sections was only one of the many objective functions considered. The possibility of using other objective functions results from the fact that the θ_i's, $\bar{E}_i$'s, β_j's, and T_j's are related, which is demonstrated by an analysis that follows the one presented by McDaniel et al. (10).

For a distillation column equipped with a total condenser at the operating condition of total reflux, the product distribution $\{b_i/d_i\}$ computed on the basis of perfect mass transfer sections ($E_{ji} = 1$ for all i and j) is given by the Fenske equation,

$$\left(\frac{b_i}{d_i}\right)_{\text{ca}} = \frac{\dfrac{B}{D}}{\displaystyle\prod_{j=1}^{N+1} K_{ji}} \qquad (10\text{-}3)$$

which is readily obtained from Equation (3-98). When vaporization efficiencies are determined such that the additional specifications are satisfied, the corresponding expression for the product distribution at total reflux is given by

$$\left(\frac{b_i}{d_i}\right)_{\exp} = \frac{\dfrac{B}{D}}{\displaystyle\prod_{j=1}^{N+1} E_{ji}(K_{ji})_{\exp}} = \frac{\dfrac{B}{D}}{\bar{E}_i^{N+1}\displaystyle\prod_{j=1}^{N+1}(K_{ji})_{\exp}} \tag{10-4}$$

The final expression on the right-hand side of Equation (10-4) follows as a result of the normalization procedure which requires that the product of the β_j's over all j be equal to unity [see Equations (9-27) and (9-28)]. The subscript "exp" on the K_{ji}'s is used to denote the fact that the K_{ji}'s are to be evaluated at the experimentally determined temperatures where a complete set of β_j's is computed. If only one or two β's are determined, then the subscript "exp" on the K_{ji}'s refers to the values of these variables as determined by the proposed calculational procedures [see Modification 1 of the procedure for the determination of the E_{ji}'s, Chapter 9].

For any one run, let the members of Equation (10-4) be divided by the corresponding members of Equation (10-3). The result so obtained may be stated in logarithmic form as follows:

$$\log_e \theta_i = -(N+1)\log_e \bar{E}_i + \log_e \prod_{j=1}^{N+1}\frac{(K_{ji})}{(K_{ji})_{\exp}} \tag{10-5}$$

As outlined in Problems 10-2 and 10-3, Equation (10-5) may be used to relate 0_1 and $\bar{0}_1$ to other potential objective functions, namely,

$$\bar{0}_2 = \frac{1}{Rc}\sum_{i=1}^{c}\left|\sum_{r=1}^{R}\log_e \theta_{ir}\right| \tag{10-6}$$

$$\bar{0}_3 = \left(\frac{N+1}{Rc}\right)\sum_{i=1}^{c}\sum_{r=1}^{R}|\log_e \bar{E}_{ir}| \tag{10-7}$$

$$\bar{0}_4 = \left(\frac{N+1}{Rc}\right)\sum_{i=1}^{c}\left|\sum_{r=1}^{R}\log_e \bar{E}_{ir}\right| \tag{10-8}$$

The expressions for $\bar{0}_2$ and $\bar{0}_4$ may be restated in terms of the geometric means as follows:

$$\bar{0}_2 = \frac{1}{c}\sum_{i=1}^{c}|\log_e \theta_{im}| \tag{10-9}$$

$$\bar{0}_4 = \frac{N+1}{c}\sum_{i=1}^{c}|\log_e \bar{E}_{im}| \tag{10-10}$$

where

$$\theta_{im} = \sqrt[R]{\theta_{i,1}\theta_{i,2}\ldots\theta_{i,R}}$$

$$\bar{E}_{im} = \sqrt[R]{\bar{E}_{i,1}\bar{E}_{i,2}\ldots\bar{E}_{i,R}}$$

The objective function $\bar{0}_1$ was used in the modeling procedure proposed in the next section. In order to obtain an average value of $\bar{E}_i$ over all runs for a given component i, the geometric mean implied by the function $\bar{0}_4$ was used in the modeling procedure described below. However, the use of $\bar{0}_3$ and $\bar{0}_4$

as objective functions in the modeling procedure was ruled out because their use led to the subsequent multimodal behavior in 0_1 for the absorber tests (9).

In addition to the functions $\bar{0}_3$ and $\bar{0}_4$, several other objective functions involving either the efficiencies or the temperatures were investigated (1, 9, 10, 13). The functions based on the efficiencies exhibited a single minimum for each of the field tests made on the packed distillation column (1). However, since this class of functions failed to exhibit unimodal behavior for the absorber runs (9), they were discarded as potential functions for the general modeling of packed columns. Although the objective functions based on the deviation of the experimental from the calculated temperatures exhibited unimodal behavior for all runs for both the packed distillation column and packed absorber (13), they were discarded for general modeling purposes because temperature profiles are not generally available for commercial columns.

Of the possible methods considered for modeling packed columns, the following one appeared the most promising.

PROCEDURE USED TO MODEL THE PACKED DISTILLATION COLUMN. The first step in the proposed modeling procedure used for packed distillation columns consisted of a logical extension of the concept of the "height equivalent to a theoretical plate" (called HETP), proposed by Peters (12), to columns in the process of separating multicomponent mixtures. For such a column, a set $\{f, N\}$ of perfect mass transfer sections does not necessarily exist such that all calculated and observed product distributions may be placed in a one-to-one correspondence. For any given set f and N, the objective functions 0_1 and $\bar{0}_1$ are seen to given an adequate measure of the deviations of the calculated from the experimental product distributions for all components. The first step of the modeling procedure for a packed distillation column is to find that pair of values f and N for which $\bar{0}_1 (f, N, \{E_{ji} = 1\})$ takes on its minimum value. Both of the function 0_1 and $\bar{0}_1$ exhibited unimodal behavior with well-defined minima as shown in Figure 10-5 for packed distillation columns and in Figure 10-10 for packed absorbers. The K-values given in Table C-1 and the enthalpies given in Table C-2 of Appendix C were used in the modeling of the packed distillation column.

Values of $\bar{0}_1 (f, N, \{E_{ji} = 1\})$ in a small neighborhood containing the optimum solution set ($f = 6, N = 11$) are presented in Table 10-3. In this work, a simultaneous search as indicated in Table 10-7 was employed in order to determine the general behavior of the objective functions over the domain of the independent variables f and N. However, since the functions were observed to be unimodal, many techniques (16) for searching $\bar{0}_1 (f, N, \{E_{ji}=1\})$ exist. For example, the method of Hooke and Jeeves (7) may be applied in a manner similar to that described by Srygley et al. (15).

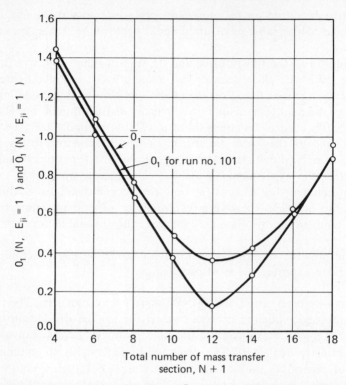

Figure 10-5. Variation of the 0_1 and $\bar{0}_1$ with the total number of perfect mass transfer sections and at equal numbers of packed sections above and below the feed-distributor plate of the distillation column. (Taken from R. McDaniel, A. A. Bassyoni, and C. D. Holland, "Use of the Results of Field Tests in the Modeling of Packed Distillation Columns and Packed Absorbers III," 645.)

The second step of the modeling procedure consisted of the determination of the vaporization efficiencies for each run at the optimum set $\{f, N\}$, found in step one, by use of any one of the procedures described in Chapter 9 for the determination of the E_{ji}'s. Since the E_{ji}'s place the calculated and observed product distributions in a one-to-one correspondence, the second step has the effect of driving the objective function $0_1 (f, N)$ to zero for each run. However, because of the large number of parameters $\{E_{jir}\}$ resulting from the second step, a procedure was needed for averaging the efficiencies in some appropriate manner.

The third step of the modeling procedure consisted of averaging of the $\bar{E}_i$'s as well as the β_j's and checking for the dependency of the efficiencies upon the operating conditions. The geometric mean was used to compute the average of the $\bar{E}_i$'s as indicated below Equation (10-10). The set $\{\bar{E}_{im}\}$ of geometric mean efficiencies over all runs R for the packed distillation column are presented in Table 10-4. The geometric mean values of the β's are also

TABLE 10-3

VALUES OF THE OBJECTIVE FUNCTION $\bar{0}_1(f, N, \{E_{ji} = 1\})$
IN THE NEIGHBORHOOD OF THE SOLUTION SET $\{f = 6, N = 11\}$ (1, 10)

Feed Plate f	Number* of Sections N	$\bar{0}_1(f, N, \{E_{ji} = 1\})$
2	3	1.4544
3	5	1.0878
4	7	0.7675
5	9	0.4884
6	11	0.3459†
7	13	0.4223
8	15	0.6212
9	17	0.8798
7	12	0.3749
6	10	0.4652
6	12	0.3832
5	10	0.3769

*N = number of packed sections plus the feed distributor plate.
†Minimum value of the objective function.

TABLE 10-4

GEOMETRIC MEAN VALUES OF THE EFFICIENCIES OVER ALL
RUNS MADE ON THE PACKED DISTILLATION COLUMN (1, 10)

Component	Modification 1 $(f = 6, N + 1 = 12)$ $\bar{E}_{im}$
C_2H_6	1.0043
C_3H_6	0.9869
$i\text{-}C_4H_{10}$	1.0216
$n\text{-}C_4H_{10}$	1.1693
$i\text{-}C_5H_{12}$	1.1972
$n\text{-}C_5H_{12}$	1.1908
C_6H_{14}	1.2169
C_7H_{16}	1.0504
C_8H_{18}	1.0754
C_9H_{20}	1.0099
$C_{10}H_{22}$	1.0448

$$\beta_1 = 1.1077$$
$$\beta_{N+1} = 0.9028$$

presented in Table 10-4. When the β's are averaged in this particular way, they are consistent with the geometric average of the $\bar{E}_i$'s.

To check for the significance of the $\bar{E}_{im}$'s (averaged over all runs R) in the reduction of 0_1, the product distributions for each run were recomputed by use of the $\bar{E}_{im}$'s and at the optimum set $\{f, N\}$ found in step one. These product distributions were then used in the reevaluation of $\bar{0}_1$, and the result obtained was

$$\bar{0}_1(6, 11, \{\bar{E}_{im}\}, \{\beta_{jm}\}) = 0.1618$$

where $\beta_1 = 1.1077$, $\beta_{N+1} = 0.9028$, and the set $\{\bar{E}_{im}\}$ is given in Table 10-3.

The minimum value obtained in step one for perfect mass transfer sections was

$$\bar{0}_1(6, 11, \{E_{ji} = 1\}) = 0.3459$$

Since the set of $\bar{E}_{im}$'s gave approximately a 50% reduction in 0_1, it is evident that the set of mean E_{im}'s and β_{jm}'s over all runs R provided a significant improvement in the calculated values of the product distributions over those predicted on the basis of perfect plates.

Next, a test for the possible effect of the operating conditions on the $\bar{E}_i$'s was made in the following manner. After the runs had been divided into three subsets in accordance with the reflux ratios employed, sets of geometric mean values of $\bar{E}_i$'s and β_j's were computed for each subset of runs. Product distributions were redetermined for each subset of runs on the basis of the corresponding set of efficiencies ($\{\bar{E}_{im}\}$, $\{\beta_{1m}, \beta_{N+1, m}\}$), and the product distributions so obtained were used to evaluate $\bar{0}_1$ for each subset of runs. The results given in Table 10-5 show that no significant improvement in the calculated product distribution was achieved by using a different set of efficiencies for each of the three sets of runs. This result implies that over the range of refluxes investigated, the efficiencies were independent of reflux ratio. This same conclusion was reached by Bassyoni (1) upon making an F-test on the $\bar{E}_i$'s.

A comparison of the calculated and observed product distributions is presented in Figure 10-6. The calculated product distributions were based on the mean $\bar{E}_i$'s and β_j's given in Table 10-4 under Modification 1. A comparison of the average deviation of the calculated from the experimental values for the terminal temperatures for the distillation column are presented in Table 10-6. The β_{jm}'s for the terminal sections ($j = 1, N + 1$) together with the $\bar{E}_{im}$'s reduced the deviations obtained for perfect sections by factors ranging from about 2 to 6.

Although the proposed modeling procedure may not yield the global

TABLE 10-5

VALUES OBTAINED FOR THE OBJECTIVE FUNCTION $\bar{0}_1$ FOR EACH OF FOUR SETS OF RUNS BY USE OF THE CORRESPONDING SETS OF $\bar{E}_i$'S AND β_j'S DETERMINED BY MODIFICATION 1 (1, 10)

Reflux Ratio L_0/D	No. Runs in Set	No. Runs Used to Compute $\{\bar{E}_{im}\}$ and $\{\beta_{1m}, \beta_{N+1, m}\}$	$\bar{0}_1$
—	17	17	0.1618
0.8	5	5	0.3155
1.2	6	6	0.0280
1.4	6	6	0.1675
—	Average*	—	0.1563

*In the computation of the average of $\bar{0}_1$ for the three subsets of runs (at $L_0/D = 0.8, 1.2,$ and 1.4), the value of $\bar{0}_1$ for each subset was weighted according to the number of runs in the subset.

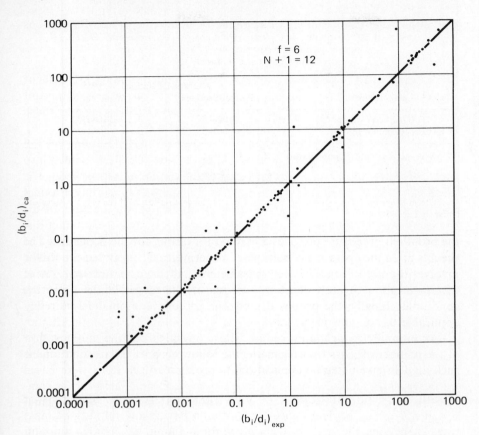

Figure 10-6. Comparison of the product distributions computed on the basis of mean values $\bar{E}_{i,m}$ and $\beta_{j,m}$ [listed under modification (1) Table 10-4] with the experimental product distributions for the packed distillation column. (Taken from R. McDaniel, A. A. Bassyoni, and C. D. Holland, "Use of the Results of Field Tests in the Modeling of Packed Distillation Columns and Packed Absorbers III," 648.)

minimum in $\bar{0}_1$, this deficiency is counterbalanced many times by the very definite unimodal behavior by 0_1 for each run and by $\bar{0}_1$ for all runs for both packed distillation columns and packed absorbers (see Part 2) when perfect mass transfer sections are employed as described in step one. Brosilow *et al.* (3) proposed a method, somewhat similar to the one described in step one except that instead of perfect mass transfer sections an approximation of the mass transfer rate expression was employed.

In summary, the proposed modeling procedure makes use of objective functions 0_1 $(f, N, \{E_{ji} = 1\})$ and $\bar{0}_1$ $(f, N, \{E_{ji} = 1\})$ which appear to always exhibit unimodal behavior. A suitable method for averaging the $\bar{E}_i$'s and β_j's over each subset or over all runs is provided by the proposed model. Also,

TABLE 10-6
AVERAGE DEVIATION OF THE CALCULATED* FROM THE EXPERIMENTAL VALUES
OF THE TERMINAL TEMPERATURES T_1 AND T_{N+1} OF THE
DISTILLATION COLUMN (1, 10)

Temperature	Average Deviation*	
	Perfect Sections	Modification 1†
T_1 (Overhead vapor)	0.02695	0.00792
T_{N+1} (Reboiler)	0.00785	0.00129

*Average deviation of temperature T_j ($j = 1, N + 1$):

$$\frac{1}{R} \sum_{r=1}^{R} \left| 1 - \frac{(T_{jr})_{ca}}{(T_{jr})_{exp}} \right|$$

†The efficiencies ($\{\bar{E}_{im}\}$ and β_{1m}, $\beta_{N+1,m}$) used in the calculation of these temperatures are listed in Table 10-4.

the proposed procedure provides a method for testing for the dependency of the $\bar{E}_i$'s upon the operating conditions. Furthermore, the proposed procedure involves the use of relatively few parameters, and these parameters appear to remain approximately constant over reasonable ranges of the operating conditions. Finally, the proposed modeling procedure is obviously directly applicable to columns with plates.

An alternate modeling procedure whose objective was the minimization of the number of mass transfer merits the following brief discussion because such models have potential applications in process control.

MINIMIZATION OF THE NUMBER OF MASS TRANSFER SECTIONS. Since sets of $\bar{E}_i$'s and β_j's may be determined for a column having any number of mass transfer sections, they may be determined for a column having the minimum number of packed sections, one packed section above and one packed section below the feed distribution plate. Such a column consists of a condenser ($j = 0$), one rectifying mass transfer section ($j = 1$), the feed distributor plate ($j = 2$), a stripping mass transfer section ($j = 3$), and a reboiler ($j = 4$). The corresponding E_i's and β_j's for Run 102 [obtained by Rubac (14)] are presented in Table 10-7. The quantities were computed by use of the general procedure described in Chapter 9.

Part 2: Modeling of a Packed Absorber

A sketch of the packed absorber used to make the field tests is shown in Figure 10-7. This absorber consisted of the following components: A Metex demister, a liquid distributor, a packing hold-down grating, two-in. metallic Pall Rings, a packing hold-up grating, and a liquid draw-off tray. The inside diameter of the absorber was 36 in., and the net length of the shell was 38 ft and 4 in. It was designed by the Stearns-Roger Manufacturing Company.

TABLE 10-7
COMPONENT EFFICIENCIES AND PLATE FACTORS FOR A COLUMN
WITH THE MINIMUM NUMBER OF MASS TRANSFER SECTIONS FOR RUN 102 (1)

Component	$\bar{E}_i{}^*$	Section Number	$\beta_j{}^*$
C_2H_6	1.002†	1	0.276
C_3H_8	2.922†	2	0.998
$i\text{-}C_4H_{10}$	2.525†	3	1.539
$n\text{-}C_4H_{10}$	3.219	4	2.357
$i\text{-}C_5H_{12}$	3.085		
$n\text{-}C_5H_{12}$	2.479		
C_6H_{14}	1.476		
C_7H_{16}	0.911		
C_8H_{18}	0.299		
C_9H_{20}	0.216†		
$C_{10}H_{22}$	0.228†		

*These values were obtained for the following conditions $N + 1 = 4$, $f = 2$, and T_1 and T_3 were taken equal to those at the midpoints of the two sections.

†Denotes components with unknown product distributions.

The 23-ft section between the packing hold-down and the packing hold-up gratings contained 4,900 lb of metallic Pall Rings.

Originally, the bottom section of the absorber in Figure 10-7 was used as a conventional one-stage separator for the purpose of separating the rich gas stream from a mixture of the rich gas, condensate, and ethylene glycol. Since the rich gas stream was presumed to have achieved a state of equilibrium with the condensate phase in the separator, the liquid-liquid separation in the bottom of the absorber was assumed to proceed independently without interaction with the rich gas above the condensate phase. Thus, in the analysis of the absorber results, the presence of the liquid-liquid separator in the bottom of the absorber was disregarded.

The rich gas from the separator entered the absorber through a pipe which was 8 in. in diameter, and it then entered the packing through three chimneys or risers. The risers had 10-in. diameters and heights of 1, 2, and 3 ft. The tops of the risers were covered with plates. On the sides of the risers were 23 equally spaced slots ($\frac{3}{4}$ in. by 12 in.) through which the gas passed before entering the packing.

The flow rates of the lean oil, lean gas, and rich oil were obtained by means of orifice meters. Samples were withdrawn from the lean gas, lean oil, and the plant inlet gas (a composite of the rich gas and the condensate streams). These samples were analyzed by means of a very sensitive gas chromatograph as described by Bassyoni (1) and McDaniel (9). The composition of the rich gas (the gas entering the bottom of the packing) was computed by means of a simple flash calculation on the basis of the analysis of the plant inlet gas. These analyses, total flow rates, and calculated results were used to determine the composition of the rich oil by material balances. This procedure

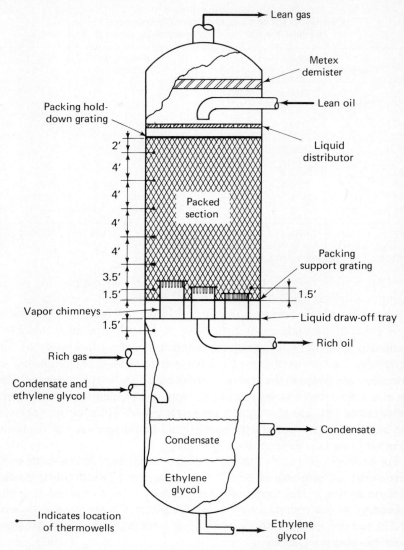

Figure 10-7. The absorber of the Zoller Gas Plant. [Taken from R. Mc-Daniel, A. A. Bassyoni, and C. D. Holland, "Use of the Results of Field Tests in the Modeling of Packed Distillation Columns and Packed Absorbers III," 636.]

was checked for several runs by analyzing the rich oil and condensate streams. The computed compositions were found to be in good agreement with the analyses. A summary of the operating conditions for the 18 field tests made on the absorbers by McDaniel (10) is given in Table 10-8. The complete set of experimental results have been presented elsewhere (9).

TABLE 10-8
SUMMARY OF THE OPERATING CONDITIONS FOR THE
FIELD TESTS MADE ON THE PACKED ABSORBER (9)

Field Test No.	Flow Rates (lb-moles/hr)			Temperature (°F)				Pressure (psia)
	Lean Oil	Rich Gas	Lean Gas	Lean Oil	Rich Oil	Rich Gas	Lean Gas	
101	188.9	2994.1	2721.1	2.9	18.6	0.0	24.0	807
102	190.5	2995.5	2726.0	2.9	18.6	0.0	24.0	807
103	188.7	2971.4	2701.2	2.9	18.6	0.0	24.0	793
201	192.1	2977.8	2700.7	1.5	18.0	−1.0	23.0	780
202	162.8	2902.0	2644.4	0.0	16.3	−1.5	20.5	784
203	165.6	2731 2	2474.6	0.4	18.0	−2.0	22.5	778
204	205 4	2787.5	2494.9	7.0	20.0	−2.0	23.0	792
205	195.9	3106.5	2824.7	0.3	18.5	2.2	24.0	786
301	197.5	2769.0	2486.2	−2.0	16.0	−1.5	23.0	794
302	184.1	2769.5	2493.2	−3.0	16.0	−2.1	25.0	794
303	166.8	2792.4	2528.9	−2.5	16.0	−2.5	25.0	792
304	214.3	2786.1	2504.7	1.0	17.0	−1.0	25.0	790
305	227.3	2785.4	2492.9	3.0	17.0	0.0	25.0	790
401	226.7	2788.2	2491.8	3.0	18.0	0.0	25.0	781
402	214.5	2799.3	2496.7	2.5	18.0	0.0	25.0	780
403	189.0	2635.2	2364.9	−1.0	18.0	−2.0	25.0	780
404	172.0	2665.0	2411.2	−1.0	17.0	−2.5	25.0	782
405	159.6	2675.7	2435.1	−1.5	16.0	−2.0	25.0	784

Since absorption is a heat liberating process, it is customary to introduce the rich gas and lean oil into the column at relatively low temperatures in order that the heat of absorption may be taken up by these streams. From the measured temperatures of the entering and leaving streams of the absorber studied by Rubac et al. (13, 14) and displayed in Figure 10-8, it follows that a *crossover* ($T^V = T^L$) of the vapor and liquid temperature profiles occurred at some z between the top of the packing ($z = z_1 = 0$) and the bottom of the packing ($z = z_{N+1} = 23$ ft). By the second law of thermodynamics, a crossover of temperature profiles cannot occur by the single mechanism of convective heat transfer alone, but a crossover is permissible when heat generation by absorption is also involved.

The temperature profiles in the packed section of the column were measured by means of seven copper-constantan thermocouples that were 18 in. in length and extended about 15 in. into the packing. Since the temperatures of the terminal streams were measured by thermocouples inserted in the lines leading to and from the column, they were taken to be representative of the average temperatures of these respective vapor and liquid streams. However, the temperatures measured in the packing required further interpretation because the thermowells were in contact with both the vapor and liquid phases as well as the packing. Except for the temperatures of the lean oil and rich gas, all other points falling on the solid curve in Figure 10-8 represent temperatures measured in the packing. The interpretation that the

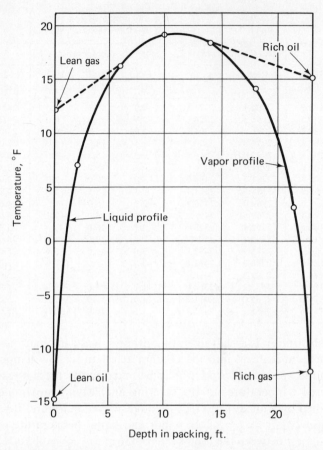

Figure 10-8. Observed temperature profiles in the packed absorber. [Taken from R. E. Rubac, R. McDaniel, and C. D. Holland, "Packed Distillation Columns and Absorbers at Steady State Operation," *A. I. Ch. E. Journal*, **15**, (1969), 573.]

vapor profile in the direction of flow of the vapor begins with an inlet temperature of about −12°F, passes through a maximum, and terminates with an exit temperature of 12°F while the liquid profile in the direction of flow of the liquid begins with a lean oil temperature of about −14°F, passes through a maximum and terminates with an exit temperature of 15°F is based in part on the following analysis.

First, in view of the reliability of the temperatures of the terminal streams, the vapor and liquid temperature profiles must crossover. The occurrence of the crossover at the point at which each profile passes through its maximum simultaneously is suggested by the experimental results and is permitted, though not required, by the equations for simultaneous mass and heat transfer in a packed absorber. These equations are readily obtained in the usual way from the integral-difference equations given by Equations (9-10), through (9-13), and (9-7), namely,

$$\frac{d(VH^V)}{dz} + \sum_{i=1}^{c} N_i H_i^V - q_G = 0$$

$$-\frac{d(Lh^L)}{dz} - \sum_{i=1}^{c} N_i H_i^V + q_G = 0$$

$$\frac{d(VH^V)}{dz} - \frac{d(Lh^L)}{dz} = 0 \qquad (10\text{-}11)$$

$$-\frac{d(Lh^L)}{dz} - \sum_{i=1}^{c} N_i h_i^L + q_L = 0$$

and

$$\frac{dv_i}{dz} + N_i = 0$$

$$\frac{dl_i}{dz} + N_i = 0 \qquad (10\text{-}12)$$

The expressions given by Equation (10-11) may be restated explicitly in terms of the temperatures T^V, T^L, and T^I in the following manner. Consider the first term of the first expression of Equation (10-11), namely,

$$\frac{d(VH^V)}{dz} = \frac{d\sum\limits_{i=1}^{c} v_i H_i^V}{dz} = \sum_{i=1}^{c} v_i \frac{dH_i^V}{dz} + \sum_{i=1}^{c} H_i^V \frac{dv_i}{dz} \qquad (10\text{-}13)$$

Since there is little change in pressure with z, it follows that

$$\frac{dH_i^V}{dz} = \frac{dH_i^V}{dT^V} \frac{dT^V}{dz} = C_{pi}^V \frac{dT^V}{dz} \qquad (10\text{-}14)$$

From the first expression given by Equation (10-12), it is seen that

$$H_i^V \frac{dv_i}{dz} = -N_i H_i^V \qquad (10\text{-}15)$$

and that

$$\sum_{i=1}^{c} H_i^V \frac{dv_i}{dz} = -\sum_{i=1}^{c} N_i H_i^V \qquad (10\text{-}16)$$

In view of the relationships given by Equations (10-13), (10-14), and (10-16) it is seen that the first expression given by Equation (10-11) reduces to

$$\left(\sum_{i=1}^{c} v_i C_{pi}^V\right) \frac{dT^V}{dz} - q_G = 0 \qquad (10\text{-}17)$$

Similarly, the second expression of Equation (10-11) may be reduced to

$$-\left(\sum_{i=1}^{c} l_i C_{pi}^L\right) \frac{dT^L}{dz} - \sum_{i=1}^{c} N_i(H_i^V - h_i^L) + q_G = 0 \qquad (10\text{-}18)$$

and the last expression of Equation (10-11) to

$$-\left(\sum_{i=1}^{c} l_i C_{pi}^L\right) \frac{dT^L}{dz} + q_L = 0 \qquad (10\text{-}19)$$

[Note that Equation (10-19) may be obtained by subtracting Equation (9-14) from (10-18).] The forms of Equations (10-17), (10-18), and (10-19) make them more convenient to use than the original expressions given by Equations (10-11) and (10-12) in the examination of the relationships that exist between the temperatures T^V, T^L, and T^I.

The above equations may be used to examine the relationships existing between the variables at the maxima as well as the intersection of the temperature profiles. First, suppose that at some z

$$\frac{dT^L}{dz} = 0 \tag{10-20}$$

Then it follows from Equation (10-19) that

$$q_L = 0 \quad \text{or} \quad T^I = T^L \tag{10-21}$$

and from Equation (10-18) that at this z

$$q_G = \sum_{i=1}^{c} N_i(H_i^V - h_i^L) \tag{10-22}$$

Conversely, if it is supposed that $T^I = T^L$, then the results given by Equations (10-20) and (10-22) follows as consequences from Equations (10-18) and (10-19).

The result given by Equation (10-22) has the interpretation that at any point at which $dT^L/dz = 0$, the rate of heat absorption by evaporation at the interface is equal to the rate of transfer of sensible heat from the vapor phase to the interface (see Figure 9-4). Similarly, for a condensation process, the rate of heat liberation by condensation at the interface is equal to the rate of transfer of sensible heat from the interface to the vapor phase.

Next, if it is supposed that at some z

$$\frac{dT^V}{dz} = 0 \tag{10-23}$$

then it follows from Equation (10-17) that at this z

$$q_G = 0 \quad \text{or} \quad T^I = T^V \tag{10-24}$$

and by Equation (9-14)

$$q_L = -\sum_{i=1}^{c} N_i(H_i^V - h_i^L) \tag{10-25}$$

Conversely, if it is supposed that $T^I = T^V$, then the results given by Equations (10-23) and (10-25) follow from Equations (10-17) and (9-14).

In terms of a vaporization process, Equation (10-25) has the interpretation that at $dT^V/dz = 0$, the rate of heat absorption at the interface by evaporation is equal to the rate of heat transfer from the liquid phase to the interface.

One of the most significant relationships resulting from this analysis is the fact that if

$$\frac{dT^V}{dz} = 0 \quad \text{and} \quad \frac{dT^L}{dz} = 0 \tag{10-26}$$

simultaneously, then at this same z

$$T^I = T^V = T^L \qquad (10\text{-}27)$$

and conversely. First, suppose that at some z, Equation (10-26) is true. Then, the result given by Equation (10-27) follows from Equations (10-17) and (10-19). Hence, if the conditions given by Equation (10-26) are true, then the crossover occurs at that z where both profiles pass through their maximum values simultaneously. Also, it follows from Equation (10-18) that at this same z

$$\sum_{i=1}^{c} N_i(H_i^V - h_i^L) = 0 \qquad (10\text{-}28)$$

Conversely, if it is supposed that the relationship given by Equation (10-27) is true, then the results given by Equations (10-26) and (10-28) follow immediately from Equations (10-17) through (10-19).

It is to be observed that the equations describing the simultaneous mass and heat transfer process do not require that the interface temperature T^I be equal to T^V and T^L at a point of crossover of the temperature profiles. At the point of crossover ($T^V = T^L$), however, Equations (10-17) and (10-19) do require that the slopes be related as follows:

$$\left(\frac{\sum_{i=1}^{c} v_i C_{pi}^V}{h_G a} \right) \frac{dT^V}{dz} = -\left(\frac{\sum_{i=1}^{c} l_i C_{pi}^L}{h_L a} \right) \frac{dT^L}{dz} \qquad (10\text{-}29)$$

Thus, from this relationship, it follows that the slopes must be either zero or of opposite sign at the point of crossover. Hence, if the crossover occurs at a point where either profile passes through a maximum, then the other profile must also pass through a maximum, and $T^I = T^V = T^L$ [see Equations (10-26) and (10-27).]

An expression for the interface temperature T^I which holds at the crossover as well as any other z within the packed section is obtained by solving Equation (9-14) for T^I in terms of the vapor and liquid film coefficients, $h_G a$ and $h_L a$, by use of the definitions of q_G and q_L that follow Equations (9-10) and (9-13), respectively, to give

$$T^I = \frac{-\sum_{i=1}^{c} N_i(H_i^V - h_i^L) + h_G a T^V + h_L a T^L}{h_G a + h_L a} \qquad (10\text{-}30)$$

When this expression is used to eliminate T^I from the definition for q_G that follows Equation (9-10), the following result is obtained:

$$q_G = Ua(T^V - T^L) + \frac{Ua}{h_L a} \sum_{i=1}^{c} N_i(H_i^V - h_i^L) \qquad (10\text{-}31)$$

where

$$Ua = \frac{1}{\dfrac{1}{h_G a} + \dfrac{1}{h_L a}}$$

The form of Equation (10-31) is particularly revealing. The first term on the right-hand side represents the overall rate of heat transfer for the case in which the heat transfer processes (or resistances) occur in series while the second term represents the contribution to the overall rate of heat transfer that results from the simultaneous (or parallel) mechanism of mass transfer across the boundary separating the vapor and liquid phases. From Equation (10-31), it follows that at the crossover

$$q_G = \frac{Ua}{h_L a} \sum_{i=1}^{c} N_i (H_i^V - h_i^L) \qquad (10\text{-}32)$$

If it is supposed as implied by Figure 10-8 that the maxima for both temperature profiles occur at the point of crossover, then by Equation (10-27), $q_G = 0$, and thus Equation (10-32) reduces again to the result given by Equation (10-28). The latter may be satisfied by either one of two conditions:

(1) $N_i = 0$ (for all i)

or

(2) $\underbrace{\sum N_i (H_i^V - h_i^L)}_{\text{positive terms}} = -\underbrace{\sum N_i (H_i^V - h_i^L)}_{\text{negative terms}}$

With regard to the first condition, there is no assurance that the rate of mass transfer N_i is zero for each component at the point where T^L and T^V pass through maxima.

In setting up the material balances involving N_i in Chapter 9, the direction of transfer from the liquid to the vapor phase was arbitrarily selected. In the following analysis of N_i in the neighborhood of the maximum value of T^L, it is supposed that $T^V \cong T^L$. Thus, when T^L is a maximum, the driving force $(K_i x_i - y_i)$ for the transfer of the lean oil from the liquid phase (where $K_i x_i > y_i$) to the vapor phase tends to increase. On the other hand for the transfer of heavy components from the gas phase (where $y_i > K_i x_i$) to the liquid phase, the driving force $(y_i - K_i x_i)$ tends toward a minimum as T^L tends toward its maximum value. However, the operating conditions of an absorber are generally selected so that the vaporization of lean oil is kept at a relatively low level. Thus, one could expect the rate of heat absorption and heat liberation to be small at the point of crossover, or Equation (10-28) represents a reasonably good approximation.

The above analysis of the experimental results serves to elucidate the mechanisms of simultaneous mass and heat transfer in a packed absorber. However, the conclusions pertaining to the behavior of the temperature profiles of the vapor and liquid phases are not directly applicable to the modeling of packed absorbers because of the lack of reliable correlations for the direct transfer coefficients ($h_G a$ and $h_L a$). In order to predict temperature profiles for both the vapor and liquid phases, reliable values for the heat transfer coefficients would be required. Thus, in the modeling of packed absorbers, the heat transfer sections are assumed to be perfect ($T^V = T^L$ at each z within the packing). As may be seen from the solutions of examples presented in Chapter 4, the calculated temperature profiles do pass through maxima as would be expected from the above analysis.

PROCEDURE USED TO MODEL THE PACKED ABSORBER. In order to place all of the absorber runs in approximate energy balance, the enthalpies were adjusted. In particular, errors in the vapor and liquid enthalpies for all components were regarded as being equally likely, and thus the enthalpies were adjusted by use of the following expressions

$$(h_i)_{co} = h_i + \epsilon C_i^L (T - T_D) \tag{10-33}$$

$$(H_i)_{co} = H_i + \epsilon C_i^V (T - T_D) \tag{10-34}$$

where T is in °R. The curve fits for h_i and H_i are presented in Table C-4 of Appendix C. The value of ϵ was determined by minimizing the sum of the squares of the heat losses over all 18 runs (see Problem 10-1). On the basis of the values of C_i^L and C_i^V given in Table 10-9 and a datum temperature $T_D = 0$, it was found that

$$\epsilon = 0.2561$$

These enthalpies as corrected by Equations (10-33) and (10-34) were used in all of the following analyses of the absorber runs.

The modeling procedure described in Chapter 9 for packed distillation columns was also used in the modeling of packed absorbers. The packed portion of the column was first divided into N elements $\Delta z_1, \Delta z_2, \ldots, \Delta z_n$ of equal size. The streams entering and leaving each element were labeled as shown in Figure 10-9. The mass transfer relationship given by Equation (9-4) was used to relate the vapor and liquid flow rates of each component leaving a given element. As in the case of a distillation column, the equations required to describe this model are of precisely the same form as those required to describe an absorber with plates.

The data obtained from the field tests consisted of the total flow rates V_1, V_{N+1}, L_0, and L_N as well as the temperatures and compositions of these streams. The temperatures for the terminal elements Δz_1 and Δz_N were taken to be equal to the observed temperatures T_1 of the lean gas stream and T_N of the rich oil stream, respectively. The experimental values of the other temperatures within the packing were not utilized directly in the modeling procedure. Instead, these temperatures were computed on the basis of perfect heat transfer sections ($T_j^V = T_j^L$, that is, the temperature of the vapor leaving each element of packing Δz_j was taken to be equal to the temperature of the liquid leaving). Thus, the proposed model for the absorber is the same as the one utilized for the packed distillation column except for the fact that the temperature T_N of the lean oil is utilized in the determination of the vapor-

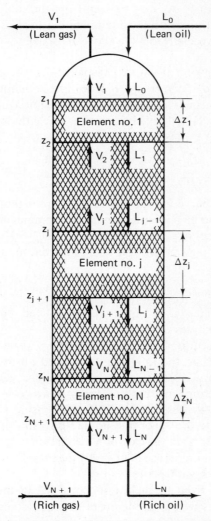

Figure 10-9. Sketch of a typical packed absorber. (Taken from R. E. Rubac, R. McDaniel, and C. D. Holland, "Packed Distillation Columns and Absorbers at Steady State Operation," 572.)

iation efficiencies by Modification 1 of Chapter 9 rather than the temperature of the reboiler T_{N+1}.

The procedure described for the determination of the mass transfer sections for packed distillation columns was also utilized for absorbers. For each run the function 0_1 $(N, \{E_{ji} = 1\})$ possessed a well-defined minimum, and the objective function over all runs $\bar{0}_1$ $(N, \{E_{ji} = 1\})$ exhibited a single minimum as shown in Figure 10-10. From this figure, it is seen that at $N = 8$,

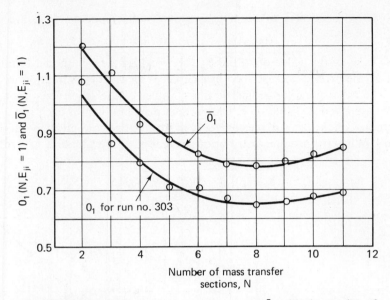

Figure 10-10. Variation of the functions 0_1 and $\bar{0}_1$ with the number of perfect mass transfer sections in the absorber. (Taken from R. McDaniel, A. A. Bassyoni, and C. D. Holland, "Use of the Results of Field Tests in the Modeling of Packed Distillation Columns and Packed Absorbers III," 645.]

TABLE 10-9
ENTHALPY CORRECTION FACTORS AND THE GEOMETRIC MEAN
VALUES OF THE EFFICIENCIES FOR THE PACKED ABSORBER (9)

Component	Enthalpy Correction Factors*		Modification 1† $(N = 8)$ $\bar{E}_{im}$
	$C_i^V \left(\dfrac{\text{Btu}}{\text{lb-mole }^\circ\text{R}}\right)$	$C_i^L \left(\dfrac{\text{Btu}}{\text{lb-mole }^\circ\text{R}}\right)$	
CO_2	8.461	19.219	2.2551
N_2	6.836	10.611	0.0435
CH_4	8.272	16.019	1.4671
C_2H_6	11.698	18.601	0.9256
$i\text{-}C_4H_{10}$	21.106	31.605	0.9323
$n\text{-}C_4H_{10}$	21.511	38.813	1.0220
$i\text{-}C_5H_{12}$	25.451	38.321	0.9016
$n\text{-}C_5H_{12}$	26.044	38.589	0.9295
$n\text{-}C_6H_{14}$	31.074	46.835	0.5302
$n\text{-}C_7H_{14}$	36.287	51.030	0.4405
$n\text{-}C_8H_{18}$	41.326	56.341	0.6342
$n\text{-}C_9H_{20}$	47.374	62.409	0.5529
$n\text{-}C_{10}H_{22}$	52.643	68.895	0.4046

$$\beta_1 = 1.0854$$
$$\beta_8 = 0.9213$$

*Calculated using the enthalpy data given in Table C-4 on the basis of $T_D = 0^\circ\text{R}$.
†See "Modifications of the Calculational Procedure for the Determination of the Vaporization Efficiencies" in Chapter 9.

the objective function $\bar{0}_1$ takes on its minimum value, namely,

$$\bar{0}_1(8, \{E_{ji} = 1\}) = 0.7834$$

On the basis of this value of N, the E_{ji}'s were calculated by use of Modification 1 of the procedure described in Chapter 9 for the determination of the E_{ji}'s. The corresponding geometric mean values of the efficiencies over all 18 runs are given in Table 10-9. The product distributions for each run were recomputed on the basis of the geometric mean efficiencies as given by Modification 1. When the product distributions were used to reevaluate $\bar{0}_1$, the result was

$$\bar{0}_1(8, \{\bar{E}_{im}\}, \beta_{1,m}, \beta_{N,m}) = 0.3316$$

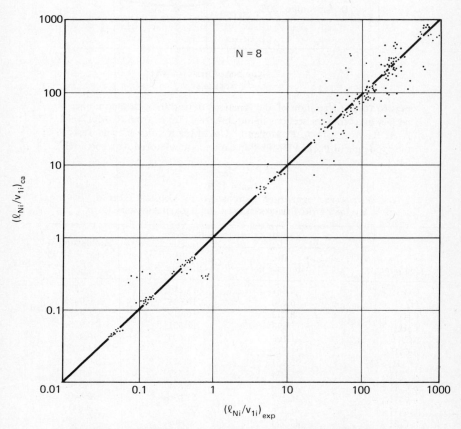

Figure 10-11. Comparison of the product distributions computed on the basis of the mean values $\bar{E}_{im}$ and β_{jm} (listed under modification 1 Table 10-4) with the experimental distributors for the packed absorber. (Taken from R. McDaniel, A. A. Bassyoni, and C. D. Holland, "Use of the Results of Field Tests in the Modeling of Packed Distillation Columns and Packed Absorbers III," 649.)

Thus, the geometric mean efficiencies are seen to have improved the product distributions given by perfect plates by about 50% as measured by the objective function $\bar{0}_1$. A comparison of the calculated and experimental product distributions is presented in Figure 10-11. A comparison of the average deviation of the calculated from the experimental values for the terminal temperatures for the packed absorber is presented in Table 10-10. Again as in the case of packed distillation columns, the β_{jm}'s for the terminal sections ($j = 1, N$) together with the $\bar{E}_{im}$'s reduce the deviations obtained for perfect sections by factors ranging from about two to five. A comparison of the calculated and experimental values of the component flow rates in the lean gas and rich oil is presented in Table 10-11. All of these calculations were based on

TABLE 10-10

AVERAGE DEVIATION OF THE CALCULATED FROM THE
EXPERIMENTAL VALUES OF THE TERMINAL TEMPERATURES
T_1 AND T_N OF THE ABSORBER (9)

Temperatures	Average Deviation* Perfect Sections	Average Deviation Modification 1
T_1 (Lean gas)	0.01756	0.00917
T_N (Rich oil)	0.02944	0.00606

*Average deviation of temperature:

$$T_j \, (j = 1, N+1) = \frac{1}{R} \sum_{r=1}^{R} \left| 1 - \frac{(T_{jr})_{ca}}{(T_{jr})_{exp}} \right|$$

TABLE 10-11

COMPARISON OF EXPERIMENTAL AND CALCULATED FLOW RATES* OF THE LEAN
GAS AND RICH OIL AND A STATEMENT OF THE CALCULATED TEMPERATURES (9)

Component	$(v_{1i})_{exp}$	$(v_{1i})_{ca}$	$(l_{Ni})_{exp}$	$(l_{Ni})_{ca}$	Section Number	Calculated Temperature °F
CO_2	17.2179	16.9905	2.0294	2.2569	1	23.009
N_2	7.1379	6.0901	0.6792	1.7270	2	28.290
CH_4	2371.548	2368.752	106.3683	109.1680	3	28.737
C_2H_6	117.7054	120.936	57.1794	57.9564	4	27.705
C_3H_8	13.2227	13.0375	63.7103	63.8971	5	25.955
$i\text{-}C_9H_{10}$	0.5847	0.5757	18.5022	18.5119	6	23.592
$n\text{-}C_9H_{10}$	0.1012	0.1002	11.7623	11.7638	7	20.297
$i\text{-}C_5H_{12}$	0.0157	0.0266	3.0375	3.0268	8	15.037
$n\text{-}C_5H_{12}$	0.0139	0.0120	1.4384	1.4404		
C_6H_{14}	0.0094	0.0098	0.6460	0.6456		
C_7H_{16}	0.1631	0.2160	10.9615	10.9086		
C_8H_{18}	0.8444	0.8034	64.5626	64.6035		
C_9H_{20}	0.2878	0.2793	55.5551	55.5634		
$C_{10}H_{22}$	0.0665	0.0603	33.8413	33.8473		

*These calculations were made on the basis of the geometric mean values of the $\bar{E}_i$'s and the β_j's (β_1 and β_8), the K values given in Table C-3 of Appendix C, and the enthalpies in Table C-4 of Appendix C and corrected as indicated by Equations (10-33), (10-34), and Table 10-9.

the K data presented in Table C-3 of Appendix C and the enthalpies presented in Table C-4 of Appendix C and corrected as indicated by Equations (10-33), (10-34), and Table 10-9.

PROBLEMS

10-1. Let the unbalance in an overall energy balance as calculated on the basis of the enthalpy curve fits of Appendix C and the experimental results be denoted by Q, that is,

$$Q = L_0 h_0 + V_{N+1} H_{N+1} - L_N h_N - V_1 H_1$$

where h_0, H_{N+1}, h_N, and H_1 are evaluated by use of the expressions for an ideal solution. On the basis of the corrected enthalpies given by Equations (10-33) and (10-34), an ϵ may be picked for any one run such that

$$L_0(h_0)_{co} + V_{N+1}(H_{N+1})_{co} - L_N(h_N)_{co} - V_1(H_1)_{co} = 0 \tag{A}$$

(a). Show that Equation (A) reduces to

$$Q + \epsilon S = 0 \tag{B}$$

where

$$S = \sum_{i=1}^{c} [l_{0i} C_i^L T_0 + v_{N+1,i} C_i^V T_{N+1} - l_{Ni} C_i^L T_N - v_{1i} C_i^V T_1];$$

$$T_D = 0.$$

(b). For any number of runs R, it is desired to find an ϵ such that the sum of the squares of the deviations [the left-hand of Equation (A)] is minimized. That is, it is desired to find that ϵ that minimizes $0(\epsilon)$, where

$$0(\epsilon) = \sum_{r=1}^{R} (Q_r + \epsilon S_r)^2$$

Show that when $\dfrac{d0(\epsilon)}{d\epsilon} = 0$,

$$\epsilon = -\frac{\sum\limits_{r=1}^{R} Q_r S_r}{\sum\limits_{r=1}^{R} S_r^2}$$

10-2. Let the variation of the K_{ji}'s with temperature be accounted for by use of the relationship suggested by Winn (17) for hydrocarbons,

$$K_{ji} = a_i(K_{jb})^{c_i} \tag{A}$$

where K_{jb} is an arbitrarily selected base component and a_i and c_i are constants for any component i. To account for the effect of operating refluxes other than total, let the function $\log_e \Omega_i$ be added to the right-hand side of Equation (10-5). (Observe that at total reflux, $\Omega_i = 1$.) On the basis of these conditions, obtain expressions for 0_1 and $\bar{0}_1$ in terms of the $\bar{E}_i$'s and other quantities; see Reference (10).

10-3. If the order of operations used to obtain the expression for $\bar{0}_1$ in Problem

10-2 is reversed, show that the functions given by Equations (10-6) and (10-8) are obtained.

NOTATION*

f = number of the feed distributor plate in Model 1 of the feed mixing section.

N = number of packed sections plus the feed distributor plate (Model 1); $N + 1$ = total number of mass transfer sections + feed distribution plate + reboiler (for a column with a total condenser).

0_1 = objective function for any one run; defined by Equation (10-1).

$\bar{0}_1$ = objective function for any number of runs R; defined by Equation (10-2).

$\bar{0}_2, \bar{0}_3, \bar{0}_4$ = objective functions; defined by Equations (10-6), (10-7), and (10-8), respectively.

r = run number.

R = total number of runs.

Subscripts

b = base component.

exp = experimental value.

m = geometric mean value.

r = run number.

Mathematical symbols

$|x|$ = absolute value of x.

$\{\bar{E}_i\}$ = set of E_i's for $1 \leq i \leq c$.

$\{\beta_j\}$ = set of β_j's for $1 \leq j \leq N + 1$.

REFERENCES

1. Bassyoni, A. A., "Formulation of a Model for Packed Distillation Columns on the Basis of Field Tests," Ph. D. dissertation, Texas A&M University, College Station, Texas, 1969.

2. ———, R. McDaniel, and C. D. Holland, "Examination of the Use of Mass Transfer Rate Expressions in the Description of Packed Distillation Columns-II," *Chem. Eng. Sci.*, **25**, (1970), 437.

*See also Chapter 9 notations.

3. Brosilow, C., R. Tanner, and H. Tureff, "Optimization of Staged Counter Current Processes," paper presented at the 63rd National Meeting of the A. I. Ch. E. in St. Louis, Missouri, February 18, 1968.

4. Fenske, M. R., "Fractionation of Straight-Run Pennsylvania Gasoline," *Ind. Eng. Chem.*, **24**, (1932), 482.

5. Holland, C. D., *Multicomponent Distillation*. Englewood Cliffs, N. J.: Pretice-Hall, Inc., 1968.

6. ———, and K. S. McMahon, "Comparison of Vaporization Efficiencies with Murphree-Type Efficiencies in Distillation-I," *Chem. Eng. Sci.*, **25**, (1970), 431.

7. Hooke, R., and T. A. Jeeves, "Direct Search Solution of Numerical and Statistical Problems," *J. Assoc. Comp. Mach.*, **8**, (1961), 2.

8. McDaniel, R., "Determination of Vaporization Efficiencies and Overall Mass Transfer Coefficients for a Packed Absorber at Steady State Operation," M. S. Thesis, Texas A&M University, College Station, Texas, 1968.

9. McDaniel, R., "Packed Absorbers at Steady State and Unsteady State Operation," Ph. D. Dissertation, Texas A&M University, College Station, Texas, 1969.

10. ———, A. A. Bassyoni, and C. D. Holland, "Use of the Results of Field Tests in the Modeling of Packed Distillation Columns and Packed Absorbers-III," *Chem. Eng. Sci.*, **25**, (1970), 633.

11. ———, R. E. Rubac, and C. D. Holland, "Packed Distillation Columns and Absorbers at Steady State Operation," presented at the 63rd National Meeting of A. I. Ch. E. at St. Louis, Missouri, February 18, 1968.

12. Peters, W. A., Jr., "The Efficiency and Capacity of Fractionating Columns," *Ind. Eng. Chem.*, **14**, (1922), 476.

13. Rubac, R. E., R. McDaniel, and C. D. Holland, "Packed Distillation Columns and Absorbers at Steady State Operation," *A. I. Ch. E. Journal*, **15**, (1969), 569.

14. ———, "Determination of Vaporization Efficiencies for Packed Columns at Steady State Operation," Ph. D. Dissertation, Texas A&M University, College Station, Texas, 1968.

15. Srygley, J. M., and C. D. Holland, "Optimum Disign of Conventional and Complex Distillation Columns," *A. I. Ch. E. Journal*, **11**, (1965), 695.

16. Wilde, D. J., *Optimum Seeking Methods*. Englewood Cliffs, N. J.: Prentice-Hall, Inc., 1964.

17. Winn, F. W., "New Relative Volatility Method for Distillation Calculations," *Hydrocarbon Processing and Petroleum Refiner*, **37**, 5 (1958), 246.

Use of Predicted
Vaporization Efficiencies
in the Modeling of
Distillation Columns
with Plates

11

The development of the expressions for the vaporization plate efficiencies, E_{ji}, for a variety of mixing models is presented in Part 1 of this chapter. Of the models presented, Model III appears to be the most promising one for the description of the mass transfer between multicomponent mixtures in the vapor and liquid phases and the liquid mixing on the plate of a distillation column. Model III makes use of the concepts of vaporization plate efficiencies, vaporization point efficiencies, a mixing parameter, and the correlations given in the *Bubble-Tray Design Manual* (4) for the number of overall mass transfer units.

The use of Model III in the modeling of distillation columns with the aid of field tests is presented in Part 2. Procedures for estimating the values of the parameters appearing in Model III such as the number of overall mass transfer units is presented in Part 3.

Part 1 : Development and Analysis of the Expressions for the Vaporization Plate Efficiencies for Various Mixing Models

Recall that the vaporization efficiency for plate j and component i is defined by

$$y_{ji} = E_{ji}\gamma_{ji}^L K_{ji}x_{ji} \tag{9-4}$$

where it is supposed that the vapor forms an ideal solution. In all of the models presented, it is assumed that the vapor and liquid leaving each plate j are in thermal equilibrium: $T_j^V = T_j^L = T_j$. The superscript L on the γ_{ji}^L is used to emphasize that the activity coefficient is to be evaluated on the basis of the mole fractions $\{x_{ji}\}$ of the liquid leaving plate j.

Since the sum of the y_{ji}'s over all components i is unity, it is evident that the following relationship must be satisfied by this choice of T_j, namely,

$$1 = \sum_{i=1}^{c} E_{ji}\gamma_{ji}^L K_{ji}x_{ji} \tag{11-1}$$

The temperature that satisfies this equation is generally different from the one obtained by use of the Murphree convention in which the temperature is taken equal to the one found by making a bubble point calculation on the liquid leaving plate j, namely,

$$1 = \sum_{i=1}^{c} \gamma_{ji}^L K_{Mji}x_{ji} \tag{11-2}$$

Note that it is a consequence of the temperature convention given by Equation (11-2) that the Murphree efficiencies for the components of a binary mixture are equal and not because of some inherent characteristic of binary mixtures. The proposed temperature convention given by Equation (11-1) avoids certain difficulties that result from the use of the Murphree-temperature convention [Equation (11-2)] in the analysis of multicomponent problems (15). Briefly, these difficulties arise because the specification that the temperature be given by the bubble point expression [Equation (11-2)] uses up one degree of freedom with the consequence that only $c - 1$ efficiencies may be specified independently, but c efficiencies may be specified independently when the proposed temperature convention [Equation (11-1)] is used. In the interest of simplicity, the activity coefficients in Equation (11-1) and (11-2) are hereafter omitted in this section.

Model I: Simple product model

The simple product model which was introduced in Chapter 9 may be regarded as a first order approximation of E_{ji}; namely,

Model I: $E_{ji} = \bar{E}_i\beta_j$ (11-3)

Although this model has been successfully employed in the modeling of existing columns as demonstrated in References (17) and (21), it is not of a predictive nature. Since it was desired to develop models which could be used eventually for design purposes, this chapter is devoted to the development of predictive models.

Next, there follows the development of the model for the case in which the liquid phase is assumed to be perfectly mixed.

Model II: Perfectly mixed liquid phase

The description and development of this model are based on the temperature convention given by Equation (11-1) rather than the Murphree temperature convention [Equation (11-2)] upon which the Murphree plate efficiency (22) is based. The relationships between the mass transfer coefficients and the number of transfer units that are consistent with the temperature convention given by Equation (11-1) are also presented. This part is concluded by the presentation of numerical examples of binary mixtures which demonstrate that essentially the same compositions of the vapor leaving a plate are predicted by use of vaporization efficiencies as predicted by use of Murphree efficiencies. Thus, the correlations of the film coefficients developed by others (4) for use with Murphree efficiencies may be used in the prediction of vaporization efficiencies for multicomponent systems.

The development of the expression for E_{ji} for this model is initiated by the enumeration of the assumptions upon which Model II is based, namely,

1. The liquid phase is perfectly mixed.
2. No mixing of the vapor occurs as it passes up through the liquid phase, but upon leaving the surface of the liquid, the vapor becomes perfectly mixed before entering the next plate.
3. The flow rate of the vapor through the liquid on plate j is given by V, where V is an appropriately selected value that lies between V_j and V_{j+1}.
4. The vapor and liquid phases are in thermal equilibrium at the temperature T_j of the liquid leaving plate j which satisfies Equation (11-1).
5. The overall mass transfer coefficient and the interfacial area are finite and positive constants for each plate.

In the following developments, the number of mass transfer units are taken to be the independent variables and the compositions the dependent variables, which is in effect the reverse of the choice made in the analyses in Reference (1) and in Chapter 9. The equation for mass transfer on the plate of a distillation column are of the same form as those for a packed column. Thus, on the basis of the above assumptions, the following differential equation which represents a component-material balance at any z for plate j is readily obtained by combining Equation (9-35) with the first expression of

Equation (10-12) and assuming that the variation of V in the direction of z is negligible.

$$V_j \frac{dy_i}{dz} + (K_G af^V)_{ji} S(K_{ji} x_{ji} - y_i) = 0 \qquad (11\text{-}4)$$

Also, in the interest of simplicity, the activity coefficient appearing in Equation (9-35) was omitted in Equation (11-4). In the initial part of the development that follows, it is supposed that $K_{ji} x_{ji} > y_i$ over the interval of integration from $z = 0$ to $z = z_T$, the total depth of liquid on plate j. Under these conditions Equation (11-4) may be solved by separation of the variables, namely,

$$\int_{y_{ji}}^{y_{j+1,i}} \frac{dy_i}{K_{ji} x_{ji} - y_i} = -\int_0^{z_T} \frac{(K_G af^V)_{ji}}{V_j} S\, dz \qquad (11\text{-}5)$$

After the implied integration has been carried out, the result so obtained may be rearranged to give the following expression for the modified Murphree plate efficiency,

$$E_{ji}^M = 1 - \exp(-n_{0Gji}) \qquad (11\text{-}6)$$

where the modified Murphree plate efficiency is defined

$$E_{ji}^M = \frac{y_{ji} - y_{j+1,i}}{K_{ji} x_{ji} - y_{j+1,i}} \qquad (11\text{-}7)$$

and the number of overall mass transfer units by

$$n_{0Gji} = K_G a_{ji} f_{ji}^V \left(\frac{S z_T}{V_j} \right) \qquad (11\text{-}8)$$

When both the numerator and denominator of the expression for E_{ji}^M given by Equation (11-7) are divided by $K_{ji} x_{ji}$ and the expression so obtained is solved for E_{ji}, the following formula is obtained:

Model II: $E_{ji} = \dfrac{y_{j+1\,i}}{K_{ji} x_{ji}} + E_{ji}^M \left[1 - \dfrac{y_{j+1,i}}{K_{ji} x_{ji}} \right]$ \qquad (11\text{-}9)

Replacement of E_{ji}^M in this expression by its equivalent as given by Equation (11-6) yields

Model II: $E_{ji} = 1 - e^{-n_{0Gji}} \left[1 - \dfrac{y_{j+1,i}}{K_{ji} x_{ji}} \right]$ \qquad (11\text{-}10)

Now, if it is supposed that the number of transfer units, n_{0Gji}, is finite and positive and that $y_{j+1,i}$ is allowed to approach $K_{ji} x_{ji}$, one obtains

$$\lim_{y_{j+1,\,i} \to K_{ji} x_{ji}} E_{ji} = 1 \qquad (11\text{-}11)$$

[Observe that the requirement that n_{0Gji} be finite and positive is equivalent to the requirement that $0 < E_{ji}^M < 1$; see Equation (11-6).] Then in view of the definition of E_{ji} given by Equation (9-4) and the result given by Equation

(11-11), it follows that

$$\lim_{\substack{y_{j+1,i} \to K_{ji}x_{ji}}} y_{ji} = \lim_{\substack{y_{j+1,i} \to K_{ji}x_{ji}}} E_{ji}K_{ji}x_{ji} = K_{ji}x_{ji}$$

Thus, if

$$y_{j+1,i} \longrightarrow K_{ji}x_{ji} \tag{11-12}$$

then

$$y_{ji} \longrightarrow K_{ji}x_{ji}$$

However, the results obtained by this limiting process are not necessarily valid because Equation (11-10) was obtained on the basis that $K_{ji}x_{ji} > y_i$ over the closed interval of integration from $z = 0$ (where $y_i = y_{ji}$) to $z = z_T$ (where $y_i = y_{j+1,i}$). On the other hand, it is permissible to reexamine Equation (11-5) when these limiting conditions [Equation (11-12)] are imposed on it. Consider first the evaluation of the limit of the left-hand side of Equation (11-5) namely,

$$\lim_{\substack{y_{j+1,i} \to K_{ji}x_{ji} \\ y_{ji} \to K_{ji}x_{ji}}} \int_{y_{ji}}^{y_{j+1,i}} \frac{dy_i}{K_{ji}x_{ji} - y_i} \tag{11-13}$$

In the limit, this integral is seen to have an infinite discontinuity at both its upper and lower limits of integration. Furthermore, in the same limit, the upper and lower limits of integration approach each other. It is shown below that the evaluation of the integral given by Equation (11-13) reduces to the evaluation of an integral of the following form:

$$\lim_{b \to 0} \int_0^{b-\epsilon} \frac{d\omega}{b - \omega} \qquad (b > \epsilon > 0) \tag{11-14}$$

Since ϵ is always smaller than b, it is evident that ϵ depends on b. Then for convenience, take

$$\epsilon = \frac{b}{e^M}$$

where M is any positive number. Thus,

$$\lim_{b \to 0} \int_0^{b-\epsilon} \frac{d\omega}{b - \omega} = -\lim_{b \to 0} \log_e \frac{\epsilon}{b} = -\lim_{b \to 0} \log_e \frac{b}{be^M} = M \tag{11-15}$$

Since M was any arbitrarily selected positive number, it follows that the value of the integral may be set equal to any positive number.

It will now be shown that Equation (11-13) may be reduced to the form given by Equation (11-14). For the sake of definiteness, consider the case in which $K_{ji}x_{ji} > y_i$ over the closed interval of integration $(0 \leq z \leq z_T)$ and further suppose that $y_{ji} > y_{j+1,i}$ prior to taking the limit. Consider first the left-hand side of Equation (11-5), and let the following change in variable be made:

$$\omega = \frac{y_i}{y_{ji}} - \frac{y_{j+1,i}}{y_{ji}} \tag{11-16}$$

Then, when

$$y_i = y_{ji}, \qquad \omega = 1 - \frac{y_{j+1,i}}{y_{ji}} = \omega_{ji}$$

$$y_i = y_{j+1,1} \qquad \omega = 0$$

Thus,

$$\int_{y_{ji}}^{y_{j+1,i}} \frac{dy_i}{K_{ji}x_{ji} - y_i} = -\int_0^b \frac{d\omega}{b - \omega} \qquad (11\text{-}17)$$

where

$$b = \frac{K_{ji}x_{ji}}{y_{ji}} - \frac{y_{j+1,i}}{y_{ji}}.$$

Since $K_{ji}x_{ji} > y_{ji}$ and $y_{ji} > y_{j+1,i}$, it follows that $b > \omega_{ji} > 0$. Thus, ω_{ji} may be set equal to $b - \epsilon$, where ϵ is a positive number which is less than b. Thus,

$$\lim_{\substack{y_{j+1,i} \to K_{ji}x_{ji} \\ y_{ji} \to k_{ji}x_{ji}}} \int_{y_{ji}}^{y_{j+1,i}} \frac{dy_i}{K_{ji}x_{ji} - y_i} = -\lim_{b \to 0} \int_0^{b-\epsilon} \frac{d\omega}{b - \omega} = -M \qquad (11\text{-}18)$$

where M is any positive number. Hence, in the limit, Equation (11-5) reduces to

$$M = n_{0Gji} \qquad (11\text{-}19)$$

On the basis of any $n_{0Gji} = M > 0$, a corresponding value of E_{ji}^M may be computed by use of Equation (11-6).

These results suggest the proposal of the following model for the case in which the experimental compositions are to be used to compute the number of overall transfer units, that is, the compositions are regarded as the independent variables and the number of overall mass transfer units as the dependent variable, which is the same choice of independent and dependent variables made in Chapter 9. If for any plate j, an infinite discontinuity arises as a result of the condition that $y_{j+1,i} \to K_{ji}x_{ji}$ at $z = z_T$, then require that the model have a second infinite discontinuity resulting from the imposed condition that y_{ji} approaches $K_{ji}x_{ji}$ at $z = 0$, the other end of the interval. Physically, this amounts to saying that if the driving force for mass transfer $(K_{ji}x_{ji} - y_{j+1,i})$ approaches zero at the floor of the plate $(z = z_T)$, then the driving force $(K_{ji}x_{ji} - y_{ji})$ for mass transfer will also approach zero at the surface of the liquid $(z = 0)$. Since the number of transfer units for plate j of this model is arbitrary $(n_{0Gji} = M)$, then the value of n_{0Gji} for plate j may be set equal to its value for any other plate for which the integral on the left-hand side of Equation (11-5) possessed no infinite discontinuity. By imposing the second condition $(y_{ji} \to K_{ji}x_{ji})$ when the first condition exists, definite meanings for n_{0Gji} and E_{ji}^M are obtained. Otherwise, if the discontinuity corresponding to the condition $y_{j+1,i} \to K_{ji}x_{ji}$ arises and the second condition is not imposed on the model, then n_{0Gji} may become infinite as shown in Chapter 9.

Further observe that for $n_{oGji} \geqq 0$, the value of E_{ji} given by Equation (11-10) is finite and positive for all finite and positive values of $y_{j+1,i}/K_{ji}x_{ji}$. For example, if $n_{oGji} \geqq 0$, then it follows from Equation (11-10) that

(a) if $0 < \dfrac{y_{j+1,i}}{K_{ji}x_{ji}} < 1$, then $0 < E_{ji} < 1$;

(b) if $\quad \dfrac{y_{j+1,i}}{K_{ji}x_{ji}} = 1$, then $E_{ji} = 1$;

(c) if $\quad \dfrac{y_{j+1,i}}{K_{ji}x_{ji}} > 1$, then $E_{ji} > 1$.

[Observe that the total reflux example that was considered in Chapter 9 wherein $y_{j+1,i} = x_{ji}$ and $K_{ji} = 1$ is represented by the second of the above cases.] Thus, after any given trial through the column whereby new sets of compositions $\{x_{ji}\}$ and $\{y_{ji}\}$) and K values $\{K_{ji}\}$ are obtained, the corresponding set of E_{ji}'s computed by use of Equation (11-10) will always be finite and positive. Since the E_{ji}'s always appear in the equations as multipliers of K_{ji} in the absorption factors $[A_{ji} = L_j/(E_{ji}K_{ji}V_j)]$ they do not introduce any discontinuities of any type in the component-material balances which describe the column [see, for example, Equation (9-19)]. Also, since E_{ji}'s are always finite and positive, their appearance as a multiplier in the temperature function [Equation (11-1)] leads to no discontinuities. Consequently, the zeroes that may appear in the numerator and denominator of the expression for E_{ji}^M never appear in the working equations of a calculational procedure that utilizes the E_{ji}'s [computed by use of Equation (11-10)] rather than the Murphree-type efficiencies.

Relationships between the number of vapor and liquid transfer units and the vaporization and Murphree plate efficiencies entrainment

First the mass transfer relationships for the temperature convention given by Equation (11-1) are presented and then those for the Murphree temperature are presented. In these developments it is assumed that the liquid phase is perfectly mixed and that the vapor entering the given plate under consideration is also perfectly mixed.

MASS TRANSFER RELATIONSHIPS FOR THE TEMPERATURE CONVENTION GIVEN BY EQUATION (11-1). The assumption given by Equation (8-6) that $T_j^V = T_j^L = T_j^{\rho} = T_j^{\varrho} = T_j$ is consistent with Equation (11-1). Since the following developments are based on the assumptions of Equation (8-6), the relationships given in Item I of Table 8-1 apply. The supposition that the temperature of the vapor and liquid phases are equal does not imply that the liquid leaves plate j at its bubble point temperature. The fact that the resistance to transfer across the interface was taken equal to zero in the development of the relationships in Table 8-1 does imply, however, that a dynamic equilibrium exists at the interface. That is, at the interface, the vapor and liquid phases are at their dew point and bubble point temperatures, which are, of course, equal to each other and to the temperature of the vapor and liquid phases for the given plate. However, the interface is the only point or position where it is supposed that a state of mass equilibrium exists.

From the results given in Item I of Table 8-1, the following relationship (deduced in Chapter 8) is needed in the present analysis, namely,

$$\frac{1}{n_{0Gi}} = \frac{1}{n_{Gi}} + \left(\frac{\gamma_i^L K_i V}{\gamma_i^V L}\right)\frac{1}{n_{Li}} = \frac{1}{n'_{Gi}} + \left(\frac{\gamma_i^L K_i V}{\gamma_i^V L}\right)\frac{1}{n'_{Li}} = \frac{1}{n'_{0Gi}} \qquad (11\text{-}20)$$

On the basis of the assumptions [Equation (8-6)] used to obtain the above expression, it follows that $n_{Gi} = n'_{Gi}$, $n_{Li} = n'_{Li}$, and $n_{0Gi} = n'_{0Gi}$. The most restrictive assumption in the development of these relationships was perhaps $\gamma_i^L = \gamma_i^c$, and this restriction is removed, of course, for the case of ideal solutions.

MASS TRANSFER RELATIONSHIPS FOR THE MURPHREE TEMPERATURE CONVENTION. In the case of the Murphree temperature convention, the liquid is assumed to be at its bubble point temperature at each point in the liquid phase on plate j. Thus, the liquid leaving is assumed to be at its bubble point temperature as implied by Equation (11-2). For each x_i on plate j, the corresponding equilibrium value of y_i (denoted by y_i^*) is approximated by use of the tangent line to the equilibrium curve, namely,

$$y_i^* = m_i x_i + b_i \qquad (11\text{-}21)$$

where, of course, it is supposed that a binary mixture is under consideration. Also, the temperature that satisfies the Murphree temperature convention [Equation (11-2)] also gives the set of hypothetical mole fractions y_{ji}^*; namely,

$$y_{ji}^* = \gamma_{ji}^L K_{Mji} x_{ji} \qquad (11\text{-}22)$$

[Note in the above equations and those that follow the superscripts V and L used in Chapter 8 to identify the bulk conditions for the vapor and liquid phases have been omitted.]

Since the mass transfer relationships given in Item II of Table 8-1 are based on the use of Equation (11-21), they are applicable for the case in which the Murphree temperature convention is employed. Consequently, the number of overall mass transfer units n'_{0GM_i} (where the subscript M has been added to emphasize the use of the Murphree temperature convention) is given by

$$\frac{1}{n'_{0GM_i}} = \frac{1}{n'_{Gi}} + \left(\frac{m_i V}{L}\right)\frac{1}{n'_{Li}} \qquad (11\text{-}23)$$

To obtain an expression for the Murphree plate efficiency for the case in which it is assumed that the liquid phase is perfectly mixed, y_i^* in the expression for n'_{0Gi} in Item II of Table 8-1 is replaced by its equivalent $\gamma_{ji}^L K_{Mji} x_{ji}$, and the indicated integration carried out to give

$$E_{Mji} = \frac{y_{ji} - y_{j+1,i}}{\gamma_{ji}^L K_{Mji} x_{ji} - y_{j+1,i}} = 1 - \exp(-n'_{0GMji}) \tag{11-24}$$

COMPARISON OF THE MASS TRANSFER RELATIONSHIPS AND CORRESPONDING PLATE EFFICIENCIES FOR THE TWO TEMPERATURE CONVENTIONS. Comparison of the Murphree efficiency [Equation (11-24)] with the modified Murphree efficiency [Equations (11-6) and (11-7)] shows that the Murphree efficiency depends on n'_{0GMi} [Equation (11-23)] while the modified Murphree efficiency depends on n'_{0Gi} [Equations (11-20)]. The number of overall transfer units, n'_{0GMi} and n'_{0Gi}, for the two efficiencies depend on the same set of transfer units (n'_G and n'_L) and are the same except for the fact that the quantity $\gamma_i^L K_i / \gamma_i^V$ appears in n'_{0Gi} in the same position that m_i appears in n'_{0GMi} [see Equations (11-20) and (11-23)].

Both models assume that a state of dynamic equilibrium exists at the interface ($\bar{f}_i^v = \bar{f}_i^c$) and have this one point (x_{ji}^c, y_{ji}^c) at the interface in common. In the model for the vaporization efficiencies, the T_j found by Equation (11-1) is the same throughout the vapor and liquid phases on plate j. Since the interface is assumed to be at equilibrium, it follows that the vaporization efficiency must be unity at the interface and, consequently, the K_{ji}'s found by use of Equation (11-1) must satisfy the equilibrium relationship $y_{ji}^v = \gamma_{ji}^c K_{ji} x_{ji}^c$. Since it is supposed in the Murphree model that the liquid is at its bubble point temperature throughout the liquid phase it follows that $y_{ji}^v = \gamma_{ji}^c K_{Mji} x_{ji}^c = \gamma_{ji}^c K_{ji}^c x_{ji}^c$ or K_{Mji}^c is equal to K_{ji}, that is, the two models have the same interface temperatures. Moreover, the temperature computed by use of vaporization efficiencies [Equation (11-1)] is seen to be the bubble point temperature of the interface (see Figure 11-1).

By use of the two rate expressions given by Equation (8-7) and the corresponding definitions of n'_{Li} and n'_{Gi} from Item I of Table 8-1, it is readily shown that

$$\frac{Ln'_L}{Vn'_G} = \frac{y_i^v - y_i}{x_i - x_i^c} \tag{11-25}$$

which is seen to be the negative of the slope of the line that connects the two common points, $(x_{j1}, y_{j+1,1})$ and $(x_{j1}^c$ and $y_{j1}^v)$, of the two models as shown in Figure 11-1. However, after the interface point has been reached, the paths for the models differ. In the case of vaporization (or modified Murphree) efficiency, a projection is made from the point (x_{j1}^c, y_{j1}^v) to the point (x_{j1}, y_{j1}) with the slope K_{j1} that satisfies Equation (11-1) (see Figure 11-1). In the case of the Murphree efficiency, a projection is made to the point (x_{j1}, y_{j1}^*) along the "tangent line" $y_1^* = m_1 x_1 + b_1$. Since the equilibrium line is not generally

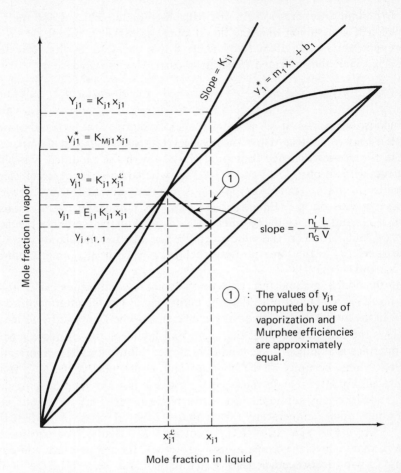

Figure 11-1. Comparison of vaporization and Murphree efficiencies [Taken from C. D. Holland, A. E. Hutton, and G. P. Pendon, "Reduction of Vaporization Efficiencies for Multicomponent Mixtures by Use of Existing Correlations for Vapor and Liquid Film Coefficients," *Chem. Eng. Sci.*, **26**, (1971), 1729.]

straight, an appropriate m_1 must be picked which is essentially equivalent to passing a straight line through the two points $(x_{j1}^{\ell}, y_{j1}^{\upsilon})$ and $(x_{j1}, y_{j+1,1})$.

The y_{j1} predicted by use of vaporization efficiencies is calculated by use of Equations (11-1), (11-10), and (11-20). [Actually, in practice, y_{j1} will have been determined once the T_j that satisfies Equations (11-2), (11-6), (11-7), (11-20) has been found.] The y_{j1} predicted by the Murphree model is found by use of Equations (11-2), (11-23) and (11-24).

Although the assumptions upon which the models are based do differ slightly, the corresponding values of y_{j1} predicted by both models on the basis of the same sets of film coefficients $\{n'_L, n'_G\}$ and points $\{x_{j1}, y_{j+1,1}\}$ appear to be in almost perfect agreement for the examples considered (see Tables 11-1 and 11-2). The number of transfer units for each film in these examples was taken to be independent of component identity just as they are for the existing correlations (4). In Illustrative Example 11-1 (the benzene-toluene system), the vapor and liquid phases closely approximate ideal solutions, but, in Illustrative Example 11-2 (the ethanol-water system), the liquid phase forms a highly nonideal solution.

The following calculational procedure was utilized to compute the y_{j1}'s by use of vaporization efficiencies:

1. Choose n'_L, n'_G, and x_{j1}, and compute $y_{j+1,1}$ from the operating line.
2. Find the T_j that satisfies the temperature convention, Equation (11-1). For each assumed value of T_j, each E_{ji} must be evaluated by use of Equations (11-10) and (11-20). [Note that if the liquid phase is highly nonideal, K_{ji} should be preceded in all equations by the γ^L_{ji}, which was omitted wherever possible in the interest of simplicity.]

TABLE 11-1

STATEMENT AND SOLUTION OF ILLUSTRATIVE EXAMPLE 11-1 (17)

I. *STATEMENT OF ILLUSTRATIVE EXAMPLE 11-1*

A benzene-toluene mixture is to be separated in a column operating at 760 mm pressure. A total condenser is to be used, $L_0 = L_j = 100$ moles/hr, $D = 50$ moles/hr, and the mole fraction of benzene (component No. 1) in the distillate is 0.95. Find y_{j1} by use of vaporization efficiencies (Model II) and by use of Murphree efficiencies (Model II) for the set $\{n_L', n_G'\}$ of film coefficients enumerated below at $x_{j1} = 0.4, 0.6$, and 0.9. (The operating line for the rectifying section intersects the equilibrium curve at an x_{j1} slightly greater than 0.3 but less than 0.4. Use the following curvefits which are based on data taken from page 578 of *Chemical Engineers Handbook* (3rd ed.), J. H. Perry, Ed.

$$P_1 \text{ (benzene)} = 36.3 \times 10^6 \exp\left[-\frac{6852.0}{T}\right], (P_1 \text{ in mm and } T \text{ in } °R)$$

$$P_2 \text{ (toluene)} = 48.15 + 10^6 \exp\left[-\frac{7636.36}{T}\right], (P_2 \text{ in mm and } T \text{ in } °R)$$

Compute m_1 by use of Equation (11-26).

II. *SOLUTION OF EXAMPLE 11-1*

Specifications			Vaporization Plate Efficiency (Model II)				Murphree Efficiency		
n'_G	n'_L	x_{j1}	$T_j(°F)$	$E_{j,1}$	$E^M_{j,1}$	$y_{j,1}$	$T_{Mj}(°F)$	$E_{Mj,1}$	$y_{j,1}$
1	1	0.4	204.4310	0.9433	0.2569	0.5957	204.3829	0.3329	0.5960
1	1	0.6	194.9532	0.9126	0.2804	0.7445	192.9477	0.3879	0.7451
1	1	0.9	181.1069	0.9585	0.3166	0.9355	180.0270	0.4521	0.9356
0.2	10	0.4	203.4859	0.9483	0.1739	0.5901	203.3829	0.1766	0.5900
0.2	10	0.6	193.0473	0.9222	0.1750	0.7297	192.9477	0.1779	0.7297
0.2	10	0.9	180.1058	0.9630	0.1762	0.9243	180.0270	0.1791	0.9242
10	0.2	0.4	205.3324	0.9180	0.0793	0.5879	203.3829	0.1256	0.5881
10	0.2	0.6	197.2668	0.8604	0.0894	0.7283	192.9477	0.1724	0.7293
10	0.2	0.9	183.0296	0.9197	0.1110	0.9268	180.0270	0.2541	0.9273

TABLE 11-2
STATEMENT AND SOLUTION OF ILLUSTRATIVE EXAMPLE 11-2 (17)

I. STATEMENT OF ILLUSTRATIVE EXAMPLE 11-2

An ethanol-water·mixture is to be separated in a column operating at total reflux ($y_{j+1,i} = x_{ji}$, $L_j/V_{j+1} = 1$) at a total pressure of 760 mm. Find y_{j1} by use of vaporization efficiencies (Model II) and by use of Murphree efficiencies (Model II) for the set of transfer units $\{n_L', n_G'\}$ enumerated below for $x_{j1} = 0.2, 0.3, 0.4$, and 0.9. Use the following curvefits for the activity coefficients that are based on the experimental results of J. S. Carey (Sc. D. Thesis in Chemical Engineering, M.I.T., 1929).

$$\frac{\log_e \gamma_1}{x_2^2} = 0.2924 + 1.1939 x_2$$

$$\frac{\log_e \gamma_2}{x_1^2} = 0.2924 + 1.1939 \, (0.5 + x_2)$$

Use the following curvefits for the vapor pressures of the pure components:

1. P_1 (ethanol) $= 745 \times 10^6 \exp(-8728.11/T)$
2. P_2 (water) $= 530.5 \times 10^6 \exp(-9038.0/T)$

where P is in mm and T is in °R. These curvefits are based on data presented in the *International Critical Tables* Vol. III, pages 212–17, McGraw-Hill Book Co. Inc., New York, 1928.

Compute m_1 by use of Equation (11-27) and also take $L_j/V_j = 1$ in the calculation of n_{0Gi}' and n_{0GMi}'.

II. SOLUTION OF EXAMPLE 11-2

Specifications			Vaporization Efficiency (Model II)				Murphree Efficiency		
n_G'	n_L'	x_{j1}	$T_j(°F)$	$E_{j,1}$	$E_{j,1}^M$	$y_{j,1}$	$T_{Mj}(°F)$	$E_{Mj,1}$	$y_{j,1}$
1	1	0.2	192.7814	0.4400	0.2043	0.2971	182.2920	0.0960	0.2329 (0.3087)
1	1	0.3	185.8769	0.5846	0.2622	0.4014	179.3242	0.1778	0.3529 (0.3978)
1	1	0.4	181.0562	0.7129	0.3086	0.4876	177.6956	0.2791	0.4660
1	1	0.9	172.8911	0.9990	0.3932	0.9006	173.2978	0.4183	0.9040
0.2	10.0	0.2	182.6709	0.4752	0.1727	0.2599	182.2920	0.1561	0.2535
0.2	10.0	0.3	179.2672	0.5896	0.1750	0.3520	179.3242	0.1687	0.3502
0.2	10.0	0.4	177.2978	0.6983	0.1762	0.4407	177.6956	0.1748	0.4414
0.2	10.0	0.9	172.8906	0.9987	0.1780	0.9003	173.2978	0.1785	0.9017
10.0	0.2	0.2	199.2087	0.2973	0.0504	0.2287	182.2920	0.0221	0.2076 (0.2337)
10.0	0.2	0 3	192.0522	0.4296	0.0735	0.3353	179.3242	0.0473	0.3141 (0.3471)
10.0	0.2	0.4	185.5680	0.5784	0.0999	0.4352	177.6956	0.0918	0.4217
10.0	0.2	0.9	172.8916	0.9986	0.1778	0.9003	173.2978	0.2062	0.9020

The values of y_{j1} corresponding to the Murphree temperature convention were calculated by the following procedure:

1. Same as above.
2. Find the T_{Mj} which satisfies Equation (11-2).
3. Evaluate m_1 by use of the appropriate formula given below.
4. Compute n_{0GM1}' by use of Equation (11-23) and y_{j1} by use of Equation (11-24).

For definiteness, the value of m_1 corresponding to the outlet liquid composition $\{x_{ji}\}$ was used. For Illustrative Example 11-1 (Table 11-1), the following formula for m_1 as suggested in Reference (4) was used:

$$m_1 = \frac{dy_1^*}{dx_1} = \frac{d\left[\alpha_1 x_1 \left(\sum_{i=1}^{2} \alpha_i x_i\right)^{-1}\right]}{dx_1} = \frac{\alpha_1}{[1 + x_1(\alpha_1 - 1)]^2} \qquad (11\text{-}26)$$

For Illustrative Example 11-2 (Table 11-2) whose liquid phase was highly nonideal, the following formula for m_1 was used:

$$m_1 = \frac{d\left[\gamma_1 \alpha_1 x_1 \left(\sum_{i=1}^{2} \gamma_i \alpha_i x_i\right)^{-1}\right]}{dx_1} = \frac{\gamma_1 \alpha_1 + \alpha_1 x_1 \dfrac{d\gamma_1}{dx_1}}{\sum_{i=1}^{2} \gamma_i \alpha_i x_i}$$

$$- \frac{\gamma_1 \alpha_1 x_1 \left[\gamma_1 \alpha_1 + \alpha_1 x_1 \dfrac{d\gamma_1}{dx_1} - \gamma_2 + (1 - x_1)\dfrac{d\gamma_2}{dx_1}\right]}{\left(\sum_{i=1}^{2} \gamma_i \alpha_i x_i\right)^2} \qquad (11\text{-}27)$$

where component 2 (the least volatile) was taken as the base component in Equations (11-26) and (11-27), that is, $\alpha_2 = K_2/K_2 = 1$. At large values of n_G' relative to n_L' (and/or large values of V relative to L), it can be seen from Figure 11-1 that the value of m_1 at x_{j1} may differ appreciably from its value at $x_{j1}^{\mathfrak{e}}$. The precise variation of m_1 depends, of course, on the shape of the given equilibrium curve. Better values of m_1 for those cases in which $n_L' < n_G'$ may be obtained for Illustrative Example 11-2 by locating the point $(x_{j1}^{\mathfrak{e}}, y_{j1}^{\mathfrak{v}})$ graphically by use of the slope $(-n_L'/n_G')$ and then taking m_1 to be equal to the slope of the straight line that passes through the points $(x_{j1}^{\mathfrak{e}}, y_{j1}^{\mathfrak{v}})$ and (x_{j1}, y_{j1}^*) as may be visualized by use of Figure 11-1. Corresponding values of y_{j1} which were obtained by use of these values of m_1 are shown in parentheses in Table 11-2 for selected values of x_{j1}. An improvement in the value of y_{j1} was obtained, that is, the agreement between the y_{j1}'s predicted by use of Murphree and vaporization efficiencies is seen to have been improved by use of the graphically determined values of m_1.

For the case in which n_G' (and/or V) is small relative to n_L' (and/or L), it may be seen from Figure 11-1 that the value of m_1 at $x_{j1}^{\mathfrak{e}}$ approaches its value at x_{j1}, and the value of m_1 given by Equation (11-27) approaches the correct value of m_1. For those entries in Table 11-2 where n_G' is small relative to n_L', it is seen that the y_{j1} computed by use of the Murphree efficiency approaches the value of y_{j1} computed by use of the vaporization efficiency.

Although some disagreement between the y_{j1}'s predicted by use of the two types of efficiencies might be expected for Illustrative Example 11-2 because the assumption given by Equation (8-6) becomes an approximation for the nonideal solution, ethanol-water, and thus n_G and n_L become only approximately equal to n_G' and n_L', respectively. The error involved in the approximations $n_G' = n_G$ and $n_L' = n_L$ appears to be small, however, because it is evident from the results of Table 11-2 that most of the deviation between the y_{j1}'s computed by use of the two types of efficiencies may be attributed to the precise method used to calculate m_1.

Since it has been shown that for the same sets of film coefficients (n'_L, n'_G) and compositions $\{x_{j1}, y_{j+1,1}\}$ approximately the same value for y_{j1} is obtained by use of vaporization efficiencies as is obtained by use of Murphree efficiencies for binary mixtures, the existing correlations for n'_L and n'_G given in the *Bubble-Tray Design Manual* (4) may be used to estimate the E_{ji}'s (for Model II) for multicomponent mixtures. This statement is based on the fact that the formulas for the vaporization efficiencies that were used in the computations for the binary mixtures are precisely the same ones that are also applicable for multicomponent mixtures. The use of vaporization efficiencies avoids the uncertainty encountered in the determination of m_1 when the Murphree equations are applied to multicomponent mixtures. Also avoided is the further apparent complication that results in the solution of the equations for multicomponent mixtures as a consequence of the fact that only $(c - 1)$ Murphree efficiencies may be fixed independently for each plate j.

Use of vaporization efficiencies as given by Model II [Equation (11-10)] and the corresponding temperature convention given by Equation (11-1) has already been demonstrated for multicomponent distillation problems (15). Neither convergence nor numerical problems were encountered provided as supposed in the development that n_{0Gji} is any finite number equal to or greater than zero.

Thus, in summary, vaporization efficiencies provide a simple and accurate method for making use of the vast body of existing data in the form of Murphree efficiencies for binary mixtures for the prediction of the efficiencies for multicomponent mixtures.

Unfortunately, as will be discussed in Chapter 12, a limination of Equation (11-10) in the description of certain experimental results was discovered wherein the product rates required that E_{ji} be less than unity while the calculated values of $y_{j+1,i}/K_{ji}x_{ji}$ were greater than unity. If $y_{j+1,i}/K_{ji}x_{ji} > 1$, it is evident from Equation (11-10) that no $n_{0Gji} > 0$ exists for which Model II gives $E_{ji} > 1$. This limitation is removed, however, by Model III, which follows. Also, for purposes of comparison, the efficiency expressions for Model II for each of the two temperature conventions are presented in Table 11-3.

Model III: Partially mixed liquid phase

There follows an extension of Model II that permits the liquid on a plate to take on a different temperature at each point along the plate. This model makes use of the concept of a vaporization efficiency at a point which con-

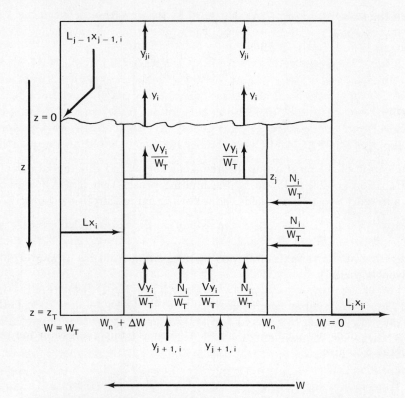

Figure 11-2. Sketch of the behavior assumed for any plate j. [Taken from J. P. Graham, J. W. Fulton, M. S. Kuk, and C. D. Holland, "Predictive Methods for the Determination of Vaporization Efficiencies," *Chem. Eng. Sci.*, **28**, (1973), 475.]

stitutes an extension of the concept of a vaporization efficiency for a plate. Hereafter, the vaporization efficiency defined by Equation (9-4) is referred to as the *vaporization plate efficiency* to distinguish it from the *vaporization point efficiency* [Equation (9-1)] which is used in the formulation of Model III. The assumptions upon which Model III are based are as follows:

1. The liquid phase is perfectly mixed in the direction z (see Figure 11-2).
2. Same as Model II.
3. Same as Model II.
4. The liquid at each location W has a different temperature (see Figure 11-2) which is equal to the temperature of the vapor leaving the surface of the liquid at this same W.)
5. The product $K_G a_i S f_i^V$ remains constant in the direction z at each W.

There follows a material balance on component i (in the vapor phase) over the element, $z_j \leqq z \leqq z_j + \Delta z$, $W_n \leqq W \leqq W_n + \Delta W$

$$\int_{W_n}^{W_n + \Delta W} \frac{V y_i}{W_T}\bigg|_{z_{j+1},W} dW - \int_{W_n}^{W_n + \Delta W} \frac{V y_i}{W_T}\bigg|_{z_j,W} dW$$

$$+ \int_{W_n}^{W_n + \Delta W} \int_{z_j}^{z_j + \Delta z} \frac{N_i}{W_T}\bigg|_{z,W} dz\, dW = 0 \qquad (11\text{-}28)$$

where

$$\frac{N_i}{W_T}\bigg|_{z,W} = K_G a_i \frac{S}{W_T}[f_i^V(K_i x_i|_{z,w} - y_i|_{z,w}].$$

Application of the mean value theorems of integral and differential calculus to Equation (11-28) followed by the limiting process wherein ΔW and Δz are allowed to go to zero yields the following expression in the limit:

$$V \frac{dy_i}{dz} + N_i = 0 \qquad (0 < z < z_T,\, 0 < W < W_T)$$

Separation of the variables followed by integration from $z = 0$ to $z = z_T$ at a given W yields

$$\int_{y_i}^{y_{j+1,i}} \frac{dy_i}{K_i x_i - y_i} = -\int_{z=0}^{z=z_T} \frac{K_G a_i S f_i^V}{V} dz \qquad (11\text{-}29)$$

Since $K_i x_i$ does not change with z at a given W, this expression may be integrated to give

$$\log_e \frac{K_i x_i - y_i}{K_i x_i - y_{j+1,i}} = -\frac{K_G a_i S f_i^V}{V} z_T \qquad (11\text{-}30)$$

For convenience, let

$$F_i(W) = K_i x_i$$

$$n_i(W) = \frac{K_G a_i S f_i^V z_T}{V} \quad \text{or} \quad \frac{1}{n_i} = \frac{1}{n_{Gi}} + \left(\frac{K_i V}{L}\right) \frac{1}{n_{Li}} \qquad (11\text{-}31)$$

Equation (11-30) may be rearranged to give the expression for the modified Murphree point efficiency, E_{Pji}^M, namely,

$$E_{Pji}^M = \frac{y_i - y_{j+1,i}}{F_i - y_{j+1,i}} = 1 - e^{-n_i} \qquad (11\text{-}32)$$

The *vaporization point efficiency* at any W at the surface of the liquid is defined by Equation (9-1), namely,

$$\tilde{f}_i^V = E_i \tilde{f}_i^L \qquad (9\text{-}1)$$

When assumptions stated above are imposed on this expression, the following definition for the vaporization point efficiency is obtained:

$$y_i = E_i F_i \qquad (11\text{-}33)$$

where y_i, E_i, and F_i all depend on W surface of the liquid, and the y_i's have been omitted in the interest of simplicity. Use of this expression to eliminate

y_i from Equation (11-30) yields the following formula for the vaporization point efficiency:

$$E_i = 1 - e^{-n_i}\left(1 - \frac{y_{j+1,i}}{F_i}\right) \tag{11-34}$$

Similarly, in the present notation, the defining equation for the vaporization plate efficiency (Equation 9-4) becomes

$$y_{ji} = E_{ji}F_i(0) \tag{9-4}$$

The formula for the vaporization plate efficiency for Model III is found by first observing that the composition of the vapor entering plate $j - 1$ is given by

$$y_{ji} = \int_0^1 y_i \, d\omega = \int_0^1 E_i F_i \, d\omega \tag{11-35}$$

where the change in variable

$$\omega = \frac{W}{W_T} \qquad (0 \leqq \omega \leqq 1)$$

has been made. When the integral appearing in Equation (11-35) is approximated by use of the well known numerical method called the "implicit method" that was applied to distillation problems in Reference (16), one obtains

$$y_{ji} = \mu y_i(1) + (1 - \mu)y_i(0) = \mu E_i(1)F_i(1) + (1 - \mu)[E_i(0)F_i(0)] \tag{11-36}$$

where $0 \leqq \mu \leqq 1$. [Although a single value of μ is implied by Equation (11-36) for all components, it is evident that the exact evaluation of the integral would generally require a different μ for each component i. However, for a wide variety of physical situations, it can be shown that the corresponding equations are satisfied by using the same μ for each component i (see Problem 11-6 for an outline of this proof).] When the expression given by Equation (11-34) for the vaporization point efficiency is substituted into Equation (11-36), the following result is obtained:

$$\begin{aligned} y_{ji} = \mu\{F_i(1) - e^{-n_i(1)}[F_i(1) - y_{j+1,i}]\} \\ + (1 - \mu)\{F_i(0) - e^{-n_i(0)}[F_i(0) - y_{j+1,i}]\} \end{aligned}$$

where $n_i(0)$ and $n_i(1)$ are the values of n_{0Gi} at $\omega = 0$ and $\omega = 1$, respectively. When Equation (9-4) is used to eliminate y_{ji} from this expression, the following formula for the vaporization plate efficiency for Model III results:

$$\begin{aligned} Model\ III:\ E_{ji} = \frac{\mu \bar{K}_{j-1,i}x_{j-1,i}}{K_{ji}x_{ji}}\left[1 - e^{-n_i(1)}\left(1 - \frac{y_{j+1,i}}{\bar{K}_{j-1,i}x_{j-1,i}}\right)\right] \\ + (1 - \mu)\left[1 - e^{-n_i(0)}\left(1 - \frac{y_{j+1,i}}{K_{ji}x_{ji}}\right)\right] \end{aligned} \tag{11-37}$$

The symbol $\bar{K}_{j-1,i}$ in Equation (11-37) is used to denote the fact that this particular K value is to be evaluated at that temperature $\bar{T}_{j-1}$ which is required to make the sum of the point values of y_i at the inlet ($\omega = 1$) of plate j equal to unity, that is, $\bar{T}_{j-1}$ is that temperature which satisfies

$$1 = \sum_{i=1}^{c} y_i(1) = \sum_{i=1}^{c} E_i(1)F_i(1)$$

$$= \sum_{i=1}^{c} [\bar{K}_{j-1,i}x_{j-1,i} - e^{-n_i(1)}(\bar{K}_{j-1,i}x_{j-1,i} - y_{j+1,i})] \qquad (11\text{-}38)$$

The temperature of the liquid leaving plate j, T_j, is that temperature required to make the sum of the point values of y_i at the outlet ($\omega = 0$) of plate j equal to unity. Thus, T_j is that temperature which satisfies

$$1 = \sum_{i=1}^{c} y_i(0) = \sum_{i=1}^{c} E_i(0)F_i(0)$$

$$= \sum_{i=1}^{c} [K_{ji}x_{ji} - e^{-n_i(0)}(K_{ji}x_{ji} - y_{j+1,i})] \qquad (11\text{-}39)$$

Observe that since $\bar{T}_{j-1}$ and T_j are to be selected such that Equations (11-38) and (11-39), respectively, are satisfied, then it follows that the general temperature convention given by Equation (11-1) is also satisfied (see Problem 11-7).

To apply Model III, it is necessary to compute two temperatures for each plate. The fact that the temperature $\bar{T}_{j-1}$ may differ from the temperature T_{j-1} of the liquid leaving plate $j - 1$ is easily demonstrated by considering the case in which plate $j - 1$ is perfect and plate j is not. For perfect plate $j - 1$, $\sum_{i=1}^{c} K_{j-1,i}x_{j-1,i} = 1$, and obviously the temperature required to satisfy this relationship may differ from the one required to satisfy Equation (11-38).

EXAMINATION OF THE FORMULA FOR E_{ji} FOR MODEL III. The choice of μ can be interpreted as the specification of the degree of backmixing of the liquid. For example, when $\mu = 0$, Model III [Equation (11-37)] reduces to Model II [Equation (11-10)] which assumes the liquid phase is perfectly mixed. As μ approaches unity, it is seen from Equation (11-36) that the y_{ji} entering plate $j - 1$ approaches $y_i(1)$ (at the inlet of plate j where the driving force for mass transfer is the greatest), and consequently the separation which may be achieved tends toward a maximum as demonstrated in Part 2. Thus, it appears reasonable to interpret μ as a mixing parameter.

If for each plate j, $n_i(0) > 0$ and $n_i(1) > 0$, then for all μ in the closed interval $0 \leq \mu \leq 1$, the E_{ji} computed by use of Equation (11-37) has the property that it is always nonzero, finite, and positive, or $0 < E_{ji} < \infty$.

Next consider the case for which Model II failed, namely, $y_{j+1,i} > K_{ji}x_{ji}$, and an E_{ji} lying between zero and unity is needed to satisfy the experimental results. Before the range of the E_{ji}'s given by Equation (11-37) may be established, it is necessary to specify the relationship between $F_i(1)$ and $F_i(0)$ and the relationship between $n_i(1)$ and $n_i(0)$. When the direction of mass transfer is from the vapor to the liquid ($y_{j+1,i} > K_{ji}x_{ji}$), then it is reasonable to suppose that $\bar{K}_{j-1,i}x_{j-1,i} < K_{ji}x_{ji}$ or $F_i(1) < F_i(0)$. Since the temperature T_j of the liquid leaving plate j of a distillation column is generally greater than the temperature $\bar{T}_{j-1}$ of the liquid at the inlet on plate j, it follows from Equation (11-20) that $n_i(1)$ generally increases monotonically with $n_i(0)$. Thus, if

1. $y_{j+1,i} > F_i(0) > 0$
2. $0 < F_i(1) < F_i(0)$
3. $n_i(1)$ increases monotonically with $n_i(0)$ and also $n_i(1) > 0$ and $n_i(0) > 0$

then there exist positive values for $n_i(1)$ and $n_i(0)$ such that $0 < E_{ji} < 1$ for all $\mu > 0$ $(0 < \mu \leq 1)$. On the basis of these conditions, it follows that the upper bound of E_{ji} is $y_{j+1,i}/F_i(0) > 1$, which occurs at the origin $[n_i(0) = 0, n_i(1) = 0]$. The lower bound is given by

$$\lim_{n_i(1) \to \infty} E_{ji} = \mu \frac{F_i(1)}{F_i(0)} + (1 - \mu) = 1 - \mu \left[1 - \frac{F_i(1)}{F_i(0)} \right]$$

From condition (2), it follows that the E_{ji} given by Equation (11-37) is less than unity for all $\mu > 0$ $(0 < \mu \leq 1)$. Since it can be shown that $dE_{ji}/dn_i(1) < 0$ for all $n_i(1) > 0$, it follows that E_{ji} decreases monotonically from its upper bound (which is greater than unity) to its lower bound (which is less than unity) as $n_i(1)$ is varied from zero to infinity. Hence, it may be concluded that for each choice of $\mu (0 < \mu \leq 1)$, there exists some positive number N such that for all $n_i(1) > N$, the values of E_{ji} lie in the open interval $0 < E_{ji} < 1$. If conditions (2) and (3) are removed, the generality of the result is lost.

Also, for the special case in which the compositions of the vapor and liquid approach that of pure component $(y_{ji} \rightarrow y_{j+1,i} \rightarrow x_{ji}, T_j \rightarrow T_{j-1})$ at one end of the column, it is readily demonstrated by use of Equations (11-37) through (11-39) that $E_{ji} \rightarrow 1$.

The use of this model for the representation of the backmixing on the plate of a distillation column is demonstrated by use of a field test in Part 2.

Model IV: Plug flow of the liquid phase

The descriptions of Models II, IV, and V presented herein are based on the temperature conventions given by Equation (11-1) while the descriptions of these models in the literature are based on the Murphree temperature convention [Equation (11-2)]. For all but the smallest of plates wherein the liquid traverses the length of the plate, the assumption of a perfectly mixed liquid phase is not valid and leads to separations which are poorer than those obtained (4). Model IV was first formulated by Lewis (20) who utilized the Murphree temperature convention.

The assumptions upon which Model IV is based are as follows:

1. The liquid is perfectly mixed in the vertical direction, but no mixing occurs in the radial direction.
2. Same as Model II.
3. The total flow rates L and V are constant at appropriately selected values.
4. Same as Model II.
5. The product $K_G a_i S f_i^V$ remains constant with respect to both z and W. This amounts to assuming that the modified Murphree point efficiency E_{Pji}^M is equal to a constant for all z and W, that is,

$$E_{Pji}^M = \frac{y_i - y_{j+1,i}}{K_{ji}x_i - y_{j+1,i}} = 1 - e^{-n_{0Gji}} \tag{11-40}$$

Both x_i and y_i depend, of course, on ω. On the basis of these assumptions, the formulas for Model IV are developed in a manner analogous to that demonstrated by Lewis (20) for the case in which the Murphree temperature convention was employed (see Problems 11-7 and 11-9). The results obtained for both of the temperature conventions are summarized in Table 11-3.

EXAMINATION OF THE FORMULA FOR E_{ji} FOR MODEL IV. For all values

$$\frac{y_{j+1\,i}}{K_{ji}x_{ji}} \leqq 1, \qquad 0 < (1 - e^{-n_{0Gji}}) < 1, \qquad 0 < A_{ji} < \infty$$

the formula for E_{ji} for Model IV (see Table 11-3) gives finite and positive numbers. If, however, $y_{j+1,i}/K_{ji}x_{ji} > 1$, then there exist infinitely many values of $A_{ji} > 0$ and $[\exp(E_{Pji}^M/A_{ji}) - 1] > 0$ for which Model IV gives negative values for E_{ji}. For example, suppose that $(1 - e^{-n_{0Gji}}) = 0.9$, $y_{j+1,i}/K_{ji}x_{ji} = 10$, and $A_{ji} = 1$. Then $E_{ji} = -3.14$.

In any iterative calculational procedure wherein new values of E_{ji} are computed on the basis of the most recent sets of compositions, temperatures, and total flow rates, the existence of sets of values of these variables that produce negative values for one or more E_{ji}'s appears to be quite likely.

That negative values of E_{ji} may be obtained for reasonable choices of values for the other variables is a consequence of the fact that in the integration of y_i over all ω, no restriction requiring each y_i to be positive or for their sum to be equal to unity over all ω was imposed. In fact, the expression for $y(\omega)$ for Model IV gives negative values for y_i for all positive values of ω less than 0.234. In other words, the difficulty arises because the conditions (or assumptions) of the model permit the behavior of the multicomponent mixture to be described by solving the differential equation for each component independent of those for the other components and thus there is no assurance that each y_i is positive and that the sum of y_i's is equal to unity. Thus, negative values of E_{ji} result from an inadequacy in the model and not from the definition of E_{ji}, for by Equation (9-4), E_{ji} is always nonzero, finite, and positive.

Model V: Partially mixing of the liquid phase by eddy diffusion

Generally, the liquid on a plate is neither perfectly mixed nor is it represented by plug flow in which no horizontal mixing occurs (4). Instead, the actual situation lies somewhere between these two extremes. To account for this situation, Gerster (11) proposed a model which is the same as the one that follows except for the temperature convention. The assumptions for Model V follows:

1. The liquid is perfectly mixed in the vertical direction (the z direction and the partial mixing in the positive direction of W is given by

$$\mathcal{I}_i = -\mathcal{D}_E S \frac{dC_i}{dW} = -\left(\frac{\mathcal{D}_E S \rho_L}{W_T}\right)\frac{dx_i}{d\omega}$$

where $\mathcal{D}_E$ is the effective diffusion coefficient for component i in the liquid.

2. Same as Model IV.

When the mathematical development of the model is carried out in a manner analogous to that demonstrated by Gerster (11) for the case in which the Murphree temperature convention is used, the results so obtained are summarized in Table 11-3 (see Problem 11-8). For purposes of comparison, those obtained by Gerster (11) are also listed.

For all $y_{j+1,i}/K_{ji}x_{ji} \leq 1$ and all positive ϕ, it is evident that the expression for E_{ji} for Model V (see Table 11-3) gives finite and positive values for E_{ji}. Unfortunately, for $y_{j+1,i}/K_{ji}x_{ji} > 1$, there exists the possiblity of computing a negative number for E_{ji}. For example, suppose

$$(1 - e^{-n_{0Gji}}) = 0.9, \qquad N_{Pe} = 10, \qquad \frac{y_{j+1,i}}{K_{ji}x_{ji}} = 10, \qquad A_{ji} = 0.3$$

For these conditions, $\phi \cong 3.54$ and $E_{ji} = -18.6$. A detailed analysis of the

TABLE 11-3
SUMMARY OF THE RESULTS FOR MODELS II, IV, AND V
FOR THE TWO TEMPERATURE CONVENTIONS (17)

MODEL II

Temperature Convention: Equation (11-1)	Temperature Convention: Equation (11-2)
Model II: $E_{ji}^M = \dfrac{y_{ji} - y_{j+1,i}}{K_{ji}x_{ji} - y_{j+1,i}}$ $= 1 - \exp(-n_{0Gji})$ where n_{0Gji} is given by Equation (11-20).	Model II: $E_{Mji} = \dfrac{y_{ji} - y_{j+1,i}}{K_{Mji}x_{ji} - y_{j+1,j}}$ $= 1 - \exp(-n'_{0GMji})$ where n_{0GMji} is given by Equation (11-23).
Model II: $E_{ji} = 1 - [\exp(-n_{0Gji})] \times \left[1 - \dfrac{y_{j+1,i}}{K_{ji}x_{ji}}\right]$	

TABLE 11-3 *(Cont.)*

MODEL IV

Temperature Convention: Equation (11-1)	Temperature Convention: Equation (11-2)

Model IV: $E_{ji}^M = \dfrac{y_{ji} - y_{j+1,i}}{K_{ji}x_{ji} - y_{j+1,i}}$

$= A_{ji}\left[\exp\left(\dfrac{E_{Pji}^M}{A_{ji}}\right) - 1\right]$

where E_{Pji}^M is defined by Equation (11-40) and where

$A_{ji} = \dfrac{L_j}{K_{ji}V_j}$

Model IV: $E_{ji} = \dfrac{y_{j+1,i}}{K_{ji}x_{ji}}$

$+ A_{ji}\left[1 - \dfrac{y_{j+1,i}}{K_{ji}x_{ji}}\right] \times$

$\left[\exp\left(\dfrac{E_{Pji}^M}{A_{ji}}\right) - 1\right]$

Model IV: $E_{Mji} = \dfrac{y_{ji} - y_{j+1,i}}{K_{Mji}x_{ji} - y_{j+1,i}}$

$= \dfrac{1}{\lambda_{ji}}[\exp(\lambda_{ji}E_{PMji}) - 1]$

where

$\lambda_{ji} = \dfrac{L_j}{m_{ji}V_j}$

$E_{PMji} = \dfrac{y_i - y_{j+1,i}}{y_i^* - y_{j+1,i}}$

$= 1 - e^{-n'_{0GMji}}$

$\dfrac{1}{n'_{0GMji}} = \dfrac{1}{n'_{Gji}}$

$+ \left(\dfrac{m_{ji}V_j}{L_j}\right)\dfrac{1}{n'_{Lji}}$

MODEL V

Temperature Convention: Equation (11-1)	Temperature Convention: Equation (11-2)

Model V: $E_{ji}^M = \dfrac{y_{ji} - y_{j+1,i}}{K_{ji}x_{ji} - y_{j+1,i}}$

$= E_{ji}^M \phi$

where E_{Pji}^M is defined by Equation (11-40) and where

$\phi = \left[\dfrac{1 - \exp(-\eta - N_{Pe})}{(\eta + N_{Pe})\left(1 + \dfrac{\eta + N_{Pe}}{\eta}\right)}\right.$

$\left. + \dfrac{\exp(\eta) - 1}{\eta\left(1 + \dfrac{\eta}{\eta + N_{Pe}}\right)}\right]$

$\eta = \dfrac{N_{Pe}}{2}\left[\sqrt{1 + \dfrac{4E_{Pji}^M}{A_{ji}N_{Pe}}} - 1\right]$

$N_{Pe} = \dfrac{W_T L_j}{\mathfrak{D}_E S\rho^L}$

$A_{ji} = \dfrac{L_j}{K_{ji}V_j}$

Model V: $E_{ji} = \dfrac{y_{j+1,i}}{K_{ji}x_{ji}} + (1 - e^{-n_{0Gji}}) \times$

$\left[1 - \dfrac{y_{j+1,i}}{K_{ji}x_{ji}}\right]\phi$

Model V: $E_{Mji} = \dfrac{y_{ji} - y_{j+1,i}}{K_{Mji}x_{ji} - y_{j+1,i}}$

$= E_{Mji}\Phi$

where the definition for Φ is obtained for the one stated for ϕ by replacing E_{Pji}^M/A_{ji} in the definition of η by $\lambda_{ji}E_{PMji}$.

---

equation for this model shows that the negative values of E_{ji} for this example may be attributed to the same causes enumerated for Model IV.

Entrainment

Instead of including the effect of entrainment in the formula for the plate efficiency as proposed by Colburn (7), this effect is readily included in the statement of the material balances. The actual entrainment may be computed however, by use of the correlations such as those proposed by Hunt *et al.* (19) and Fair (8, 9). The correlations proposed by Fair make use of the fractional entrainment ψ, which is defined as follows:

$$\psi = \frac{e}{L + e} \qquad (11\text{-}41)$$

where

$L =$ flow rate of the "dry" liquid leaving the given plate under consideration; lb/hr.

$e =$ flow rate of the liquid entrained in the vapor leaving the given plate under consideration; lb/hr.

Since it is further assumed that the streams e and L have the same composition, it follows that the flow rates appearing in the defining expression for ψ [Equation (11-41)] may be expressed in moles per unit time. Thus, the effect of entrainment may be visualized as shown in Figure 11-3. The component-material balances for plate j are as follows:

$$(1 - \psi)l_{j-1,i} - \psi l_{ji} - v_{ji} - l_{ji} + v_{j+1,i} + \psi l_{j+1,i} = 0$$

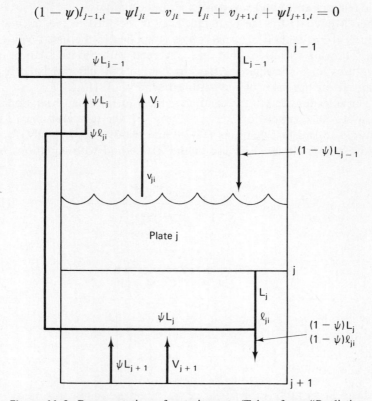

Figure 11-3. Representation of entrainment. (Taken from "Predictive Methods for the Determination of Vaporization Efficiencies," 480.)

Then, in terms of the component-flow rates of the vapor, one obtains

$$(1 - \psi)A_{j-1,i} - [1 + (1 + \psi)A_{ji}]v_{ji} + [(1 + \psi)A_{j+1,i}]v_{j+1,i} = 0 \qquad (11\text{-}42)$$

where $A_{ji} = L_j/E_{ji}K_{ji}V_j$. Thus, the matrix equation for the component-material balances remains tridiagonal in form.

Part 2: Analysis of the Experimental Results

Although the ultimate goal is obviously the correlation of the mixing parameter μ as a function of the operating conditions and geometry of a plate, the present effort was limited to the demonstration of the capacity of Model III for the representation of actual columns in the process of separating multicomponent mixtures.

In the analysis of the results of the field tests, it was supposed that the correlations given in the *Bubble-Tray Design Manual* (4) and summarized in Part 3 could be used to compute the number of gas phase and liquid phase transfer units. Entrainment was not regarded as being significant in the field test presented and was neglected in the analysis.

When Model III for the plate efficiency is included in the calculational procedure, the θ method of convergence is applied in the usual way except for the following variations that are concerned with the calculation of the temperatures and efficiencies. After the component-material balances had been solved on the basis of the assumed sets, $\{T_{jn}\}$, $\{V_{jn}\}$, $\{E_{jin}\}$, of values of the variables for the nth trial and after the θ method had been applied to give new sets of compositions, $\{x_{ji,n+1}\}$, $\{y_{ji,n+1}\}$, the temperatures $\bar{T}_{j-1}$ and T_j required to satisfy Equations (11-38) and (11-39), respectively, for each plate j were approximated by use of the K_b method with equations of the form:

$$\bar{K}_{j-1,b,n+1} = \frac{1 - \sum_{i=1}^{c} e^{-n_i(1)}y_{j+1,i,n+1}}{\sum_{i=1}^{c}(1 - e^{-n_i(1)})\bar{\alpha}_{j-1,i,n}x_{j-1,i,n+1}} \qquad (11\text{-}43)$$

$$K_{jb,n+1} = \frac{1 - \sum_{i=1}^{c} e^{-n(0)}y_{j+1,i,n+1}}{\sum_{i=1}^{c}(1 - e^{-n_i(0)})\bar{\alpha}_{jin}x_{ji,n+1}} \qquad (11\text{-}44)$$

where

$$\bar{\alpha}_{j-1,i,n} = \bar{K}_{j-1,i,n}/\bar{K}_{j-1,b,n};$$

$$\frac{1}{n_i(0)} = \frac{1}{n_{Gi}|_{T_{jn}}} + \left(\frac{K_{ji,n}V_{jn}}{L_{jn}}\right)\frac{1}{n_{Li}|_{T_{jn}}};$$

$$\frac{1}{n_i(1)} = \frac{1}{n_{Gi}|_{\bar{T}_{j-1,n}}} + \left(\frac{\bar{K}_{j-1,i,n}V_{jn}}{L_{jn}}\right)\frac{1}{n_{Li}|_{\bar{T}_{j-1,n}}}.$$

Then the new sets of temperatures $\{\bar{T}_{j-1,n+1}\}$ and $\{T_{j,n+1}\}$ corresponding to the above values of K_b were computed [see Equation (3-73)]. Of these two sets of temperatures, only the set $\{T_{j,n+1}\}$ was used in the enthalpy balances. The enthalpy balances were stated in the form called the *constant-composition method* [see Equations (3-77) through (3-81)] and solved for the total flow rates to be assumed for the next trial. At this point, the vaporization plate efficiencies to be assumed for the next trial were computed by use of Equation (11-37) and the most recent values of the variables.

Also, it should be mentioned that except for the feed plate, the composition at the inlet of each plate j was taken equal to the composition of the liquid leaving the plate above (plate $j - 1$). In the case of the feed plate f, the composition of the liquid at the inlet ($\omega = 1$) was computed as follows for the field tests presented below:

$$\bar{x}_{f-1,i} = \frac{FX_i + l_{f-1,i}}{F + L_{f-1}}$$

where F is the total flow rate of the feed and L_{f-1} is the total flow rate of the liquid entering the feed plate f from the plate above. This formula implies that the feed stream mixes with the liquid stream L_{f-1} at the entrance to the feed plate prior to flashing and that any vapor formed by the flash contacts the liquid above it. It should be noted that other models for the feed plate behavior may be employed. For example, if it is supposed that the feed flashes prior to contacting the liquid on the feed plate and forms the streams V_F and L_F, then $\bar{x}_{f-1,i} = (l_{Fi} + l_{f-1,i})/(L_F + L_{f-1})$.

The results of a field test made on an *o*-xylene column are presented in Table 11-4. The sources of all of the physical properties data used in the analysis of this column are also presented in Table 11-4. Since the column was operated at a relatively low pressure, the variation of the total pressure was taken into account.

The effect of μ on the distillate and bottoms compositions is demonstrated in Table 11-5. Observe that as μ increases, the separation achieved by the column improves. In fact for values of μ greater than about 0.38, the resulting mixing effects achieved by μ counterbalance the mass transfer effects imposed by the n_{oG}'s to the extent that the separations are better than those achieved by perfect plates. Thus, from the results presented in Table 11-5, it is evident that μ has the capacity for changing the compositions of the distillation and bottoms over relatively wide ranges.

The effect of μ on the temperature at the inlet and outlet, respectively, of each plate j is demonstrated by the results presented in Table 11-6. At $\mu = 0$, the mixing is perfect, but the plates are not perfect because the number of overall mass transfer units, n_{oGji}, are finite, and in this case the deviations between $\bar{T}_{j-1}$ and T_j are caused by mass transfer effects alone. Perfect plates correspond, of course, to the case in which $\mu = 0$ and $n_{oGji} \rightarrow \infty$ for

TABLE 11-4

OPERATING CONDITIONS, GEOMETRY, AND SOURCES OF THE
PHYSICAL PROPERTIES DATA FOR THE o-XYLENE COLUMN (12)

I. *Operating Conditions and Geometry*

$L_0/D = 13.35$	$h_w = 3$ in.
$f = 15$ (feed plate)	$Z_l = 4$ ft and 1 in. (distance between
$N = 40$	weirs)
Total condenser at 19.7 psia	ΔP (across condenser) $= 2$ psi
I.D. of column $= 5.5$ ft	ΔP (per plate) $= 0.225$ psi
Plates = Koch Flexitray — AF,	Pressure in reboiler $= 26.7$ psia
singlepass flow	Thermal condition of feed: liquid at
	148.9°F and 54.7 psia

Component	Comp. No.	Feed lb/hr	Feed mole/hr	Distillate lb/hr	Distillate mole/hr	Bottoms lb/hr	Bottoms mole/hr
o-Xylene	1	4,360.5	41.07479	4,280.5	40.32121	—	0.75358
m-Xylene	2	} 170.0*	0.80068	85.0	0.80068	—	—
p-Xylene	3		0.80068	85.0	0.80068	—	—
Ethylbenzene	4	8.5	0.80068	85.0	0.80068	—	—
C_9+Aromatics†	5	3,935.5	32.7399	15.5	0.12896	3,920.0	32.61503
Non-Aromatics†	6	25.5	0.22723	25.5	0.22723	—	—

*The experimental value for both m-xylene and p-xylene was 170 lb/hr, and an equal distribution (85 lb/hr) was assumed for computational purposes.

†C_9 + aromatics were taken to be cumene, and the nonaromatics were taken to be ethylcyclohexane. (These choices were based on experimental evidence which was available.)

II. *Sources of the Physical Properties Data*

1. Vapor and liquid enthalpies except for the liquid enthalpy of ethylcyclohexane were taken from API Research Project 44. [Selected Values of Properties of Hydrocarbons and Related Compounds, American Research Project 44, Thermodynamics Research Center, Texas A&M University]. The liquid enthalpy of ehycyclohexane was estimated by use of Theisen's correlation.*
2. The vapor pressures were computed by use of the Antoine equation, and the constants for this equation were taken from *Lange's Handbook of Chemistry* 10th ed. by N. A. Lange, Handbook Publishers, Inc., McGraw-Hill Book Company, New York, N.Y., 1967.
3. Liquid viscosities for the pure components were taken from API Research Project 44.
4. Critical properties for all components except ethylcyclohexane were taken from AP Research Project 44. The critical properties for ethycyclohexane were estimated by use of Lydersen's method.*
5. Atomic diffusion volumes were taken from Reference (10).
6. Molal volumes were taken from Reference (33).

The Properties of Gases and Liquids Their Estimation and Correlation, by R. C. Reid and T. K. Sherwood, McGraw-Hill Book Company, New York, N.Y., 2nd Edition, 1966.

all j and i. Again, it is evident that as μ is increased, the range of the corresponding temperature profiles increases which implies progressively better separations. For a given set of overall mass transfer units, the best separation that may be achieved by a given column is approached as μ approaches unity.

The effect of μ on the vaporization plate efficiencies is demonstrated in

TABLE 11-5
EFFECT OF THE MIXING PARAMETER μ ON DISTILLATE AND BOTTOMS COMPOSITIONS (12)

x_{Di}, Mole Fraction in Distillate at μ Indicated

Component	Experimental	Perfect Plates*	$\mu = 0.0$	$\mu = 0.25$	$\mu = 0.5$	$\mu = 0.75$	$\mu = 0.939$†	$\mu = 0.975$	$\mu = 1.0$
o-Xylene	0.95190	0.92217	0.88555	0.90910	0.93073	0.94592	0.95189	0.95261	0.95305
m-Xylene	} 0.03781	0.01889	0.01886	0.01889	0.01890	0.01890	0.01890	0.01890	0.01802
p-Xylene		0.01889	0.01884	0.01888	0.01890	0.01890	0.01890	0.01890	0.01802
Ethylbenzene	0.00189	0.00189	0.00189	0.00189	0.00189	0.00189	0.00189	0.00189	0.00189
C$_9$ + aromatics	0.00304	0.03280	0.06949	0.04587	0.02422	0.00903	0.00305	0.00233	0.00190
Nonaromatics	0.00536	0.00536	0.00536	0.00536	0.00536	0.00536	0.00536	0.00536	0.00536

x_{Bi}, Mole Fraction in Bottoms at μ Indicated

Component	Experimental	Perfect Plates*	$\mu = 0.0$	$\mu = 0.25$	$\mu = 0.5$	$\mu = 0.75$	$\mu = 0.939$†	$\mu = 0.975$	$\mu = 1.0$
o-Xylene	0.02258	0.06032	0.01680	0.07690	0.04945	0.03018	0.02259	0.02167	0.02112
m-Xylene	0.0	0.00001	0.00005	0.00002	0.0	0.0	0.0	0.0	0.0
p-Xylene	0.0	0.00002	0.00008	0.00003	0.0	0.0	0.0	0.0	0.0
Ethylbenzene	0.0	0.0	0.0	0.0	0.0	0.0	0.0	0.0	0.0
C$_9$ + aromatics	0.97742	0.93965	0.89306	0.923061	0.95054	0.96982	0.97741	0.97832	0.97888
Nonaromatics	0.0	0.0	0.0	0.0	0.0	0.0	0.0	0.0	0.0

*For this case $\mu = 0$ and $n_{OGji} \longrightarrow \infty$ for all j and i.
†This μ minimizes the objective function $0(\mu)$.

TABLE 11-6

EFFECT OF THE MIXING PARAMETER μ ON THE CALCULATED
TEMPERATURE FOR THE o-XYLENE COLUMN (12)

Plate No.	Perfect Plates $T_j(°C)$	$\mu = 0$		$\mu = 0.5$		$\mu = 1.0$	
		$\bar{T}_{j-1}(°C)$	$T_j(°C)$	$\bar{T}_{j-1}(°C)$	$T_j(°C)$	$\bar{T}_{j-1}(°C)$	$T_j(°C)$
0	93.33		93.33		93.33		93.33
1	151.54	151.72	151.84	151.38	151.48	151.23	151.37
2	152.15	152.35	152.46	151.99	152.08	151.87	151.92
4	153.35	153.51	153.68	153.16	153.24	152.94	152.97
6	154.55	154.79	154.91	154.31	154.40	153.96	153.99
8	155.77	156.01	156.14	155.47	155.57	154.98	155.02
10	157.03	157.24	157.38	156.65	156.78	156.03	156.11
12	158.32	158.48	158.63	157.88	158.03	157.16	157.30
14	159.64	159.72	159.88	159.15	159.34	158.43	158.65
15	160.32	160.33	160.52	159.89	160.05	159.28	159.59
16	160.90	160.98	161.10	160.50	160.60	160.03	160.07
18	162.09	162.13	162.26	161.61	161.73	160.96	160.97
20	163.33	163.31	163.46	162.76	162.91	161.84	161.86
22	164.63	164.51	164.67	163.97	164.16	162.72	162.74
24	165.97	165.73	165.91	165.26	165.49	163.60	163.63
26	167.34	166.97	167.16	166.62	166.89	164.51	164.57
28	168.72	168.22	168.42	168.05	168.34	165.49	164.61
30	170.08	169.47	169.67	169.51	169.80	166.62	166.84
32	171.39	170.70	170.90	170.95	171.23	168.01	166.40
34	172.63	171.91	172.09	172.34	172.58	169.82	170.43
36	173.79	173.08	172.34	173.63	173.83	172.07	172.80
38	174.87	174.20	173.25	174.82	174.97	174.37	174.91
40	175.88	175.27	174.35	175.89	176.01	176.10	176.32
41	176.36		175.93		176.46		176.73

Table 11-7 for o-xylene and the C_9 + aromatics. Observe that as μ is varied from $\mu = 0$ to $\mu = 1$, the values of E_{ji} for o-xylene (a relatively light component) varies from numbers less than unity to numbers approximately equal to or greater than unity while the converse is true for the E_{ji} for the C_9 + aromatics (a relatively heavy component). The nature of the variations of the E_{ji}'s is seen to account for the improvement in the separations achieved as μ is increased from $\mu = 0$ to $\mu = 1$.

The μ which best described the experimental results of the o-xylene column was taken to be equal to that μ which minimized the following objective function:

$$0(\mu) = \frac{1}{5}\left[\left|\log_e \frac{(X_{D1})_{\exp}}{(X_{D1})_{ca}}\right| + \left|\log_e \frac{(X_{D2} + X_{D3})_{\exp}}{(X_{D2} + X_{D3})_{ca}}\right| \right.$$
$$+ \left|\log_e \frac{(X_{D4} + X_{D5} + X_{D6})_{\exp}}{(X_{D4} + X_{D5} + X_{D6})_{ca}}\right| + \left|\log_e \frac{(x_{B1})_{\exp}}{(x_{B1})_{ca}}\right| \quad (11\text{-}45)$$
$$\left. + \left|\log_e \frac{(x_{B5})_{\exp}}{(x_{B5})_{ca}}\right|\right]$$

The variation of $0(\mu)$ with μ is presented in Table 11-8, and it is seen that a $\mu = 0.939$, the function $0(\mu)$ takes on a minimum value of 0.000186.

Model III may be used to describe not only columns in the process of separating mixtures which very nearly obey the ideal solution law (such as the o-xylene column) but also columns in the process of separating highly nonideal solutions. For example, the objective function of the same form as Equation (11-45) for a field test on a methanol column in the process of

TABLE 11-7

EFFECT OF THE MIXING PARAMETER μ ON THE
VAPORIZATION PLATE EFFICIENCIES (12)

Plate No.	E_{ji} at $\mu = 0$		E_{ji} at $\mu = 0.5$		E_{ji} at $\mu = 0.939$		E_{ji} at $\mu = 1.0$	
	o-Xylene	C$_9$ + aromatics	o-Xylene	C$_9$ + aromatics	o-Xylene	C$_9$ + aromatics	o-Xylene	C$_9$ + aromatics
0*	1.000	1.00	1.000	1.000	1.000	1.000	1.000	1.000
1	0.998	1.06	0.998	0.989	0.989	0.845	0.985	0.811
2	0.997	1.06	0.999	0.997	0.995	0.876	0.995	0.848
4	0.995	1.05	0.999	0.996	0.998	0.885	0.999	0.860
6	0.994	1.05	1.00	0.995	1.00	0.890	1.00	0.865
8	0.992	1.04	1.00	0.995	1.00	0.892	1.00	0.868
10	0.991	1.04	1.00	0.994	1.01	0.894	1.01	0.869
12	0.989	1.04	1.00	0.993	1.01	0.895	1.01	0.870
14	0.988	1.03	1.00	0.992	1.01	0.896	1.02	0.871
15	0.993	1.02	1.01	0.960	0.998	0.804	1.06	0.764
16	0.992	1.02	1.00	0.996	0.999	0.996	0.996	0.997
18	0.990	1.02	1.00	0.995	1.00	0.996	0.999	0.997
20	0.988	1.02	1.00	0.994	1.00	0.993	1.00	0.996
22	0.985	1.02	1.00	0.992	1.00	0.988	1.00	0.993
24	0.983	1.02	1.00	0.991	1.01	0.981	1.00	0.988
26	0.980	1.02	1.01	0.990	1.01	0.969	1.01	0.978
28	0.978	1.02	1.01	0.989	1.02	0.953	1.01	0.962
30	0.975	1.01	1.02	0.989	1.03	0.936	1.02	0.940
32	0.973	1.01	1.02	0.990	1.07	0.923	1.05	0.916
34	0.970	1.01	1.03	0.991	1.13	0.921	1.11	0.899
36	0.968	1.01	1.04	0.993	1.23	0.936	1.24	0.904
38	0.965	1.01	1.04	0.994	1.36	0.962	1.47	0.939
40	0.963	1.01	1.05	0.996	1.50	0.983	1.76	0.976
41*	1.000	1.00	1.00	1.000	1.00	1.000	1.00	1.000

*Perfect plates were assumed for the condenser-accumulator section and the reboiler.

TABLE 11-8

VARIATION OF THE OBJECTIVE FUNCTION $0(\mu)$ WITH μ (12)

μ	$0(\mu)$	μ	$0(\mu)$
Perfect plates*	0.4825	0.925	0.009898
0.0	0.7455	0.9375	0.001351
0.25	0.5940	0.939	0.000186
0.50	0.3903	0.940	0.000490
0.75	0.1524	0.945	0.002157
0.85	0.0665	0.950	0.007099
0.90	0.0280	0.975	0.022978
		1.000	0.037547

*For this case $\mu = 0$ and $n_{0Gji} \longrightarrow \infty$ for all j and i.

separating a highly nonideal mixture of alcohol, water, and traces of other oxygenated organic compounds was reduced to 0.0000137.

It should be noted, however, that the value of μ found for each of these examples cannot be regarded as a pure mixing parameter because μ has been used to absorb all of the inaccuracies in the correlations for the point values of n_G and n_L, the different mixing effects on each plate, the experimental data, and the inaccuracies of the model itself. Actually, the proposed model is not dependent on the particular correlations used for n_G and n_L (4). If and when better ones are developed, they may be employed in the model. In any event in spite of the acknowledged limitations, it appears that Model III and the mixing parameter μ do provide a useful method for treating plate efficiencies for columns in the process of separating multicomponent mixtures. In fact, from the results and analyses of field tests, it appears that the single-parameter mixing model gives an adequate description of existing columns. Since only one mixing parameter is involved per column and all other parameters contained within the model are given by existing correlations, it is concluded that the proposed mixing model (Model III) represents a firm basis upon which to build correlations for existing columns that may be subsequently employed as a predictive method in column design. Also, the capacity of μ to produce separations that range from ones much poorer to ones far better than those which may be achieved by perfect plates suggests that it is worthwhile to reexamine the existing data for single plates in search for a correlation of μ as a function of the geometry and operating conditions of a plate.

Part 3: Summary of Existing Correlations for the Prediction of the Number of Transfer Units and Entrainment

For the prediction of the number of transfer units n_G and n_L, the correlations proposed in the *Bubble-Caps Design Manual* (4) are presented. (This book is referred to hereafter as simply the *A.I.Ch.E. Manual.*)

PREDICTION OF n_G AND n_L. For the prediction of n_G, the following correlation for n_G is given by Equation (2) on page 27 of the *A.I.Ch.E. Manual* (4):

$$n_{Gi} = \frac{0.776 + 0.116h_w - 0.290\mathfrak{F} + 0.0217\dfrac{Q^*}{l}}{(N_{Sci})^{0.5}} \qquad (11\text{-}46)$$

*This particular form of this equation is open to question because the additional term, "+0.2 times the liquid hydraulic gradient in inches," appeared in the numerator of this equation as it appeared on a loose-leaf calculational form sheet used by the authors of Reference (4).

where

h_w = outlet weir height, inches;

$\mathfrak{F} = u_G \sqrt{\rho^V}$, (ft/sec) (lb/ft³)$^{1/2}$;

Q = flow rate of clear liquid, gal/min;

l = average liquid flow width, fit;

u_G = superficial velocity in cu ft per sec per sq ft of bubbling area of the plate, ft/sec;

$N_{Sci} = \mu^V / \rho^V D_i^V$, dimensionless ($\mu^V$ is the viscosity, ρ^V is the mass density of the gas phase, and D_i^V is the diffusivity of component i of a multicomponent mixture). Any consistent set of units may be employed.

For the prediction of n_L, the correlation given by Equation (9) on page 34 of the *A.I.Ch.E. Manual* (4) is recommended:

$$n_{Li} = [(1.065 \times 10^4)D_{im}^L]^{0.5}(0.26\mathfrak{F} + 0.15)t_L \tag{11-47}$$

where D_{im}^L is the liquid diffusivity in sq ft per hr of the solute i in the solvent m. The liquid contact time t_L in sec is computed by use of Equation (9) on page 35 of the *A.I.Ch.E. Manual* as follows:

$$t_L = \frac{37.4(Z_c)(Z_l)}{\dfrac{Q}{l}} \tag{11-48}$$

where Z_l is the length of liquid travel (distance between the inlet and outlet weirs), ft. The liquid holdup on a bubble tray, cu in. per sq in. of bubbling area was correlated as follows [Equation (10) on page 35 of the *A.I.Ch.E. Manual*].

$$Z_c = 1.65 + 0.19h_w - 0.65\mathfrak{F} + 0.020\frac{Q}{l} \tag{11-49}$$

For sieve trays, which were not included in the *A.I.Ch.E. Manual*, Smith (27) gives the following expressions that were recommended by Gerster *et al.* (11).

$$n_{Li} = 100(D_{im}^L)^{0.5}(0.49\mathfrak{F} + 0.17)t_L \tag{11-50}$$

and

$$Z_c = 0.24 + 0.725h_w - 0.29h_w\mathfrak{F} + 0.01\frac{Q}{l} \tag{11-51}$$

PREDICTION OF FRACTIONAL ENTRAINMENT. Fair *et al.* (8, 9) developed generalized correlations for both bubble cap and sieve trays by use of the liquid-vapor flow parameter

$$\frac{W^V}{W^L}\sqrt{\frac{\rho^V}{\rho^L}}$$

and a capacity parameter. The Souders and Brown (28) coefficient was selected as the capacity parameter:

$$C_{SB} = u_n \sqrt{\frac{\rho^V}{\rho^L - \rho^V}} \tag{11-52}$$

where W^L and W^V are the flow rates of the liquid and vapor, respectively, lb/hr. The vapor velocity u_n is based on the *net area* of the tray available for liquid disengagement. Normally, the net area is taken equal to the total area minus the area blocked off by one downcomer. This area should be reduced to account for unusual baffling effects.

For bubble cap trays, the charts given by Fair *et al*, are recommended subject to the following limitations:

1. The system is low to nonfoaming.
2. Weir height is less than 15% of tray spacing.
3. Bubbling area occupies most of the area between weirs.

The C_{SB} corresponding to the flood point may be determined from charts as described by Fair *et al*.

The value of C_{SB} so obtained corresponds to a surface tension of 20 dynes/cm and is denoted by $(C_{SB})_{\sigma=20}$. Fair *et al*. proposed the following correlation for computing C_{SB} at the surface tension σ of the system under consideration:

$$C_{SB} = (C_{SB})_{\sigma=20}\left(\frac{\sigma}{20}\right)^{0.2} \tag{11-53}$$

This value of C_{SB} at the flood point [the one given by Equation (11-53)] is used to compute $u_{n,\text{flood}}$ from Equation (11-52). The *percent flood* is computed by use of the design or operating value u_n as follows:

$$\text{percent flood} = \frac{u_n}{u_{n,\text{flood}}} \times 100 \tag{11-54}$$

Then the percent flood parameter is used to determine the fractional entrainment ψ. Similar correlations are given by Fair *et al*. for sieve trays (8, 9).

Prediction of certain physical properties which appear in the correlations

There follows a selected set of formulas for the prediction of diffusivities and viscosities for both vapor and liquid mixtures. In the interest of simplicity, the superscripts V and L have been omitted below because the ordering of the treatments serves to distinguish between the properties of the two phases.

PREDICTION OF DIFFUSIVITIES AND VISCOSITIES FOR VAPORS. For most systems, vapor diffusivities for binary pairs may be predicted by use of any one of several methods such as the Hirschfelder-Bird-Spotz equation (14), a modified form of this equation proposed by Slattery (26), or the equation recently proposed by Fuller *et al.* (10).

The diffusivities D_i for each component i of a multicomponent mixture may be predicted by use of the diffusivities D_{ij} for the binary pairs by use of the following relationship proposed by Wilke (32).

$$D_i = \frac{1 - y_i}{\displaystyle\sum_{j=1,\neq i}^{c} \frac{y_i}{D_{ij}}} \tag{11-55}$$

In the absence of experimental data, the viscosity of pure gaseous components at low pressures may be predicted as suggested by Bird *et al.* (2, 3, 14).

For mixtures of gases, at low densities, the semiempirical formula of Wilke (30) gives adequate results for most purposes.

For the case of elevated pressures and/or elevated temperatures, several methods that give essentially same results over the ranges of reduced temperature and pressure wherein they overlap (3) are available (5, 6, 18, 25).

PREDICTION OF THE DIFFUSIVITIES AND VISCOSITIES OF LIQUID MIXTURES. Unfortunately, methods for the prediction of the aforementioned physical properties for the liquid phases, particularly multicomponent mixtures, are far inferior to the corresponding methods for the vapor phase.

Of the methods proposed for the prediction of the diffusivities of very dilute solutions of a solute i in a solvent j, the Wilke-Chang equation (33) is generally regarded as one of the more reliable methods. Recently, Reddy *et al.* (24) proposed the following modified version of the Wilke-Chang correlation. Although the Reddy equation was shown to be superior to the Wilke-Chang equation for the prediction of the diffusion of water in an organic liquid, neither correlation holds satisfactorily for systems involving highly viscous liquids such as the diffusion of water in ethylene glycol.

Since the above correlations are applicable only for very dilute solutions of the solute i in the solvent j, the following method for predicting the effect of concentration of the solute on the diffusivity was developed by Wilke (31) on the basis of the suggestion of Powell *et al.* (23) that for an ideal solution, the product $D\mu$ at constant temperature should vary linearly with concentration. Thus,

$$\frac{D_{ij}\mu_{ij}}{T} = \left(\frac{D_{ji}^0\mu_i}{T} - \frac{D_{ij}^0\mu_j}{T}\right)x_i + \frac{D_{ij}^0\mu_j}{T} \tag{11-56}$$

where D_{ij} is the diffusivity of the solute i in the solvent j at the concentration corresponding to the mole fraction x_i. For nonideal solutions, Powell *et al.* (23) suggested that the following activity correction be included.

$$\frac{D_{ij}\mu_{ij}}{T} = \left[\left(\frac{D_{ij}^0\mu_j}{T} - \frac{D_{ij}\mu_j}{T}\right)x_i + \frac{D_{ij}^0\mu_j}{T}\right]\left[1 + \frac{\partial \log_e \gamma_i}{\partial \log_e x_i}\right] \tag{11-57}$$

were γ_i is the activity coefficient of the solute i in the solvent j at the mole fraction x_i. In the generalizations of the correlation given in the *A.I.Ch.E. Manual* for η_{Li}, the binary diffusivity was replaced by the effective binary diffusivity to give Equation (11-46). The effective binary diffusivity D_{im} represents the diffusivity of component i relative to the mixed solvent m (composed of all components except i).

Of the methods that have been proposed for predicting D_{im} for ternary mixtures, Himmelblau (13) recommends the following equation proposed by Tang (29) for the diffusion of dilute component 1 through a mixture of two solvents, components 2 and 3:

$$D_{im}^0 \mu_m^{1/2} = x_2 D_{12}^0 \mu_2^{1/2} + x_3 D_{13}^0 \mu_3^{1/2} \tag{11-58}$$

In the absence of proposed correlations for D_{im} multicomponent mixtures, the following modifications of Equation (11-58) are suggested. To account for concentrated solutions of the solute (component 1) in a mixed solvent composed of $c - 1$ components, it is suggested that Equation (11-58) be modified as follows so that it will approach the proper limits for all binary pairs of solute 1 with each of the $c - 1$ components of the solvent, that is,

$$D_{1m} \mu_m^{1/2} = \frac{1}{1 - x_1} [x_2 D_{12} \mu_{12}^{1/2} + x_3 D_{13} \mu_{13}^{1/2} + \cdots + x_c D_{1c} \mu_{1c}^{1/2}] \tag{11-59}$$

The diffusivities D_{ij} and viscosities μ_{ij} for the binary pairs are evaluated at the same relative values of mole fractions as the original mixture. For example, D_{12} and μ_{12} are to be evaluated at the mole fractions X_1 and X_2, defined as follows:

$$X_1 = \frac{x_1}{x_1 + x_2}, \qquad X_2 = \frac{x_2}{x_1 + x_2} \tag{11-60}$$

Although Equation (11-56) does approach the proper binary limits as well as the ternary limit given by Equation (11-58) (provided, of course, x_1 is very small), it should be regarded as a tentative expression because it is without experimental verification.

Actually, a formula for multicomponent mixtures that has a theoretical basis has been stated by Bird *et al.* (2), namely,

$$D_{1m} = \frac{1 - x_1}{\sum\limits_{j=1}^{c} \frac{x_j}{D_{ij}}} \tag{11-61}$$

However, Tang (29) demonstrated that the diffusivities predicted by use of Equation (11-58) gave far better agreement with the experimental results for ternary mixtures than did Equation (11-61).

Although numerous methods have been proposed for the prediction of the viscosities of binary mixtures of liquids, methods for the prediction of the viscosities of multicomponent mixtures of liquid are nonexistent. In view of this dilemma, it is suggested that the following Arrehenius equation for binary mixtures be generalized to include multicomponent mixtures:

$$\log_e \mu_m = x_1 \log_e \mu_1 + x_2 \log_e \mu_2 \tag{11-62}$$

The use of this correlation was suggested in the *A.I.Ch.E. Manual* for compounds that are similar, structurally and chemically.

Unless experimental data for multicomponent mixtures or correlations based thereon are available, it is suggested (without experimental verification) that the following generalized form of Equation (11-62) be used:

$$\log_e \mu_m = \sum_{i=1}^{c} x_i \log_e \mu_i \tag{11-63}$$

PROBLEMS

11-1. Verify the relationship given by Equation (11-20).

11-2. Verify the relationship given by Equation (11-23).

11-3. Verify the relationship given by Equation (11-25).

11-4. Verify the formulas for m_1 which are given by Equations (11-26) and (11-27).

11-5. If y_{ji} is defined by Equation (11-36), E_{ji} by Equation (9-4) [as stated below Equation (11-34)], and the temperatures $\bar{T}_{j-1}$ and T_j are determined such that Equations (11-38) and (11-39), respectively, are satisfied, then show that the temperature convention given by Equation (11-1) is satisfied.

11-6. Consider

$$y_{ji} = \int_0^1 [F_i - e^{-n_i}(F_i - y_{j+1,i})]\, d\omega = \int_0^1 f_i(\omega)\, d\omega \tag{A}$$

where the function $f_i(\omega)$ is continuous with the exception of possible finite discontinuities at the end point $\omega = 0$ and $\omega = 1$.

11-6(a). For the case where $f_i(\omega)$ is continuous over the closed interval and its mean value is bounded by $f_i(0)$ and $f_i(1)$, show that a μ_i exists such that

$$y_{ji} = \mu_i f_i(1) + (1 - \mu_i) f_i(0) \tag{B}$$

11-6(b). If $f_i(\omega)$ varies linearly with ω, show that $\mu_i = 1/2$ for each component i and thus $\mu_i/\mu_k = 1$, where k refers to any arbitrary component of the mixture.

11-6(c). If $f_i(\omega) = f_i(1) \neq f_i(0)$, $0 \leq \omega < 1$, show that $\mu_i = 1$ for each component i or $\mu_i/\mu_k = 1$ as μ_i and μ_k approach 1.

11-6(d). If $f_i(\omega) = f_i(0) \neq f_i(1)$, $0 \leq \omega < 1$, show that a necessary condition for each component i to be satisfied is that $\mu_i = 0$. In this case it will be observed that the equations are satisfied if $\mu_i/\mu_k = 1$ or any other arbitrary constant as both μ_i and μ_k go to zero. Also, observe that this case corresponds to the physical situation of a perfectly mixed liquid phase.

11-6(e). If each $f_i(\omega)$ is continuous over the closed interval $0 \leqq \omega \leqq 1$ and has a mean value lying between $f_i(0)$ and $f_i(1)$, then a sufficient condition for Equation (B) to be satisfied is that $\mu_i/\mu_k = 1$.

11-7(a). Show that the differential equation which describes Model IV is given by

$$y_{j+1,i} - y_i = \frac{-L}{V} \frac{dx_i}{d\omega} \tag{A}$$

where $\omega = W/W_T$.

11-7(b). Use Equation (11-40) to show that

$$\frac{dx_i}{d\omega} = \left(\frac{1}{K_{ji} E_{Pji}^M} \right) \frac{dy_i}{d\omega} \tag{B}$$

Then show that when $dx_i/d\omega$ is eliminated from Equations (A) and (B), and the result so obtained is integrated from $\omega = 0$ to any ω, one obtains

$$y_i(\omega) = y_{j+1,i} - [y_{j+1,i} - y_i(0)] \exp\left(\frac{E_{Pji}^M \omega}{A_{ji}} \right) \tag{C}$$

11-7(c). For the numerical example used in the text in the examination of Model IV, show that Equation (11-40) gives

$$y_i(0) = 1.9 K_{ji} x_{ji} \tag{D}$$

at $\omega = 0$.

11-7(d). Use the results given by Parts (b) and (c) to show that

$$y_i = 0 \text{ at } \omega \cong 0.234$$

and that $y_i > 0$ for $\omega < 0.233$ and that $y_i < 0$ for $\omega > 0.234$.

11-7(e). Show that the formula given in Table 11-3 for E_{ji}^M for Model IV is obtained by integrating the expression given by Equation (C) over all ω. Then use the defining equation $y_{ji} = E_{ji} K_{ji} x_{ji}$ to eliminate $y_{ji}/K_{ji} x_{ji}$ from E_{ji}^M to obtain the expression given in Table 11-3 for E_{ji} for Model IV.

11-8(a). Show that the differential equation corresponding to a material balance on component i for Model V is given by

$$\frac{1}{N_{Pe}} \frac{d^2 x_i}{d\omega^2} + \frac{dx_i}{d\omega} + \frac{V}{L} (y_{j+1,i} - y_i) = 0 \tag{A}$$

11-8(b). The boundary conditions customarily used are as follows:

$$\mathcal{I}_i \big|_{\omega=0} = 0 = -\frac{\mathcal{D}_E S \rho^L}{W_T} \frac{dx_i}{d\omega}\bigg|_{\omega=0} \tag{B}$$

and

$$x_i = x_{ji} \text{ at } \omega = 0$$

(Observe that the first of these boundary conditions is analogous to the boundary condition for a perfectly insulated wall in a heat transfer problem.) Show that the solution which satisfies the differential equation [Equation (A)] and the boundary conditions [Equation (B)] is

$$\frac{y_{j+1,i} - K_{ji} x_i}{y_{j+1,i} - K_{ji} x_{ji}} = \frac{\exp(\eta\omega)}{1 + \dfrac{\eta}{\eta + N_{Pe}}} + \frac{\exp[-(\eta + N_{Pe})\omega]}{1 + \dfrac{\eta + N_{Pe}}{\eta}} \tag{C}$$

11-8(c). By use of Equations (C), (11-40), and the values given for the numerical example presented in the text in the analysis of Model V, show that $y_i = 0$ at $\omega \cong 0.1525$.

11-8(d). Show that the formula given in Table 11-3 for E_{ji}^M for Model V may be obtained by first eliminating x_i from Equation (C) by use of Equation (11-40) and then integrating the resulting expression for y_i over all ω to give y_{ji}. Then use the defining equation $y_{ji} = E_{ji}K_{ji}x_{ji}$ to eliminate $y_{ji}/K_{ji}x_{ji}$ from E_{ji}^M to obtain the expression given in Table 11-3 for E_{ji} for Model V.

11-9. Develop the expressions given in Table 11-3 for Models II, IV, and V which are based on the Murphree temperature convention [Equation (11-2)] and the associated mass transfer relationships.

NOTATION*

C_i = concentration, lb-moles/unit volume.

$\mathfrak{D}_E$ = effective diffusivity; defined in the list of assumptions given for Model V.

e = flow rate of the liquid entrained in the vapor leaving the plate.

E_i = vaporization point efficiency.

E_{ji} = vaporization efficiency; defined by Equation (9-4).

E_{ji}^M = modified Murphree plate efficiency; defined by Equation (11-7).

E_{Mji} = Murphree plate efficiency; defined by Equation (11-24).

E_{Pji}^M = modified Murphree point efficiency for plate j.

$\bar{E}_i$ = component efficiency; defined by Equation (11-3).

f_{ji}^V, f_{ji}^L = fugacities of pure component i in the vapor and liquid states, respectively; evaluated at the total pressure and at the temperature of plate j, atm.

$\hat{f}_{ji}^V, \hat{f}_{ji}^L$ = fugacities of component i in the vapor and liquid streams leaving plate j, respectively; evaluated at the total pressure, temperature, and compositions of these respective phases, atm.

F_i = value of this product $K_i x_i$ at any point along W.

$K_G a_i$ = product of the overall mass transfer coefficient and the interfacial area; defined in Table 8-1.

K_{ji} = the ideal solution K value ($K_{ji} = f_{ji}^L/f_{ji}^V$) for component i; evaluated at the pressure of plate j and at the temperature of the liquid leaving plate j.

K_{Mji} = The K value for component i; evaluated at the pressure of plate j and at the corresponding bubble point temperature of the actual liquid leaving plate j.

L_j = total flow rate of liquid leaving plate j, lb-moles/hr.

*See also Chapter 8 notations.

$n_G,\ n'_G$ = number of gas phase transfer units; defined in Table 8-1.

$n_L,\ n'_L$ = number of liquid phase transfer units; defined in Table 8-1.

$n_{0G},\ n'_{0G}$ = number of overall gas phase transfer units for use with vaporization efficiencies; defined by Equation (11-20).

n'_{0GM} = number of overall gas phase transfer units for use with Murphree efficiencies; defined by Equation (11-23).

N_i = rate of mass transfer of component i from the liquid to the vapor phase; lb-moles/(unit time) (unit length of packing).

N_{Pe} = Peclet number.

P = total pressure for a given plate, atm.

S = cross-sectional area of the column, ft².

V_j = total flow rate of the vapor leaving plate j, lb-moles/hr.

W = distance measured from the outlet weir, $W = W_T$, to the inlet of the liquid, $W = 0$.

x_{ji} = mole fraction of component i in the actual liquid leaving plate j.

y_{ji} = mole fraction of component i in the actual vapor leaving plate j.

y^*_{ji} = mole fraction that component i would have if the vapor leaving plate j were in equilibrium with the liquid leaving.

z = depth of liquid on a given plate; measured from the surface ($z = 0$) to the floor of the plate ($z = z_T$), ft.

Greek letters

ψ = fractional entrainment; defined by Equation (11-41).

$\omega = W/W_T,\ 0 \leqq \omega \leqq 1$.

α_i = relative volatility; $\alpha_i = K_i/K_b$ where K_b is the base component, $\alpha_b = 1$.

β_j = plate factor for plate j.

γ^L_{ji} = activity coefficients for component i in the liquid phase; evaluated at the temperature, pressure, and composition of this stream as it leaves plate j.

$\rho^L,\ \rho^{\mathcal{L}}$ = total molal density of the liquid; evaluated at the bulk conditions and at the conditions at the interface, respectively.

Subscripts

i = component number.

j = plate number.

M = Murphree efficiency or temperature convention.

Superscripts

L = to be evaluated at the bulk conditions of the liquid phase.

$\mathcal{L}$ = to be evaluated at the conditions of the liquid at the interface.

M = modified Murphree efficiency.

V = to be evaluated at the bulk conditions of the vapor phase.

$\mathcal{V}$ = to be evaluated at the conditions of the vapor at the interface.

Mathematical symbols

$\sum_{i=1}^{c} x_i$ = sum over all components from $i = 1$ through $i = c$.

$\{n_i\}$ = set of all numbers $n_1, n_2, \ldots, n_c$.

REFERENCES

1. Bassyoni, A. A., R. McDaniel, and C. D. Holland, "Examination of the Use of Mass Transfer Rate Expressions in the Description of Packed Distillation Columns—II," *Chem. Eng. Sci.*, **25**, (1970), 437.

2. Bird, R. B., J. O. Hirschfelder, and C. F. Curtiss, "Theoretical Calculation of the Equation of State and Transport Properties of Gases and Liquids," *Trans. ASME*, **76**, (1954), 1011.

3. ———, W. E. Stewart, and F. N. Lightfoot, *Transport Phenomena*. New York: John Wiley & Sons, Inc., 1960.

4. *Bubble-Tray Design Manual*. American Institute of Chemical Engineers, 345 East 47 Street, New York, N. Y. 1958.

5. Carr, N. L., R. Kobayashi, and D. B. Burroughs, "Viscosity of Hydrocarbon Gases Under Pressure," *Am. Inst. Min. & Met. Engrs. Petroleum Tech.*, **6**, (1954), 264.

6. *Chemical Engineer's Handbook* (4th ed.), eds. R. H. Perry, C. H. Chilton, and S. D. Kirkpatrick. New York: McGraw-Hill Book Company 1963.

7. Colburn, A. P., "Effect of Entrainment on Plate Efficiency in Distillation," *Ind. Eng. Chem.*, **28**, (1936), 526.

8. Fair, J. R., "How to Predict Sieve Tray Entrainment and Flooding," *Petro/ Chem. Engr.*, **33**, 10 (1961), 45.

9. ———, and R. L. Mathews, "Better Estimates of Entrainment from Bubble-Cap Trays," *Petroleum Refiner*, **37**, 4 (1958), 153.

10. Fuller, E. N., P. D. Schettler, and J. C. Giddings, "A New Method for Prediction of Binary Gas-Phase Diffusion Coefficients," *Ind. Eng. Chem.*, **58**, (1966), 19.

11. Gerster, J. A., A. B. Hill, M. N. Hochgraf, and D. N. Robinson, *Tray Efficiencies in Distillation Columns—Final Report from the University of Delaware to the A.I.Ch.E. Committee*. 345 East 47 Street, New York, N. Y. American Institute of Chemical Engineers, 1958.

12. Graham, J. P., J. W. Fulton, M. S. Kuk, and C. D. Holland, "Predictive

Methods for the Determination of Vaporization Efficiencies," *Chem. Eng. Sci.*, **28**, (1973) 473.

13. Himmelblau, D. M., "Diffusion of Dissolved Gases in Liquid," *Chem. Rev.*, **64**, (1964), 527.

14. Hirschfelder, J. O., R. B. Bird, and E. L. Spotz, "The Transport Properties of Gases and Gaseous Mixtures II," *Chem. Rev.*, **44**, (1949), 205.

15. Holland, C. D., *Multicomponent Distillation*. Englewood Cliffs, N. J.: Prentice-Hall, Inc., 1963.

16. ———, *Unsteady State Processes with Applications in Multicomponent Distillation*, Prentice-Hall, Inc., Englewood Cliffs, N. J., 1966.

17. ———, A. E. Hutton, and G. P. Pendon, "Prediction of Vaporization Efficiencies for Multicomponent Mixtures by Use of Existing Correlations for Vapor and Liquid Film Coefficients," *Chem. Eng. Sci.*, **26**, (1971), 1723.

18. Hougen, O. A., K. M. Watson, and R. A. Ragatz, *C.P.P. Charts*, 2nd ed. New York.: John Wiley & Sons, Inc., 1960.

19. Hunt, C. d'A., D. N. Hanson, and C. R. Wilke, "Capacity Factors in the Performance of Perforated Plate Columns," *A. I. Ch. E. Journal*, **1**, (1955), 441.

20. Lewis, W. K., Jr., "Rectification of Binary Mixtures, "Plate Efficiency of Bubble Cap Columns," *Ind. Eng. Chem.*, **28**, (1936), 399.

21. McDaniel, R., A. A. Bassyoni, and C. D. Holland, "Use of the Results of Field Tests in the Modeling of Packed Distillation Columns and Packed Absorbers—III," *Chem. Eng. Sci.*, **25**, (1970), 633.

22. Murphree, E. V., "Rectifying Column Calculations," *Ind. Eng. Chem.*, **17**, (1925), 747.

23. Powell, R. E., W. E. Roseveare, and H. Eyring, "Diffusion, Thermal Conductivity, and Viscous Flow of Liquids," *Ind. Eng. Chem.*, **33**, (1941), 430.

24. Reddy, K. A., and L, K. Doraiswamy, "Estimating Liquid Diffusivity" *I & EC Fundamentals*, **6**, (1967), 77.

25. Reid, R. C., and T. K. Sherwood, *The Properties of Gases and Liquids*, 2nd ed. New York: McGraw-Hill Book Company, 1966.

26. Slattery, J. C., and R. B. Bird, "Calculation of the Diffusion Coefficient of Dilute Gases and the Self-diffusion Coefficient of Dense Gases," *A. I. Ch. E. Journal*, **4**, (1958), 137.

27. Smith, B. D., *Design of Equilibrium Stage Processes*. New York: McGraw-Hill Book Company, 1963.

28. Souders, M., and G. G. Brown, "Design of Fractionating Columns," *Ind. Eng. Chem.*, **26**, (1934), 98.

29. Tang, Y. P., "Diffusion of Carbon Dioxide in Liquids—Concentration Dependence of Diffusion Coefficient and Diffusion in Mixed Solvents," Ph.D. dissertation, University of Texas, 1963.

30. Wilke, C. R., "A Viscosity Equation for Gas Mixtures," *J. Chem. Phys.*, **18**, (1950), 517.

31. ———, "Estimation of Liquid Diffusion Coefficients," *Chem. Eng. Progr.*, **45**, (1949), 218.

32. ———, "Diffusional Properties of Multicomponent Gases," *Chem. Eng. Progr.*, **46**, (1950), 95.

33. ———, and P. Chang, "Correlation of Diffusion Coefficients in Dilute Solutions," *A.I.Ch.E. Journal*, **1**, (1955), 264.

Modeling of a Packed Liquid-liquid Extraction Column

12

A picture of the liquid-liquid extraction column is shown in Figure 12-1 (a simplified diagram is presented in Figure 5-1) and a detailed diagram appears in Figure 12-2. This extraction column was a part of the Number 1 SO_2 Plant at the Baytown Refinery of the Exxon Company.

In Part 1 of this chapter, the process and experimental tests are described as well as the Black and Hartwig (1, 2) method for the prediction of the phase equilibria for kerosene-sulfur dioxide systems. In Part 2, the use of the results of the field tests in the modeling of the extraction column is demonstrated.

Part 1 : Description of the Edeleanu Process, Experimental Results, and the Prediction of the Liquid-liquid Equilibria

The Edeleanu process (named after its originator, Dr. Lazar Edeleanu) was developed in an effort to make Rumanian lamp oil competitive with Pennsylvania kerosene. The Rumanian lamp oil originally possessed poor

Figure 12-1. The liquid-liquid extraction column at the Number 1 SO_2 Plant at the Baytown Refinery of the Exxon Company. [Taken from A. E. Hutton, and C. D. Holland, "Use of Field Tests in the Modeling of a Liquid-liquid Extraction Column," *Chem. Eng. Sci.*, **27**, (1972), 920.]

burning qualities, namely, excessive smoking characteristics that substantially reduced its desirability as an illuminating fuel. This poor burning quality was traced to the relatively high proportion of aromatic and unsaturated hydrocarbons in the Rumanian lamp oil. The basic principles involved in the removal of these undesirable materials were described by Edeleanu *et al.* (6) in 1907. The use of sulfuric acid (one of the earliest selective solvents)

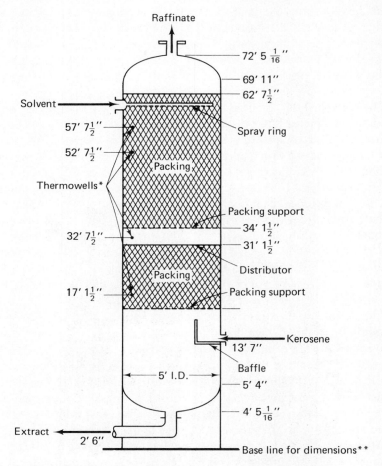

Figure 12-2. Detailed diagram of extraction column. (Taken from "Use of Field Tests in the Modeling of a Liquid-liquid Extraction Column," 920.)

was tried and did reduce the aromatic hydrocarbon content. However, it was found that polymerization, sulfonation, and oxidation took place during the acid treatment, thereby causing high acid losses, high loss of aromatic hydrocarbons in the acid layer, and the formation of large amounts of acid tar. In the search for a better process, a variety of solvents were tried. Edeleanu found that of these, liquid sulfur dioxide exhibited a remarkably high solubility for aromatic and unsaturated hydrocarbons, while paraffins and napthenes showed a very low solubility. This physical property of liquid sulfur dioxide forms the basis of the Edeleanu process (5). In addition to making "smokeless" kerosenes, the Edeleanu process has been used in plants for upgrading lubricating oils (3), producing high-octane-number blending stocks for gasolines (12), producing aviation fuels (10), upgrading the quality of diesel fuels (4), and producing improved catalytic cracking stocks (11).

The Number 1 SO_2 plant at the Baytown Refinery of the Exxon Company was designed to treat 12,000 barrels per day of kerosene feed by the Edeleanu process. The feed contained about 30% (by volume) of aromatic hydrocarbons (9). First, the feed was dried in calcium chloride driers, and then it was cooled to about 50°F by contacting it with chilled raffinate from the extractor. Next, the feed was chilled to about $-10°F$ by means of a propane refrigerant, and then it was introduced at the bottom of the liquid-liquid extractor. The raffinate stream that was withdrawn from the top of the extractor consisted of about 15% sulfur dioxide and 85% (by volume) of hydrocarbons, which were mostly paraffinic in character. The heavier extract phase, which was withdrawn from the bottom of the extractor, contained about 85% sulfur dioxide and 15% (by volume) of hydrocarbons, which consisted mostly of aromatics.

The liquid-liquid extraction column (see Figures 5-1 and 12-2) operated at essentially isothermal conditions at a temperature of about $-10°F$ and at an average pressure of about 40 psig. It was packed with $1\frac{1}{2}$ in. Pall Rings. The extract phase first exchanged heat with the sulfur dioxide stream and then it flowed through the equipment for the removal of the solvent. This extract-solvent removal equipment consisted of three strippers and a flash drum. After treatment in this equipment, the extract-product stream contained about 85% aromatics and 15% (by volume) paraffin and naphthenes.

After the raffinate left the top of the extractor, it flowed through a heat exchanger and then through the raffinate-solvent removal equipment, which consisted of two strippers and a flash drum. The raffinate product contained about 94% paraffins and naphthenes and 6% (by volume) aromatic hydrocarbons. Recoveries of from 85% to 95% (by volume) of the aromatic hydrocarbons were customarily obtained.

The sulfur dioxide recovered from both the extract and the raffinate streams was compressed, condensed, chilled, and recycled back to the top of the extractor.

Description of the plant tests and reduction of the data

Ten plant tests were made with the composition of the kerosene feed held fixed. The charge rate of the kerosene feed was varied from about 6,000 to 9,000 barrels per day, and the ratio of the flow rate of the solvent to the flow rate of the feed was varied from about 0.8 to 1.5. The average temperatures of the column were varied from $-16.5°F$ to $5°F$. A list of the complete sets of operating conditions for the plant tests is presented in Table 12-1. Prior to making a plant test, a period of at least three hours was allowed for the column to come to steady state at the specified set of operating conditions for the given plant test.

Samples of the feed, extract, raffinate, and solvent were analyzed by means of a mass spectrometer in conjunction with certain physical and chemical anaylytical techniques as described by Hutton (7).

TABLE 12-1
SUMMARY OF THE OPERATING CONDITIONS FOR THE TEN FIELD TESTS (7, 8)

Run No.	Feed Rate (Barrels/Day)	Solvent/Feed (Volume basis)	Avg. Temp. of Column (°F)
1	9129.84	1.0077	2.6
2	9129.84	0.7930	5.0
3	6035.20	1.0604	5.1
4	6035.20	1.0604	−6.7
5	6035.20	1.0604	−16.5
6	9025.14	0.8421	−1.8
7	9025.14	1.0061	−1.1
8	6110.64	0.9819	−4.6
9	6110.64	0.9819	−4.6
10	6110.64	1.4892	−5.8

Other data collected for each of the field tests consisted of the temperature, pressure, and flow rate of all terminal streams (the kerosene feed, solvent, raffinate, and extract streams). In addition to these terminal temperatures, the temperatures at four locations within the column were measured as indicated in Figures 5-1 and 12-2. Flow rates of all streams were determined by use of orifice meters. The raw data as well as the product distributions computed therefrom are presented by Hutton (7). Since all of the runs were made consecutively with the same kerosene feed, the analysis of the feed (see Table 12-2) was taken equal to the average of the feed analyses.

The results for each run were placed in component and total material balance by selecting from the total flow rates and compositions for the four terminal streams the three regarded as the most accurate and the flow rate and composition of the fourth were determined by difference. By examination of the nature of the various methods of analysis, it was concluded that the most accurate sets of analyses consisted of those for the feed, raffinate, and solvent. The analysis for the extract was found by difference. By independent checks such as tank gauging, the flow rates of the feed, the solvent sulfur dioxide, and the extract were found to be the most accurate of the four, and thus the total flow rate of the raffinate was found by difference.

TABLE 12-2

ANALYSIS OF THE KEROSENE FEED USED IN THE FIELD TESTS (Based on the molal average of the feed analyses) (7, 8)

Mole Distribution (Basis: 100 Moles of Feed)

Component No.	$Z = +2$ Paraffins	$Z = 0$ 1-ring Naphthenes	$Z = +2$ 2-ring Naphthenes	$Z = -6$ Alkyl Aromatics	$Z = -8$ Indans	$Z = -10$ Indenes	$Z = -12$ Naphthalenes
7	0.0000	0.0000	0.0000	0.1915	0.0000	0.0000	0.0000
8	0.0000	0.0000	0.0000	0.7813	0.0000	0.0000	0.0000
9	1.8356	2.1764	0.0865	1.6102	0.0000	0.0000	0.0000
10	1.8924	1.4311	0.4309	2.0380	0.6263	0.0000	1.3470
11	2.6624	1.3302	0.8496	1.8531	1.8220	0.2003	4.5991
12	4.7756	1.8107	1.7665	1.2796	2.5823	0.4132	5.5770
13	7.0785	2.0228	2.8702	0.7174	2.6410	0.7870	3.2887
14	5.6165	1.8459	2.6829	0.4234	1.7312	0.8580	1.2869
15	5.2783	1.9342	2.9175	0.3368	0.9179	0.6305	0.4225
16	3.1869	1.3355	2.0396	0.1482	0.2894	0.3012	0.1622
17	0.0000	0.0000	0.0000	0.0759	0.1085	0.0958	0.0000
Totals	32.3262	13.8868	13.6437	9.4554	10.7186	3.2860	16.6834
mole average Carbon numbers	13.0809	12.4420	13.6987	10.8054	12.6981	13.7888	12.0291

Total paraffins $= 32.3262$

Total naphthenes $= 27.5305$

Total saturates $= 59.8566$

Total aromatics $= 40.1433$

Molecular weight of saturates $= 183.78$

mole average $C_r = 13.0735$

mole average $C_a = 12.0636$

mole average $Z_r = -9.3550$

mole average $Z_a = 0.6242$

Molecular weight of aromatics $= 159.65$

C_a "effective" (for calculation of K_r) $= 12.2889$

C_a "effective" (for calculation of K_a) $= 11.7395$

Molecular weight of feed $= 174.10$

Prediction of the phase equilibria for the kerosene-sulfur dioxide system

The correlations of Black and Hartwig (1, 2) appeared to constitute the best method available for predicting the phase equilibria for the kerosene-sulfur dioxide system. In this procedure, the phase equilibria for a pseudo ternary mixture is predicted first, and then the results so obtained are used in the prediction of the phase equilibria for the multicomponent mixture. This same order is followed in the presentation of an abbreviated development of their procedure.

Consider a ternary system composed of aromatics a, saturates r, and the solvent sulfur dioxide s. Let x_i denote the mole fraction of component i $(i = a, r, s)$ in the heavier solvent phase and y_i the mole fraction of component i in the hydrocarbon phase. The distribution coefficient K_i for any component is defined in the usual way, namely,

$$K_i = \frac{y_i}{x_i} \qquad (i = a, r, s) \tag{12-1}$$

Let H denote the nonsolvent material in each phase. Then the distribution coefficient K_H for the nonsolvent material is given by

$$K_H = \frac{1 - y_s}{1 - x_s} \tag{12-2}$$

The effect of the molecular structure of the saturate r and the aromatic a constituents on the phase equilibria of the kerosene-sulfur dioxide system was correlated as functions of the carbon number C and the hydrogen deficiency Z, defined as follows:

$Z =$ number of hydrogen atoms minus 2 times the number of carbon atoms $\tag{12-3}$

The equilibrium state for the ternary mixture of the kerosene-sulfur dioxide system is uniquely determined for a given set of values for K_a, K_r, K_s, and K_H (see Problem 12-1). The prediction method consists of the development of relationships for K_a, K_r, and K_H as functions of the independent variables C_a, C_r, Z_a, and Z_r with K_s held fixed. Thus, if the $\log_{10} K_i$, where $i = a, r, H$, is expanded in a Taylor's series about a known reference state and only the linear terms are taken to be significant, one obtains

$$\log_{10} K_i = \log_{10} K_{i,\text{Ref}} + \left(\frac{\partial \log_{19} K_{i,\text{Ref}}}{\partial C_a}\right)\Delta C_a + \left(\frac{\partial \log_{10} K_{i,\text{Ref}}}{\partial C_r}\right)\Delta C_r$$

$$+ \left(\frac{\partial \log_{10} K_{i,\text{Ref}}}{\partial Z_a}\right)\Delta Z_a + \left(\frac{\partial \log_{10} K_{i,\text{Ref}}}{\partial Z_r}\right)\Delta Z_r \qquad (i = a, r, H)$$

$$\tag{12-4}$$

where,

$$\Delta C_a = (C_a - C_{a,\text{Ref}});$$
$$\Delta C_r = (C_r - C_{r,\text{Ref}});$$
$$\Delta Z_a = (Z_a - Z_{a,\text{Ref}});$$
$$\Delta Z_r = (Z_r - Z_{r,\text{Ref}}).$$

In the interest of simplicity, the partial derivatives appearing in Equation (12-4) are replaced by the coefficients B_{ai}, B_{ri}, $B_{Z_a i}$, and $B_{z,i}$ to give

$$\log_{10} K_i = \log_{10} K_{i,\text{Ref}} + B_{ai} \Delta C_a + B_{ri} \Delta C_r + B_{Z_a i} \Delta Z_a + B_{Z,i} \Delta_{Zr}$$
$$(i = a, r, H) \qquad (12\text{-}5)$$

Values of the B coefficients may be evaluated through the use of Equations (12-6) through (12-9) [proposed by Black and Hartwig (2)] and Equations (12-10) and (12-11) [developed by Hutton (7) on the basis of results presented by Black and Hartwig (2)].

$$B_{ar} = A - BZ_a \qquad (12\text{-}6)$$

$$B_{aH} = B_{aa} - \frac{C + DZ_a}{C_r} \qquad (12\text{-}7)$$

$$B_{rr} = E - F(Z_a - 6)^3 \qquad (12\text{-}8)$$

$$B_{rH} = G + \left[\frac{C + DZ_a}{128} \right] \Delta C_{a,\text{alkyl}} \qquad (12\text{-}9)$$

$$B_{Z_a r} = L - M(Z_a + 6) \qquad (12\text{-}10)$$

$$B_{Z_a H} = P + Q(C_r - 12) - R(Z_a + 6) \qquad (12\text{-}11)$$

where
$\Delta C_{a,\text{alkyl}}$ = number of alkyl carbons on the aromatic minus the number of alkyl carbons on the aromatic hydrocarbons of the reference system.

Values for the parameters A, B, C, D, E, F, G, L, M, P, Q, and R are given in Table C-5 of Appendix C as functions of K_s. [It should be mentioned that since the data (2) used by Hutton (7) in the computation of the values of $B_{Z_r a}$, $B_{Z_r r}$, and $B_{Z_r H}$ were collected at $-28.9°C$ rather than $-10°F$, it was in effect assumed that these coefficients were independent of temperature over this range.]

To account for the effect of multi-alkyl substitutions on the values of K_i $(i = a, r, H)$, Black and Hartwig (2) proposed that the following interpretation be given to C_a in Equation (12-5):

C_a = total number of carbons in the ring of the aromatic hydrocarbon
$+ C_{a,\text{alkyl}}$ \qquad (12-12)

where $C_{a,\text{alkyl}}$ is computed in a different manner for each of the K's. For the calculation of $\log_{10} K_a$:

$$C_{a,\text{alkyl}} = \text{total number of alkyl carbons} + (\lambda - 1)(\psi - 1) \quad (12\text{-}13)$$

where

λ = number of alkyl substitutions;

ψ = fraction of the total number of carbons in the aromatic compound which are paraffinic carbons.

For the calculation of $\log_{10} K_r$:

$$C_{a,\text{alkyl}} = \text{total number of alkyl carbons} + \frac{(\lambda - 1)(\psi + 1)}{3} \quad (12\text{-}14)$$

For the calculation of $\log_{10} K_H$:

$$C_{a,\text{alkyl}} = \text{total number of alkyl carbons} \quad (12\text{-}15)$$

Prediction of the phase equilibria for multicomponent mixtures

After the phase equilibria for the pseudo ternary mixture (composed of a pseudo aromatic, a pseudo saturate, and sulfur dioxide) have been predicted by the procedure described above, the K values ($K_{a,P}$ and $K_{r,P}$) obtained for the pseudo components are used in the prediction of the K values for the individual components I within the aromatic and saturate fractions. The pseudo aromatic and saturate compounds are defined as those hypothetical compounds that have carbon and hydrogen numbers equal to the respective mole averages of the carbon and hydrogen numbers of these compounds in the kerosene feed.

The term *individual component I* is used to mean either a single chemical compound or a given class of hydrocarbon compounds. For example, each class of hydrocarbons given in the analysis of the feed in Table 12-2 may be regarded as an individual component.

To predict the K value for any component I in the aromatic or saturate fraction, Black and Hartwig (2) assumed that these K values could be represented by the linear terms of a Taylor's series expansion about the K value for the pseudo components of the pseudo ternary mixture, that is,

$$\log_{10} K_{i,I} = \log_{10} K_{i,P} + \left(\frac{\partial \log_{10} K_{i,P}}{\partial C_i}\right)\Delta C_i$$

$$+ \left(\frac{\partial \log_{10} K_{i,P}}{\partial Z_i}\right)\Delta Z_i \quad (i = a, r) \quad (12\text{-}16)$$

where

$$\Delta C_i = C_{i,I} - C_{i,P} \quad (i = a, r);$$

$$\Delta Z_i = Z_{i,I} - Z_{i,P} \quad (i = a, r).$$

Observe that for $i = a$,

$$\left(\frac{\partial \log_{10} K_{a,P}}{\partial Z_a}\right)_{C_a} = \left(\frac{\partial \log_{10} K_{a,P}}{\partial \log_{10} K_{a,\text{Ref}}}\right)\left(\frac{\partial \log_{10} K_{a,\text{Ref}}}{\partial Z_a}\right)_{C_a}$$

The second term on the right-hand side of this expression is recognized as $B_{Z_a a}$, and if one lets

$$\left(\frac{\partial \log_{10} K_{a,P}}{\partial \log_{10} K_{a,\text{Ref}}}\right) = f_{Z_a a} \tag{12-17}$$

then

$$\left(\frac{\partial \log_{10} K_{a,P}}{\partial Z_a}\right)_{C_a} = f_{Z_a a} B_{Z_a a} \tag{12-18}$$

Thus, it follows that for any component I of the aromatic fraction of the original kerosene feed, Equation (12-16) may be stated as follows:

$$\log_{10} K_{a,I} = \log_{10} K_{a,P} + f_{aa} B_{aa} \Delta C_a + f_{Z_a a} B_{Z_a a} \Delta Z_a \tag{12-19}$$

and for any component I of the saturate fraction of the original kerosene feed, as

$$\log_{10} K_{r,I} = \log_{10} K_{r,P} + f_{rr} B_{rr} \Delta C_r + f_{Z_r r} B_{Z_r r} \Delta Z_r \tag{12-20}$$

where

$$f_{aa} = \left(\frac{\partial \log_{10} K_{a,P}}{\partial \log_{10} K_{a,\text{Ref}}}\right);$$

$$f_{Z_r r} = \left(\frac{\partial \log_{10} K_{r,P}}{\partial \log_{10} K_{r,\text{Ref}}}\right);$$

$$f_{rr} = \left(\frac{\partial \log_{10} K_{r,P}}{\partial \log_{10} K_{r,\text{Ref}}}\right).$$

Black and Hartwig (2) recommended that the following f's be set equal to unity, namely

$$f_{rr} = 1, \qquad f_{Z_a a} = 1, \qquad f_{A_r r} = 1 \tag{12-21}$$

On the basis of results given by Black and Hartwig (2), Hutton (7) obtained the following curvefit for f_{aa}

$$f_{aa} = 0.4899 + 16.1589 \times 10^{-4} K_r^{-3.32059} \tag{12-22}$$

which is applicable for $0.2 \leq K_s \leq 0.7$ and $4 \leq (C_{a,I} - C_{a,P}) \leq 6$.

Values of B_{aa}, $B_{Z_a a}$, $B_{Z_r r}$, and B_{rr} needed in Equations (12-19) and (12-20) were obtained from those given in Table C-5 either by Lagrangian-parabolic interpolation (for any K_s lying between any two values given in Table C-5) or by linear extrapolation for any $K_s < 0.269$. For $0.2 \leq K_s \leq 0.269$, Hutton (7) lists the following formulas for the linear extrapolations;

$$B_{aa} = 0.33349 - 0.49180 K_s$$

$$B_{Z_a a} = 0.19392 - 0.17777 K_s$$

$$B_{Z_r r} = 0.10921 - 0.12903 K_s \tag{12-23}$$

$$E = 0.24054 - 0.50871 K_s$$

$$F \times 10^4 = 0.03575 - 0.08 K_s$$

In summary, Equations (12-5), (12-19), and (12-20) may be used to predict the phase equilibria for the kerosene-sulfur dioxide system for napthenic and aromatic hydrocarbons in the C_6 to C_{15} range.

Prediction of the phase equilibria for the ten field tests

First, the K values for the pseudo ternary were computed, and then the K values for the multicomponent mixture were computed as follows.

Step 1. The analysis of the feed given in Table 12-2 was used to calculate the numbers C_r, Z_a, and Z_r for the pseudo aromatic and pseudo saturate components of the pseudo ternary mixture.

Step 2. Next, C_a and $C_{a,\text{alkyl}}$ were evaluated by use of Equations (12-12) through (12-15). [Since the number of alkyl substitutions on each aromatic hydrocarbon was not available from the laboratory analyses, it was assumed that the aromatic fraction consisted of equal parts of mono- and di-substituted alkylaromatic hydrocarbons.]

Step 3. On the basis of the results of Steps 1 and 2, and selected values from Tables C-5 and C-6, the pseudo ternary values of K_a, K_r, and K_H listed in Table C-7 were obtained. For values of K_s lying between those listed in Table C-5, the B coefficients and other parameters needed in the computations were found by Lagrangian-parabolic interpolation. The K values for $0.2 \leq K_s \leq 0.269$ were obtained by extrapolation as described by Hutton (7). [For any intermediate value of K_s which arose in subsequent calculations, the corresponding values of K_a, K_r, and K_H for the pseudo ternary were obtained from those listed in Table C-7 by Lagrangian-parabolic interpolation.]

Step 4. The results of Steps 1 through 3 were used in Equations (12-19) and (12-20) to predict the distribution coefficients for each class of components I of the aromatic and saturate fractions of the feed.

Use of the correlations of Black and Hartwig in the θ method of convergence

The use of the correlations of Black and Hartwig requires only two minor modifications in the calculational procedure described in Chapter 9. First, in the initiation of the calculational procedure, an initial value of K_{js} is assumed for each mass transfer section j or "plate j" rather than the compositions of the two streams leaving the plate. Then, the distribution coefficients $\{K_{ji}\}$ corresponding to K_{js} for each plate j were computed by the method of Black and Hartwig as described above. Then after a given trial had been made, the corrected values of K_{jH} were computed for each plate j by use of Equation (12-2) and the corrected compositions. Then, the set of distribution coefficients consistent with the K_{jH} for each plate j was computed by use of the correlations of Black and Hartwig (1, 2).

Part 2 : Use of the Results of Field Tests in the Modeling of the Extraction Column

The operating conditions for the ten field tests (see Table 12-1) were selected for the purpose of determining the effect of feed rate, solvent to the feed ratio, and temperature on the product distribution. The K values for the pseudo ternaries and for the multicomponent mixtures were computed by use of the procedures described above. The individual components I of the multicomponent mixtures consisted of the seven classes of compounds listed in the analysis of the kerosene feed (Table 12-2). These two representations of the kerosene-sulfur dioxide system are referred to hereafter as the "pseudo ternary" and the "multicomponent" mixtures. No attempt was made to include the effect of temperature on the K values.

Two potential methods for modeling the plate efficiencies for the column were first tested by use of pesudo ternary mixtures. The best of these models for the plate efficiency was then used in the analysis of the multicomponent mixtures. Each of the two methods made use of the number of perfect plates (or perfect mass transfer sections) required to minimize an objective function of the same form as the one presented in Chapter 10 for absorbers. For any one run, the objective function employed is given by

$$0_1(N, \{E_{ji} = 1\}) = \frac{1}{c} \sum_{i=1}^{c} |\log_e \theta_i| \tag{12-24}$$

and over all runs R by

$$\bar{0}_1(N, \{E_{ji} = 1\}) = \frac{1}{Rc} \sum_{r=1}^{R} \sum_{i=1}^{c} |\log_e \theta_{ir}| \tag{12-25}$$

where

$$\theta_i = \frac{\left(\frac{v_{1i}}{l_{Ni}}\right)_{\exp}}{\left(\frac{v_{1i}}{l_{Ni}}\right)_{ca}}.$$

The objective function 0_1 for each run was evaluated for different choices of N for both the pseudo ternary and for the multicomponent mixtures. The values so obtained for the objective function 0_1 for a typical run (Run 7) and the value of $\bar{0}_1$ over all ten runs for different choices of N are shown in Figures 12-3 and 12-4 for pseudo ternary and the multicomponent mixtures. Clearly, the minimum in $\bar{0}_1$ occurs at $N = 3$ for both the pseudo ternary and the multicomponent mixtures. For the pseudo ternary mixture, the minimum in 0_1 occurred at $N = 3$ for each run (7). For the multicomponent mixture, the minimum in 0_1 occurred at $N = 3$ for Runs 1, 2, 3, 4, 7, 10, and at $N = 4$ for Runs 5, 6, 8, and 9. In all subsequent calculations described, the number of mass transfer sections (or plates) was fixed at $N = 3$.

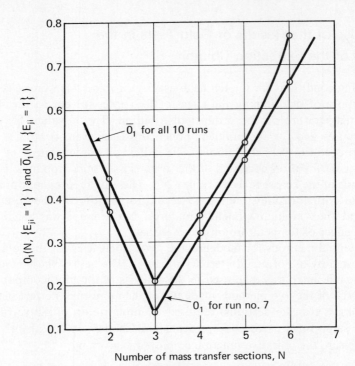

Figure 12-3. Variation of functions 0_1 and $\bar{0}_1$ (for the pseudo ternary representation of the kerosene-sulfur dioxide system) with number of perfect mass transfer sections. (Taken from "Use of Field Tests in the Modeling of a Liquid-liquid Extraction Column," 930.)

Next, the pseudo ternary mixture was used to investigate two potential models for representing the plate efficiencies, namely, the simple product model (see Chapters 9 and 11),

$$Model\ I:\ E_{ji} = \bar{E}_i \beta_j \qquad (12\text{-}26)$$

and the vaporization efficiency form of the model for the perfectly mixed liquid phase (see Chapter 11),

$$Model\ II:\ E_{ji} = 1 - [\exp(-n_{0Gji})]\left[1 - \frac{y_{j+1,i}}{K_{ji}x_{ji}}\right] \qquad (12\text{-}27)$$

Model II was discarded because it could not be used to account for experimental results of the following type. Suppose that for perfect plates, it is found that $(l_{Ni}/v_{1i})_{ca} < (l_{Ni}/v_{1i})_{exp}$. Then, on the average, it is necessary to decrease the product $E_{ji}K_{ji}$ or the E_{ji}'s must be less than unity on the average in order to increase $(l_{Ni}/v_{1i})_{ca}$. If, however, $y_{j+1,i} > K_{ji}x_{ji}$, then there exists no n_{0Gji} such that Model II will give an $E_{ji} < 1$.

All of these difficulties were overcome through the use of Model I. For

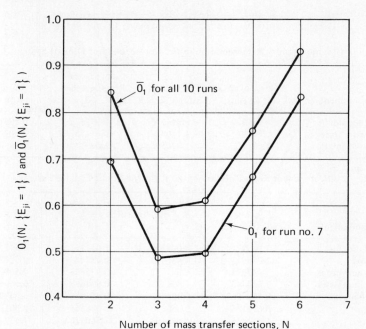

Figure 12-4. Variation of functions 0_1 and $\bar{0}_1$ (for the eight-component representation of the kerosene-sulfur dioxide system) with number of perfect mass transfer sections. (Taken from "Use of Field Tests in the Modeling of a Liquid-liquid Extraction Column," 932.)

the case of isothermal operation, the β_j's in Model I may be set equal to unity to give

$$E_{ji} = \bar{E}_i \qquad \text{(for all } j) \tag{12-28}$$

Corresponding to each $(l_{Ni}/v_{1i})_{\text{exp}}$, an $\bar{E}_i$ was obtained for the pseudo aromatic a, the pseudo saturate r, and sulfur dioxide s for each run as described in Chapter 9 such that for each run

$$0_1(3, \{\bar{E}_i\}) = 0 \tag{12-29}$$

The geometric mean values of the $\bar{E}_i$'s for each component over all ten runs are given in Table 12-3. On the basis of this set of mean values for the $\bar{E}_i$'s (denoted by $\{\bar{E}_{i,m}\}$), the product distributions for each run were recomputed. When the objective function was evaluated on the basis of the product distributions so obtained, it was found that over all ten runs

$$0_1(3, \{\bar{E}_{i,m}\}) = 0.1875$$

while the value of $\bar{0}_1$ for perfect mass transfer sections was

$$\bar{0}_i(3, \{E_{ji} = 1\}) = 0.2074$$

The reduction in the value of the objective function achieved through

TABLE 12-3
GEOMETRIC MEAN VALUES FOR THE COMPONENT EFFICIENCIES (7, 8)

I. Pseudo Ternary Representation of the Kerosene-Sulfur Dioxide System

Component	$\bar{E}_{i,m}$ over all 10 runs	$\bar{E}_{i,m}$ over Runs 1, 2, 6, 7	$\bar{E}_{i,m}$ over Runs 3, 4, 8, 9, 10	$\bar{E}_i$ for Run 5
Sulfur dioxide	0.984709	0.994578	0.995479	0.896117
Pseudo saturate	0.882947	1.120882	0.814157	0.509992
Pseudo aromatic	1.071722	1.020566	1.089738	1.199081

II. Multicomponent Representation of the Kerosene-Sulfur Dioxide System

Component	$\bar{E}_{i,m}$ over all 10 runs	$\bar{E}_{i,m}$ over Runs 1, 2, 6, 7	$\bar{E}_{i,m}$ over Runs 3, 4, 8, 9, 10	$\bar{E}_i$ for Run 5
Sulfur dioxide	1.013971	1.048254	1.041315	0.777104
Paraffins	1.056315	1.436433	0.937423	0.561201
1-ring napthenes	1.043222	1.391792	0.935373	0.568266
2-ring napthenes	1.044445	1.396327	0.934687	0.569617
Alkyl benzene	0.5509562	0.526882	0.553495	0.643792
Indans	0.6901063	0.666830	0.688345	0.801799
Indenes	1.350656	1.315164	1.340976	1.557485
Naphthalenes	0.8548068	0.786377	0.880545	1.028959

the use of this set of $\bar{E}_{i,m}$'s rather than perfect plates was not as great for the pseudo ternary mixture as it was for the case in which the system was represented by the multicomponent mixture. The mean $\bar{E}_i$'s over all ten runs are listed in Table 12-3 for the multicomponent representation of the feed. For the multicomponent mixture and the mean $\bar{E}_{i,m}$'s over all ten runs, it was found that

$$\bar{0}_1(3, \{\bar{E}_{i,m}\}) = 0.2073$$

while the corresponding value of $\bar{0}_1$ for multicomponent mixture for perfect mass transfer sections over all ten runs was

$$\bar{0}_1(3, \{E_{ji} = 1\}) = 0.5912$$

Some further reduction in $\bar{0}_1$ was achieved by use of a set of mean values of $\bar{E}_i$'s over the runs at feed rates of about 6,000 barrels per day and another set of mean values of the $\bar{E}_i$'s for the runs made at feed rates of about 9,000 barrels per day. These sets of $\bar{E}_{i,m}$'s are listed in Table 12-3 for both the pseudo ternary and the multicomponent mixtures. The $\bar{E}_i$'s for Run 5 did not appear to fit well in either group and were omitted in the analysis that follows. Actually, the temperature at which Run 5 was made was significantly lower than those for the others and because of this the $\bar{E}_i$'s for Run 5 are listed in Table 12-3. For Runs 3, 4, 8, 9, and 10, which were made at a feed rate of about 6,000 barrels per day, the values of $\bar{0}_1$ for the pseudo ternary mixture were

$$\bar{0}_1(3, \{\bar{E}_{i,m}\}) = 0.08904$$

$$\bar{0}_1(3, \{E_{ji} = 1\}) = 0.1669$$

and for the multicomponent mixture, the values of $\bar{0}_1$ were

$$\bar{0}_1(3, \{\bar{E}_{i,m}\}) = 0.08123$$

$$\bar{0}_1(3, \{E_{ji} = 1\}) = 0.5726$$

For Runs, 1, 2, 6, and 7, which were made at feed rates of about 9,000

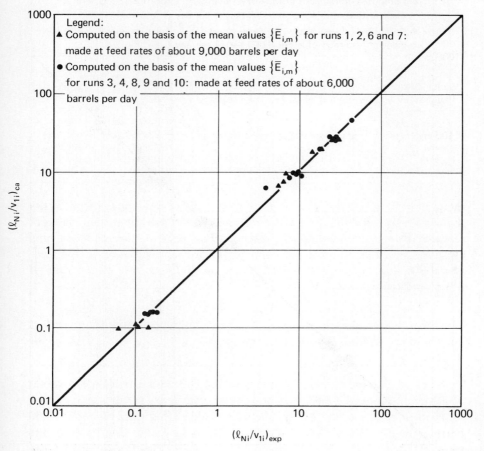

Figure 12-5. Comparison of the experimental and calculated product distributions for the pseudo ternary representation of the kerosene-sulfur dioxide system. (Taken from "Use of Field Tests in the Modeling of a Liquid-liquid Extraction Column," 932.)

barrels per day, the values of $\bar{0}_1$ for the pseudo ternary mixture were

$$\bar{0}_1(3, \{\bar{E}_{i,m}\}) = 0.1730$$

$$\bar{0}_1(3, \{E_{ji} = 1\}) = 0.1834$$

and for the multicomponent mixture, the values of $\bar{0}_1$ were

$$\bar{0}_1(3, \{\bar{E}_{i,m}\}) = 0.2007$$

$$\bar{0}_1(3, \{E_{ji} = 1\}) = 0.5110$$

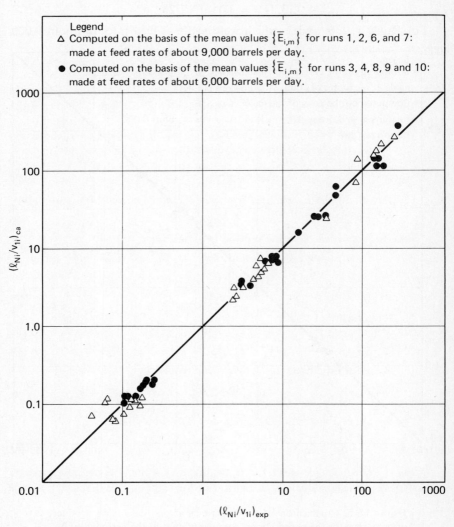

Figure 12-6. Comparison of the experimental and calculated product distributions for an eight-component representation of the kerosene-sulfur dioxide system. (Taken from "Use of Field Tests in the Modeling of a Liquid-liquid Extraction Column," 932.)

A comparison of the experimental product distributions with those calculated by use of a set of mean $\bar{E}_i$'s over Runs 1, 2, 6, and 7 and a set of mean $\bar{E}_i$'s over Runs 3, 4, 8, 9, and 10 is shown in Figure 12-5 for the pseudo ternary mixture and in Figure 12-6 for the multicomponent mixture.

In conclusion, both the pseudo ternary and the multicomponent representations of the kerosene-sulfur dioxide system gave adequate descriptions of the experimental observations. Although the component efficiencies appeared to be relatively insensitive to the operating conditions, they did exhibit some dependency on the feed rate.

PROBLEMS

12-1. For a known set of values for K_a, K_r, K_s, and K_H, show that the equilibrium state (the sets $\{x_i\}$ and $\{y_i\}$) are uniquely determined. The following order for the proof is suggested:

(a). Show that when y_s is eliminated from Equation (12-2) by use of Equation (12-1), the following formula is obtained for the computation of x_s namely,

$$x_s = \frac{1 - K_H}{K_s - K_H} \tag{A}$$

After x_s has been computed, y_s may be computed by use of Equation (12-1),

$$y_s = K_s x_s \tag{B}$$

(b). For convenience, define the two new variables

$$x'_a = \frac{x_a}{1 - x_s} \tag{C}$$

$$y'_a = \frac{y_a}{1 - y_s} \tag{D}$$

and recall that

$$x_a + x_r + x_s = 1 \tag{E}$$

$$y_a + y_r + y_s = 1 \tag{F}$$

By use of Equations (12-1), (12-2), and (C) through (F), show that

$$x'_a = \frac{K_r - K_H}{K_r - K_a} \tag{G}$$

(c). By use of Equations (12-2), (C), and (D), show that

$$y'_a = x'_a \frac{K_a}{K_H} \tag{H}$$

For a known set of values for K_a, K_r, K_s, and K_H, the above relationships may be used to evaluate each element of the sets $\{x_a, x_r, x_s\}$ and $\{y_a, y_r, y_s\}$.

NOTATION

$A, B, \ldots, F, G, L, M, P, Q, R$ = parameters appearing in Equations (12-6) through (12-11).

$B_{ai}, B_{ri}, B_{Z_a a}, B_{Z,i}$ = coefficients appearing in Equation (12-5).

C_a, C_r = carbon numbers for aromatics a and saturates r, respectively.

$f_{aa}, f_{rr}, f_{Z_a a}, f_{Z,r}$ = coefficients appearing in Equations (12-19) and (12-20).

K_i = distributions coefficient for component i; defined by Equation (12-1).

N = total number of plates or mass transfer sections.

0_1 = objective function for a given run.

$\bar{0}_1$ = objective function over a given number of runs R.

r = run number.

R = total number of runs.

Z = hydrogen deficiency; defined by (12-3).

Greek letters

λ = number of alkyl substitutions.

ψ = fraction of the total number of carbons in the aromatic compound which are paraffinic carbons.

Subscripts

a = aromatic.

H = nonsolvent material.

i = any component or pseudo component $a, r, s,$ or H.

I = individual component within the aromatic or saturate fractions of the feed.

j = plate or mass transfer section number; the sections are numbered in the same order as shown in Chapter 10.

P = a pseudo compound.

Ref = reference state.

r = saturate.

s = solvent, sulfur dioxide.

REFERENCES

1. Black, C., and G. M. Hartwig, "Liquid-liquid Phase Equilibria and Multistage Extraction: Part 1, Ternary Liquid-liquid Phase Equilibria; Part 2, Multi-

component Phase Equilibria and a Simplified Algebraic Calculation Method for Multistage Extraction," *Chem. Eng. Progr. Symp. Ser.*, **63**, 81, (1967), 65.

2. ————, "Phase Equilibria and Extraction Results for Petroleum Fractions with Sulfur Dioxide: Part 1, Binary and Ternary Equilibria; Part 2, Multicomponent Equilibria and Extraction Results for Kerosenes, *Chem. Engr. Progr. Symp. Ser.*, **64**, 88, (1968), 66.

3. Bray, U. B., "Solvent Treatment of California Lubricating Stocks, Particularly with Sulfur Dioxide, Sulfur Dioxide-Benzene," *The Science of Petroleum*, Vol. III. New York: Oxford University Press 1938, p. 1893.

4. Dickey, S. W., "Diesel Fuel of 50-Centene Value Produced in New Sulfur Dioxide Extraction Plate," *Pet. Proc.*, (June, 1938), 538.

5. Edeleanu, L., "The Refining Process with Liquid Sulfur Dioxide," *J. Inst. Petr. Techn.*, **18**, (1932), 900.

6. ————, and G. Gane, "Hydrocarbons Extraits des Goudrons Acides du Petrole," *Report of the 3rd International Petroleum Congress*, Bucarest, **2**, (1907), 665.

7. Hutton, A. E., "Liquid-Liquid Extraction in Packed Columns at Steady State Operation," Ph.D. dissertation, Texas A&M University, College Station, Texas, 1971.

8. ————, and C. D. Holland, "Use of Field Tests in the Modeling of a Liquid-Liquid Extraction Column," *Chem. Eng. Sci.* **27**, (1972), 919.

9. J. H. McClintock, Exxon Company, Baytown, Texas, personal communication.

10. Moy, J. A. E., "The Production of Aromatic Hydrocarbons for Aviation Fuels by Solvent Extraction," *Ind. Chemist*, **24**, (1948), 505.

11. Reidel, J. C., "Phillips' Borger Refinery Throughput Capacity Now Rated at 80,000 Bbl. per Day," *Oil and Gas J.*, **51**, 20 (1952), 116.

12. Saegebarth, E., A. G. Broggine, and E. Sliffen, "High-Octane-Number Blending Stocks Produced by Solvent Extractions," *Oil and Gas J.*, **36**, 3 (1937), 49.

Appendices
Selected Numerical Methods
for Solving Single Variable
and Multivariable Problems

Brief descriptions of certain of the well-known procedures for solving sets of one or more nonlinear equations by trial and error procedures are described below. The convergence characteristics of the numerical methods presented are also considered.

Solution of Single Variable Problems by Use of Direct Iteration

Any iterative procedure in which the calculated values of the variables are used without alteration to make the next trial calculation is referred to herein as *direct iteration*. A variety of names are used in the literature to describe these calculational procedures for solving sets of linear and nonlinear algebraic equations, for example, *iteration*, *successive iteration*, and *successive substitution*. To introduce this subject, the special case of a single variable function is considered first. This portion of the presentation follows closely with that presented previously (5).

The method of direct iteration is best illustrated by numerical examples. First, suppose it be desired to find the value of x which satisfies the following equation

$$x - \frac{1}{2}x - 2 = 0 \tag{A-1}$$

Although the answer, $x = 4$, is readily obtained by inspection, it is informative to solve this problem by the method of direct iteration. Equation (A-1) may be rearranged to the following form:

$$x_{k+1} = \frac{1}{2}x_k + 2 \tag{A-2}$$

The subscript k denotes the number of the trial. For the kth trial the assumed value for x (the idependent variable) is denoted by x_k and the calculated value of x (the dependent variable) by x_{k+1}. Thus the problem is to find that assumed value of x such that the value of x calculated by Equation (A-2) is equal to it. In the method of direct iteration the first assumed value for x is selected arbitrarily. Corresponding to the assumed value for x, the calculated value is obtained by Equation (A-2). In the second trial the assumed value of x is taken equal to the value calculated by the first trial. Continuation of the procedure leads to the correct value of x for some problems. The results obtained by starting with an assumed value of x equal to unity are shown in Figure A-1. Observe that the slope of the x_{k+1} versus x_k is less than unity and that the procedure of direct iteration does converge to the solution $x = 4$.

The fact that convergence was obtained when the x in Equation (A-1) with the coefficient of unity was taken as the dependent variable cannot be taken to mean that convergence will be obtained when the other x appearing in Equation (A-1) is selected as the dependent variable. That is, suppose Equation (A-1) to be rearranged to the following form:

$$x_{k+1} = 2x_k - 4 \tag{A-3}$$

When the procedure of direct iteration is initiated by use of any assumed value of x less than 4, the calculated values of x are progressively further away from the desired solution. Similarly, when calculations are initiated by use of a value of x greater than 4, the calculated values of x are again progressively further away from the solution, $x = 4$. Direct iteration fails to give the desired solution because the slope of the function is greater than unity at and in the neighborhood of the desired solution.

When the slope is a function of the assumed value of x, the procedure of direct iteration may or may not lead to a solution depending on the initial

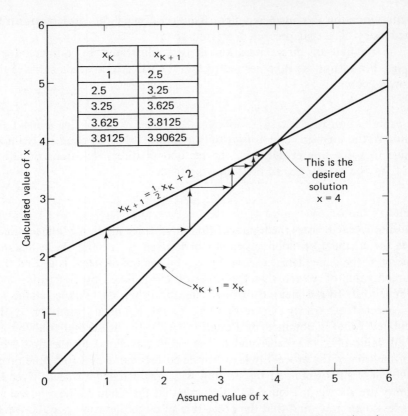

Figure A-1. When the slope of the function is greater than unity, the method of direct-iteration fails to give the solution. [Taken from Holland, C. D., *Multicomponent Distillation.* Englewood Cliffs, N.J.: Prentice-Hall, Inc., 1963, 4.]

choice for the assumed value of x. This fact is illustrated by the following example. Suppose that it is desired to find the roots of the following equation:

$$x^2 - 4x - 4 = 0 \qquad (A\text{-}4)$$

By use of the familiar quadratic formula, the roots, $x = (2 + 2\sqrt{2})$ and $x = (2 - 2\sqrt{2})$, are readily found. If the x with the coefficient of (-4) is taken to be the dependent variable, Equation (A-4) may be written as follows:

$$x_{k+1} = \frac{1}{4}x_k^2 - 1 \qquad (A\text{-}5)$$

For any initial value of x less than $(2 + 2\sqrt{2})$, the method of direct iteration converges to the root $x = (2 - 2\sqrt{2})$. For any initial value of x greater than $(2 + 2\sqrt{2})$, direct iteration fails to give the other root. In the remainder of the argument, the right-hand side of an equation such as Equation (A-5) is represented by the functional notation $F(x)$. Then Equation (A-5) can be stated as

$$x = F(x) \qquad (A\text{-}6)$$

where it is understood that

$$F(x) = \frac{1}{4}x^2 - 1$$

An examination of the results obtained for all of the functions considered shows that a sufficient condition for the method of direct iteration to converge to the solution is that

$$\left| \frac{dF(x)}{dx} \right| < 1 \qquad (A\text{-}7)$$

at and in the neighborhood of the desired solution. This is not a necessary condition because there is the chance that the correct value of x will be selected as the assumed value of x for the first trial.

Newton's Method

One systematic procedure for solving a trial and error problem is the method proposed by Newton. Suppose that it is desired to find the value of x such that $f(x) = 0$. Newton's method consists of the repeated use of the first two terms of the Taylor series expansion of $f(x)$ about some value of x, say x_k, as follows:

$$f(x) = f(x_k) + f'(x_k)(x - x_k) \qquad (A\text{-}8)$$

Since only the first two terms of the expansion are used, this is in general an approximation of the function $f(x)$. Now let $x = x_{k+1}$, the value of x that makes $f(x) = 0$. Then Equation (A-8) reduces to

$$x_{k+1} = x_k - \frac{f(x_k)}{f'(x_k)} \qquad (A\text{-}9)$$

Graphically, Newton's method may be regarded as the linear extension of the function from the point $[x_k, f(x_k)]$ to the point $(x_{k+1}, 0)$, as shown in Figure A-2. The slope of this line is $f'(x_k)$. If the approximation of the function represented by Equation (A-8) were correct (that is if $f(x)$ is a straight line), the value of x_{k+1} computed by the first trial would be the correct value. In Figure (A-2) the use of Newton's method to solve for the positive root ($x = 2 + 2\sqrt{2}$) which satisfies Equation (A-4) is shown. Although direct iteration failed for this case, it is seen that Newton's method converges rapidly to the desired solution. Also, it is to be observed that if the first assumed value of x lies to the right of $x = 2$, Newton's method converges to the root $x = (2 + 2\sqrt{2})$. If the first assumed value x lies to the left of $x = 2$, Newton's method converges to the root $x = (2 - 2\sqrt{2})$. Also, if

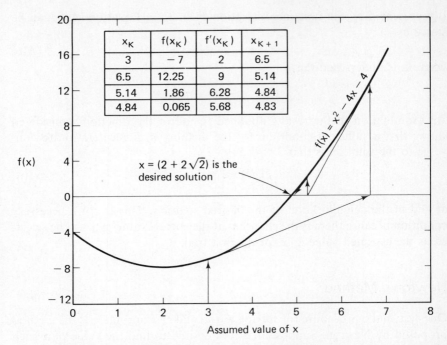

x_K	$f(x_K)$	$f'(x_K)$	x_{K+1}
3	-7	2	6.5
6.5	12.25	9	5.14
5.14	1.86	6.28	4.84
4.84	0.065	5.68	4.83

$x = (2 + 2\sqrt{2})$ is the desired solution

$f(x) = x^2 - 4x - 4$

Figure A-2. According to Newton's method, the next best value for the root is given by the straight line extrapolation with the slope $f'(x_k)$ from x_k, $f(x_k)$ to $(x_{k+1}, 0)$. (Taken from *Multicomponent Distillation*, 8.)

the value $x = 2$ is selected initially, Newton's method fails because $f'(2) = 0$; and if a point inflection occurs in the neighborhood of the root, Newton's method may fail to converge as shown by Sokolnikoff (9).

Before Newton's method should be employed, the behavior of the function in the neighborhood of the root should be established and the initial value of x selected accordingly. Certain corrective measures to be taken in the application of Newton's method to functions containing several roots are discussed as the need for them arises in the application of this method. Although Newton's method has some shortcomings, it may be successfully applied to solve many of the trial-and-error problems associated with distillation calculations.

Interpolation (*Regula Falsi*)

Some of the limitations of Newton's method are overcome by the method of interpolation *regula falsi* (9), which may be demonstrated by use of the previous example in which it was desired to find the positive root of Equation (A-4). First, for two arbitrarily selected values of x, the corresponding values of the function are calculated. Let these be $x_1 = 2$ and $x_2 = 6.5$, which give, respectively, $f(2) = -8$ and $f(6.5) = 12.25$. Now, let these two points be connected by a straight line. This line intersects the x-axis at $x = 3.777$, which is a better value than either $x = 2$ or $x = 6.5$. Corresponding to $x_3 = 3.777$, $f(x_3) = -4.86$. When the last two points $(6.5, 12.25)$ and $(3.777, -4.86)$ are connected by a straight line, a still better value for x, $x = 4.55$, is obtained. Continuation of this procedure leads to the desired root.

The interpolation formula is readily developed as follows. The equation of the straight line connecting points $[x_k, f(x_k)]$ and $[x_{k+1}, f(x_{k+1})]$ is

$$\frac{f(x) - f(x_k)}{x - x_k} = \frac{f(x_k) - f(x_{k+1})}{x_k - x_{k+1}} \tag{A-10}$$

Let the value of x at which $f(x) = 0$ be denoted by x_{k+2}. Then for $x = x_{k+2}$, Equation (A-10) is readily solved for x_{k+2} to give

$$x_{k+2} = \frac{x_k f(x_{k+1}) - x_{k+1} f(x_k)}{f(x_{k+1}) - f(x_k)} \tag{A-11}$$

It is to be observed that this method does not fail as did Newton's method when a value of x is selected for which $f'(x) = 0$. However, like Newton's method, if $f(x)$ is a straight line, the correct value for x is determined by the first trial. Also worthy of note is the fact that the derivative of the function is not involved in the calculation of the next best value of x. However, as in Newton's method, provisions must be made in the treatment of functions having several roots in order to be assured that calculations will be carried out in the interval containing the desired root. Also, it should be noted that instead of interpolating between the two most recent points, other schemes may be used. For example, one may interpolate between the two most recent points for which the functions $f(x)$ were of opposite sign.

Solution of Systems of Nonlinear Algebraic Equations by Use of Direct Iteration

Let it be required to find the desired solution to a nonlinear set of algebraic equations, which may be represented as follows:

$$f_1(x_1, x_2, \ldots, x_n) = 0$$
$$f_2(x_1, x_2, \ldots, x_n) = 0 \tag{A-12}$$
$$\vdots \qquad\qquad \vdots$$
$$f_n(x_1, x_2, \ldots, x_n) = 0$$

Let the trial vector for the kth trial be denoted by

$$\mathbf{X}_k = [x_{1k} x_{2k} \ldots x_{nk}]^T$$

For convenience let each function of Equation (A-12) be represented by

$$f_i(\mathbf{X}) = 0 \qquad (1 \leqq i \leqq n) \tag{A-13}$$

and let

$$\mathbf{f}(\mathbf{X}) = [f_1(\mathbf{X})f_2(\mathbf{X}) \dots f_n(\mathbf{X})]^T \tag{A-14}$$

Then, in matrix notation, Equation (A-12) may be represented by

$$\mathbf{f}(\mathbf{X}) = \mathbf{0} \tag{A-15}$$

Now, let the solution set of values of the variables of Equation (A-15) be denoted by $\mathbf{a}$ where,

$$\mathbf{a} = [a_1 a_2 \dots a_n]^T \tag{A-16}$$

The equations employed in the iterative procedure may be obtained by rearranging the expressions given by Equation (A-12) to the following form:

$$x_i = F_i(\mathbf{X}) \qquad (1 \leqq i \leqq n) \tag{A-17}$$

It should be observed that the functions $F_i(\mathbf{X})$ may be formed from any suitable combination of the functions $f_i(\mathbf{X})$. That is, F_i may be any function for which $F_i(\mathbf{X})$ approaches a_i as $\mathbf{X}$ approaches $\mathbf{a}$ for each i from $i = 1$ through $i = n$.

In the Jacobi method of iteration, a new set $\mathbf{X}_k$ of values of the variables is computed on the basis of an assumed set $\mathbf{X}_{k-1}$ as follows:

$$x_{ik} = F_i(\mathbf{X}_{k-1}) \qquad (1 \leqq i \leqq n) \tag{A-18}$$

The set $\mathbf{X}_k$ so obtained is used to compute the next set in a similar manner.

Now let R denote the set of all vectors $\mathbf{X}$ whose elements satisfy the condition $|x_j - a_j| \leqq h, (1 \leqq j \leqq n)$, for which

$$\sum_{j=1}^{n} \left| \frac{\partial F_i(\mathbf{X})}{\partial x_j} \right| < \mu < 1 \qquad (1 \leqq i \leqq n) \tag{A-19}$$

for each i. If the starting vector $\mathbf{X}_0$ is an element of R, then

$$\lim_{N \to \infty} \mathbf{X}_N = \mathbf{a} \tag{A-20}$$

In the procedure which is analogous to the Gauss-Seidel solution of linear algebraic equations, the most recent value of each variable is used at each point in the calculational procedure. The Gauss-Seidel type iteration differs from the Jacobi type iteration in that instead of computing each element $\mathbf{X}_k$ on the basis of an assumed vector $\mathbf{X}_{k-1}$, as indicated by Equation (A-18), the following procedure is employed:

$$\begin{aligned}
x_{1k} &= F_1(x_{1,k-1}, x_{2,k-1}, x_{3,k-1}, \dots, x_{n-1,k-1}, n_{n,k-1}) \\
x_{2k} &= F_2(x_{1k}, x_{2,k-1}, x_{3,k-1}, \dots, x_{n-1,k-1}, x_{n,k-1}) \\
x_{3k} &= F_3(x_{1k}, x_{2k}, x_{3,k-1}, \dots, x_{n-1,k-1}, x_{n,k-1}) \\
& \cdot \qquad\qquad \cdot \\
& \cdot \qquad\qquad \cdot \\
& \cdot \qquad\qquad \cdot \\
x_{nk} &= F_n(x_{1k}, x_{2k}, x_{3k}, \dots, x_{n-1,k}, x_{n,k-1})
\end{aligned} \tag{A-21}$$

When the conditions given by Equations (A-19) and (A-20) are satisfied, convergence of the Gauss-Seidel method can be assured (3).

Another sufficient condition for the convergence of the Jacobi method of iteration follows. Let R be the set of all starting vectors $\mathbf{X}_0$ for which the largest Hilbert norm (denoted by $\|\mathbf{B}\|_{\text{III}}$) of any matrix $\mathbf{B}$ generated by the iterative process has the property that

$$\|\mathbf{B}\|_{\text{III}} = \sqrt{\Lambda_1} \tag{A-22}$$

where

$$\mathbf{B} = \begin{bmatrix} \dfrac{\partial F_1}{\partial x_1} & \dfrac{\partial F_1}{\partial x_2} & \cdots & \dfrac{\partial F_1}{\partial x_n} \\[2mm] \dfrac{\partial F_2}{\partial x_1} & \dfrac{\partial F_2}{\partial x_2} & \cdots & \dfrac{\partial F_2}{\partial x_n} \\[2mm] \cdots\cdots\cdots\cdots\cdots\cdots \\[1mm] \dfrac{\partial F_n}{\partial x_1} & \dfrac{\partial F_n}{\partial x_2} & \cdots & \dfrac{\partial F_n}{\partial x_n} \end{bmatrix}$$

$\Lambda_1 =$ the largest eigenvalue of $\mathbf{B}^T\mathbf{B}$, [see References (3, 4, 6, 7, 8)].

Newton-Raphson Method

The extension of the method of Newton to functions of several variables is called the Newton-Raphson method. It is developed in a manner analogous to that shown for Newton's method. Suppose the set of values of x and y is to be found that makes $f_1(x, y) = 0$ and $f_2(x, y) = 0$, simultaneously. If each function is represented by a Taylor series expansion about the set of values (x_k, y_k), the following result is obtained when all terms that contain derivatives of higher order than the first are neglected:

$$f_1(x, y) = f_1(x_k, y_k) + \frac{\partial f_1(x_k, y_k)}{\partial x}(x - x_k) + \frac{\partial f_1(x_k, y_k)}{\partial y}(y - y_k) \tag{A-23}$$

$$f_2(x, y) = f_2(x_k, y_k) + \frac{\partial f_2(x_k, y_k)}{\partial x}(x - x_k) + \frac{\partial f_2(x_k, y_k)}{\partial y}(y - y_k) \tag{A-24}$$

For the set (x_{k+1}, y_{k+1}) of values that gives

$$f_1(x_{k+1}, y_{k+1}) = f_2(x_{k+1}, y_{k+1}) = 0$$

Equations (A-23) and (A-24) reduce to

$$0 = f_1 + \frac{\partial f_1}{\partial x}\Delta x_{k+1} + \frac{\partial f_1}{\partial y}\Delta y_{k+1} \tag{A-25}$$

and

$$0 = f_2 + \frac{\partial f_2}{\partial x}\Delta x_{k+1} + \frac{\partial f_2}{\partial y}\Delta y_{k+1} \tag{A-26}$$

respectively, where

$\Delta x_{k+1} = x_{k+1} + x_k;$

$\Delta y_{k+1} = y_{k+1} - y_k.$

Also, it is to be understood that the functions and their derivatives are evaluated at $x = x_k$ and $y = y_k$. The problem then reduces to two equations in the two unknowns, Δx_{k+1} and Δy_{k+1}. The values calculated (x_{k+1} and y_{k+1}) by the kth trial are used as the assumed values for $(k + 1)$st trial. This procedure is repeated until the desired accuracy of the roots is obtained.

For the general case of n independent equations in n unknowns, n equations of the form of Equations (A-25) and (A-26) are obtained upon application of the Newton-Raphson method. The equations so obtained may be represented by the following matrix equation.

$$\mathbf{J}_k \, \Delta \mathbf{X}_k = -\mathbf{f}_k \tag{A-27}$$

where $\mathbf{J}_k$ is the square Jacobian matrix of order n and $\Delta \mathbf{X}_k$ and $\mathbf{f}_k$ are conformable column vectors,

$$\mathbf{J}_k = \mathbf{J}(\mathbf{f}_k/\mathbf{X}_k) = \begin{bmatrix} \dfrac{\partial f_1}{\partial x_1} & \dfrac{\partial f_1}{\partial x_2} & \cdots & \dfrac{\partial f_1}{\partial x_n} \\ \cdot & \cdot & \cdots & \cdot \\ \cdot & \cdot & \cdots & \cdot \\ \cdot & \cdot & \cdots & \cdot \\ \dfrac{\partial f_n}{\partial x_1} & \dfrac{\partial f_n}{\partial x_2} & \cdots & \dfrac{\partial f_n}{\partial x_n} \end{bmatrix}$$

$$\Delta \mathbf{X}_k = [\Delta x_1 \, \Delta x_2 \ldots \Delta x_n]^T; \qquad \Delta x_j = x_{j,k+1} - x_{jk}$$

Also

$$\Delta \mathbf{X}_k = \mathbf{X}_{k+1} - \mathbf{X}_k; \qquad \mathbf{X}_k = [x_{1k} x_{2k} \ldots x_{nk}]^T$$

$$\mathbf{f}_k = [f_{1k}, f_{2k} \ldots f_{nk}]^T$$

In the application of this method, it is recommended that the convergence characteristics be checked by solving a wide variety of examples. The use of different initial sets of values for the variables should also be investigated. Also, if only positive roots of the functions are desired, provisions should be made for an alternate selection of variables for the next trial when one or more negative values are computed by an intermediate trial.

There follows a formulation of a set of sufficient conditions for the convergence of the Newton-Raphson method.

Let the solution set of the independent variables be represented by the column vector $\boldsymbol{\alpha}$,

$$\boldsymbol{\alpha} = [\alpha_1 \alpha_2 \ldots \alpha_n]^T$$

As the name of this set of values of the variables implies,

$$f_i(\boldsymbol{\alpha}) = 0 \qquad (1 \leqq i \leqq n)$$

or

$$\mathbf{f}(\boldsymbol{\alpha}) = \mathbf{0} \tag{A-28}$$

Suppose that the elements of $\mathbf{f}$ and $\mathbf{J}$ are continuous over all starting vectors $\mathbf{X}_0$ and all vectors $\mathbf{X}_k$ generated therefrom by use of the Newton-Raphson method. Further suppose that the determinate of $\mathbf{J}$ is nonzero ($|\mathbf{J}| \neq 0$) for all $\mathbf{X}_0$ and $\mathbf{X}_k$. Then it can be shown that if any starting vector $\mathbf{X}_0$ is "close enough" to the solution vector $\boldsymbol{\alpha}$, then the Newton-Raphson procedure converges to the solution vector $\boldsymbol{\alpha}$. "Close enough" means that the product of the Hilbert norms [see Carnahan *et al.* (3)] of the matrices required to relate the error vectors $\mathbf{E}_N$ and $\mathbf{E}_0$ is less than unity, where $\mathbf{E}_N = \mathbf{X}_N - \boldsymbol{\alpha}$ and $\mathbf{E}_0 = \mathbf{X}_0 - \boldsymbol{\alpha}$.

The Broyden Modification of the Newton-Raphson Method

In the class of methods proposed by Broyden (1), the partial derivatives $\partial f_i / \partial x_j$ in the Jacobian matrix $\mathbf{J}_k$ of Equation (A-27) are evaluated only once. In each successive trial, the elements of the inverse of the Jacobian matrix are corrected by use of the computed values of the function f_i. Throughout the development that follows, it is supposed that the functions f_i are real variable functions of real variables and that the functions are continuous and differentiable. If the Jacobian matrix $\mathbf{J}_k$ in the Newton-Raphson equation [Equation (A-27)] is nonsingular, then $\mathbf{J}_k^{-1}$ exists and

$$\mathbf{X}_{k+1} = \mathbf{X}_k - \mathbf{J}_k^{-1}\mathbf{f}_k \tag{A-29}$$

where $\Delta\mathbf{X}_k = \mathbf{X}_{k+1} - \mathbf{X}_k$.

Let $\mathbf{P}_k$ be defined as follows:

$$\mathbf{P}_k = -\mathbf{A}_k^{-1}\mathbf{f}_k \tag{A-30}$$

where $\mathbf{A}_k$ is some approximation of $\mathbf{J}_k$. Then a simple modification of the Newton-Raphson method consists of

$$\mathbf{X}_{k+1} = \mathbf{X}_k + s_k\mathbf{P}_k \tag{A-31}$$

where s_k is a scalar that is picked as described below.

Now let the vector $\mathbf{X}$ be defined by

$$\mathbf{X} = \mathbf{X}_k + s\mathbf{P}_k \tag{A-32}$$

where s is arbitrary. Each function f_i may now be regarded as a function of the single variable s. Since the partial derivatives $\partial f_i / \partial x_j$ are assumed to exist, then the total derivative of each f_i with respect to s is given by

$$\frac{df_i}{ds} = \sum_{j=1}^{n} \frac{\partial f_i}{\partial x_j} \frac{dx_j}{ds} \qquad (1 \leq i \leq n) \tag{A-33}$$

It will now be shown that an approximation to the derivatives given by Equation (A-33) may be used to improve the approximate matrix $\mathbf{A}_k$. By use of the relationship obtained by differentiating the members of Equation (A-32) with respect to s,

$$\frac{d\mathbf{X}}{ds} = \mathbf{P}_k$$

it is readily shown that Equation (A-33) may be stated in the form

$$\frac{d\mathbf{f}}{ds} = \mathbf{JP}_k \tag{A-34}$$

Thus, if an accurate estimate of $d\mathbf{f}/ds$ were available, it could be used with Equation (A-34) to estimate a condition that any approximation to the Jacobian matrix must satisfy. Suppose, however, that the partial derivatives are evaluated numerically. Since the value of the Jacobian matrix at the point $\mathbf{X}_{k+1}$ is desired, the derivative $d\mathbf{f}/ds$ at the point $\mathbf{X}_{k+1}$ may be evaluated numerically. The numerical approximation of the derivative df_i/ds may be computed as follows:

$$\frac{df_i}{ds} \cong \frac{f_i(\mathbf{X}_k + s_k\mathbf{P}_k) - f_i[\mathbf{X}_k + (s_k - \epsilon)\mathbf{P}_k]}{\epsilon} \tag{A-35}$$

or in simpler notation

$$-\frac{df_i}{ds} \cong \frac{f_i(s_k - \epsilon) - f_{i,k+1}}{\epsilon}$$

Thus, the complete set of derivatives may be represented as follows:

$$\mathbf{f}(s_k - \epsilon) \cong \mathbf{f}_{k+1} - \epsilon\frac{d\mathbf{f}}{ds} \tag{A-36}$$

Elimination of $d\mathbf{f}/ds$ from Equations (A-34) and (A-36) yields

$$\mathbf{f}_{k+1} - \mathbf{f}(s_k - \epsilon_k) \cong \epsilon_k\mathbf{JP}_k \tag{A-37}$$

In the class of methods proposed by Broyden, the improved approximation $\mathbf{A}_{k+1}$ of $\mathbf{A}_k$ is selected such that the following equation is satisfied:

$$\mathbf{f}_{k+1} - \mathbf{f}(s_k - \epsilon_k) = \epsilon_k\mathbf{A}_{k+1}\mathbf{P}_k \tag{A-38}$$

Next, let the square matrix $\mathbf{H}_k$ of order n be defined by

$$\mathbf{H}_k = -\mathbf{A}_k^{-1} \tag{A-39}$$

and the column vector $\mathbf{Y}_k$ of order n by

$$\mathbf{Y}_k = \mathbf{f}_{k+1} - \mathbf{f}(s_k - \epsilon_k) \tag{A-40}$$

When stated in terms of these matrices, Equations (A-30) and (A-38) become

$$\mathbf{P}_k = \mathbf{H}_k\mathbf{f}_k \tag{A-41}$$

$$\mathbf{H}_{k+1}\mathbf{Y}_k = -\epsilon_k\mathbf{P}_k \tag{A-42}$$

respectively. Also, Equations (A-38) and (A-40) may be combined to give

$$\mathbf{Y}_k = \epsilon_k\mathbf{A}_{k+1}\mathbf{P}_k \tag{A-43}$$

which relates the change $\mathbf{Y}_k$ in the vector function to a change of $\mathbf{X}$ in the direction $\mathbf{P}_k$. In the procedure called *Method 1* by Broyden, $\mathbf{A}_{k+1}$ is chosen so that the change in $\mathbf{f}$ predicted by $\mathbf{A}_{k+1}$ in a direction $\mathbf{Q}_k$ orthogonal to $\mathbf{P}_k$ is the same as would be predicted by $\mathbf{A}_k$, that is,

$$\mathbf{A}_{k+1}\mathbf{Q}_k = \mathbf{A}_k\mathbf{Q}_k; \qquad \mathbf{Q}_k^T\mathbf{P}_k = 0 \tag{A-44}$$

When these two relationships are combined with Equation (A-43), the following formula for $\mathbf{A}_{k+1}$ is obtained:

$$\mathbf{A}_{k+1} = \mathbf{A}_k + \frac{(\mathbf{Y}_k - \epsilon_k\mathbf{A}_k\mathbf{P}_k)\mathbf{P}_k^T}{\epsilon_k\mathbf{P}_k^T\mathbf{P}_k} \tag{A-45}$$

In solving problems by computer, it is preferable to store $\mathbf{H}_k$ rather than $\mathbf{A}_k$. To obtain $\mathbf{H}_{k+1}$ from $\mathbf{A}_{k+1}$, Householder's formula is used. If $\mathbf{A}$ is a nonsingular matrix and $\mathbf{X}$ and $\mathbf{Y}$ are vectors, all of order n, and if $(\mathbf{A} + \mathbf{X}\mathbf{Y}^T)$ is nonsingular, then Householder's formula

$$(\mathbf{A} + \mathbf{X}\mathbf{Y}^T)^{-1} = \mathbf{A}^{-1} - \frac{\mathbf{A}^{-1}\mathbf{X}\mathbf{Y}^T\mathbf{A}^{-1}}{1 + \mathbf{Y}^T\mathbf{A}^{-1}\mathbf{X}}$$

applies. When this formula is applied to Equation (A-45), one obtains

$$\mathbf{H}_{k+1} = \mathbf{H}_k - \frac{(\epsilon_k\mathbf{P}_k + \mathbf{H}_k\mathbf{Y}_k)\mathbf{P}_k^T\mathbf{H}_k}{\mathbf{P}_k^T\mathbf{H}_k\mathbf{Y}_k} \tag{A-46}$$

In applying these relationships, ϵ_k is set equal to s_k, which amounts to using a full step size to approximate the total derivative [see Equation (A-35)]. The numerical value of s_k is picked such that convergence is promoted, that is, for the kth trial, s_k is picked such that the Euclidean norm of $\mathbf{f}_k$ is less than the Euclidean norm of $\mathbf{f}_{k-1}$, that is,

$$\left[\sum_{i=1}^{n} f_i^2(s_k)\right]^{1/2} < \left[\sum_{i=1}^{n} f_i^2(s_{k-1})\right]^{1/2} \tag{A-47}$$

The first value of s_k (denoted by the second subscript) is taken as unity,

$$s_{k,1} = 1$$

If this value of s_k satisfied Equation (A-47), it was used; otherwise, a second value was computed by use of the following formula developed by Broyden:

$$s_{k,2} = \frac{(1 + 6\eta)^{1/2} - 1}{3\eta} \tag{A-48}$$

where

$$\eta = \frac{\left[\sum_{i=1}^{n} f_i^2(1)\right]}{\left[\sum_{i=1}^{n} f_i^2(0)\right]}.$$

As pointed out by Broyden, other methods for picking s_k may be used. For example, s_k may be picked such that the Euclidean norm of $\mathbf{f}$ is minimized.

The steps of a calculational procedure proposed by Broyden follow:

1. Assume an initial set of values of the variables, $\mathbf{X}_0$, and compute

$$\mathbf{f}_0 = \mathbf{f}(\mathbf{X}_0)$$

2. Approximate the elements of $\mathbf{H}_0$. [Broyden obtained a first approximation of the elements of $\mathbf{A}_0$ by use of the formula

$$\frac{\partial f_i}{\partial x_j} \simeq \frac{f_i(x_j + h_j) - f_i(x_j)}{h_j}$$

where h_j was taken to be roughly equal to 0.001 x_j. Next, compute $\mathbf{H}_0$ by use of Equation (A-39), namely, $\mathbf{H}_0 = -\mathbf{A}_0^{-1}$.]

3. On the basis of the most recent values of $\mathbf{H}$ and $\mathbf{f}$, say $\mathbf{H}_k$ and $\mathbf{f}_k$, compute

$$\mathbf{P}_k = \mathbf{H}_k \mathbf{f}_k$$

4. Find the s_k such that the Euclidean norm of $\mathbf{f}(\mathbf{X} + s_k\mathbf{P}_k)$ is less than of $\mathbf{f}(\mathbf{X}_k)$. [First, try $s_{k,1} = 1$ and if Equation (A-47) is satisfied, proceed to Step 5; otherwise, compute $s_{k,2}$ by use of Equation (A-48).] In the course of making these calculations, the following vectors will have been computed:

$$\mathbf{X}_{k+1} = \mathbf{X}_k + s_k\mathbf{P}_k$$

$$\mathbf{f}_{k+1} = \mathbf{f}(\mathbf{X}_{k+1})$$

5. Test $\mathbf{f}_{k+1}$ for convergence. If convergence has not been achieved, compute

$$\mathbf{Y}_k = \mathbf{f}_{k+1} - \mathbf{f}_k$$

6. Compute

$$\mathbf{H}_{k+1} = \mathbf{H}_k - \frac{(\mathbf{H}_k\mathbf{Y}_k + s_k\mathbf{P}_k)\mathbf{P}_k^T\mathbf{H}_k}{\mathbf{P}_k^T\mathbf{H}_k\mathbf{Y}_k}$$

and return to Step 3.

A Generalized Scaling Procedure

There follows the development of a generalized scaling procedure that was suggested and used by Burdett (2) in the analysis of a system of 17 evaporators at unsteady-state operation. This system of equations was solved by the Newton-Raphson method which may be represented by Equation (A-27). For large matrices $\mathbf{J}_k$ that result from problems of this type, the following generalized scaling procedure is more readily applied than the one employed in the solution of the numerical examples in Chapter 2.

Let $\mathbf{R}_k$ be a diagonal matrix whose elements r_{ii} are just greater than the absolute value of the corresponding elements of $\mathbf{X}_k$, that is,

$$r_{11} > |x_{1k}|, r_{22} > |x_{2k}|, \ldots, r_{nn} > |x_{nk}| \tag{A-49}$$

(Except for the restriction that r_{ii} must never be set equal to zero, the inequality given by Equation (A-49) need not be adhered to precisely in practice.) The elements of $\mathbf{R}_k$ should be close, however, in absolute value to the corresponding elements of $\mathbf{X}_k$. Equation (A-27) may be restated in terms of the matrix $\mathbf{R}_k$ in the following manner:

$$\mathbf{J}_k \mathbf{R}_k (\mathbf{R}_k^{-1} \Delta \mathbf{X}_k) = -\mathbf{f}_k \tag{A-50}$$

Let

$$\Delta \mathbf{Y}_k = \mathbf{R}_k^{-1} \Delta \mathbf{X}_k \tag{A-51}$$

$$\mathbf{D}_k = \mathbf{J}_k \mathbf{R}_k \tag{A-52}$$

After the multiplication implied by Equation (A-52) has been carried out, form the diagonal matrix $\mathbf{M}_k$ whose elements m_{ii} are selected such that for each row

$$m_{ii} = \text{maximum of } |d_{ij}| \text{ over all elements of row } i \tag{A-53}$$

Premultiplication of each side of Equation (A-50) by $\mathbf{M}_k^{-1}$ yields

$$(\mathbf{M}_k^{-1} \mathbf{D}_k) \Delta \mathbf{Y}_k = -(\mathbf{M}_k^{-1} \mathbf{f}_k) \tag{A-54}$$

Next, compute the values of the elements of $\mathbf{E}_k$ and $\mathbf{F}_k$, which are defined as follows:

$$\mathbf{E}_k = \mathbf{M}_k^{-1} \mathbf{D}_k, \qquad \mathbf{F}_k = \mathbf{M}_k^{-1} \mathbf{f}_k \tag{A-55}$$

Since

$$\mathbf{E}_k \Delta \mathbf{Y}_k = -\mathbf{F}_k \tag{A-56}$$

it follows that

$$\Delta \mathbf{Y}_k = -\mathbf{E}_k^{-1} \mathbf{F}_k \tag{A-57}$$

After the elements of $\Delta \mathbf{Y}_k$ have been determined, values of the variables contained in $\mathbf{X}_{k+1}$ which are to be used in the next trial are found by use of the relationship

$$\mathbf{X}_{k+1} = \mathbf{X}_k + \Delta \mathbf{X}_k = \mathbf{X}_k + \mathbf{R}_k \Delta \mathbf{Y}_k \tag{A-58}$$

which follows from Equation (A-51).

To compare the generalized scaling procedure with the scaling procedure used in the solution of the illustrative examples, reconsider the first trial of Illustrative Example 2-4. Since the scaling factors listed in items (2) through (4) of the solution of this example satisfy the inequality given by Equation (A-49), they may be taken as the elements of the diagonal matrix $\mathbf{R}_0$, that is, $r_{11} = F, r_{22} = T_0, r_{33} = F, r_{44} = T_0, r_{55} = F$, and $r_{66} = F/50$, where the definitions of the original matrices $\mathbf{X}_k$, $\mathbf{J}_k$, and $\mathbf{f}_k$ are those given by Equations (2-33), (2-34), and (2-35). Then by use of the generalized scaling procedure, the following matrices E_0 and F_0 were obtained:

$$E_0 = \begin{bmatrix} 1 & -0.2500 & 1.0 & 0.0 & 0.0 & 0.0 \\ -0.4 & -1.0 & 0.0 & 0.0 & 0.0 & 0.158 \\ 0.0 & -0.0913 & -1.0 & -0.0913 & 0.5107 & 0.0 \\ 0.0 & 1.0 & 0.667 & -1.0 & 0.0 & 0.168 \\ 0.0 & 0.0 & 0.5105 & 0.5615 & -1.0 & 0.0 \\ 0.0 & 0.0 & -1.0 & 1.0 & 0.0 & 0.164 \end{bmatrix},$$

$$F_0 = \begin{bmatrix} -0.088 \\ 0.048 \\ 0.01532 \\ -0.01867 \\ 0.02961 \\ -0.116 \end{bmatrix}$$

A comparison of the elements of J_0 (see Illustriative Example 2-4) with the corresponding elements of E_0 and the elements of f_0 (see Illustrative Example 2-4) with the corresponding elements of F_0 shows that the generalized scaling procedure is very closely related to the procedure used in the solution of the illustrative examples. In fact, for this particular choice of R_0, it is evident that each element of E_0 and F_0 may be obtained by division of each row of J_0 and f_0 of Illustrative Example 2-4 by the element of that row of J_0 which has the largest absolute value.

REFERENCES

1. Broyden, C. G., "A Class of Methods for Solving Nonlinear Simultaneous Equations," *Mathematics of Computation,* **19**, (1965), 577.

2. Burdett, J. W., "Prediction of Steady State and Unsteady State Response Behavior of a Multiple Effect Evaporator System," Ph.D. dissertation, Texas A&M University, College Station, Texas, 1969.

3. Carnahan, Brice, H. A. Luther, and J. O. Wilkes, *Applied Numerical Methods.* New York: John Wiley & Sons, Inc., 1969.

4. Faddeeva, V. N., *Computational Methods of Linear Algebra.* New York: Dover, 1959.

5. Holland, C. D., *Multicomponent Distillation.* Englewood Cliffs, N. J.: Prentice-Hall, Inc., 1963.

6. ———, *Unsteady State Processes with Applications in Multicomponent Distillation.* Englewood Cliffs, N.J.: Prentice-Hall, Inc., 1966.

7. Lapidus, Leon, *Digital Computation for Chemical Engineers*. New York: McGraw-Hill Book Company, Inc., 1963.

8. Smith, G. D., *Numerical Solution of Partial Differential Equations*. New York: Oxford University Press, 1965.

9. Sokolnikoff, I. S., and E. S. Sokolnikoff, *Higher Mathematics for Engineers and Physicists*, 2nd ed. New York: McGraw-Hill Book Company, Inc., 1941.

Theorems B

Although the following definitions and theorems are to be found in most texts dealing with functional analysis, they are repeated here for the convenience of the reader.

DEFINITION B-1. Continuity of $f(x)$ at x_0: The function of $f(x)$ is said to be continuous at the point x_0 if, for every positive number ϵ, there exists a δ_ϵ depending on ϵ such that for all x of the domain for which

$$|x - x_0| < \delta_\epsilon$$

then

$$|f(x) - f(x_0)| < \epsilon$$

DEFINITION B-2. Continuity of $f(x)$ in an interval: A function which is continuous at each point in an interval is said to be continuous in the interval.

THEOREM B-1. If the function $f(x)$ is continuous in the interval $a \leqq x \leqq b$ and $f(a) \leqq k \leqq f(b)$, then there exists a number c in the interval $a < c < b$ such that

$$f(c) = k$$

THEOREM B-2. Mean value theorem of differential calculus: If the function $f(x)$ is continuous in the interval $a \leqq x \leqq b$ and differentiable at every point

404

of the interval $a < x < b$, then there exists at least one value ξ such that

$$f(b) = f(a) + (b - a)f'[a + \xi(b - a)]$$

where $0 < \xi < 1$.

THEOREM B-3. Mean value theorem of integral calculus: If the function $f(x)$ is continuous in the interval $a \leq x \leq b$, then

$$\int_a^b f(x)\, dx = f(\xi)(b - a)$$

where $a \leq \xi \leq b$.

THEOREM B-4. Taylor's theorem: If the functions $f(x), f'(x), \ldots, f^{(n)}(x)$ are continuous for each x in the interval $a \leq x \leq b$, and $f^{(n+1)}(x)$ exists for each x in the interval $a < x < b$, then there exists a ξ in the interval $a < x < b$ such that

$$f(a + h) = f(a) + hf'(a) + \frac{h^2}{2!}f^{(2)}(a) + \frac{h^3}{3!}f^{(3)}(a) + \cdots + \frac{h_n}{n!}f^{(n)}(a) + R_n$$

where $h = b - a$, and the remainder R_n is given by the formula

$$R_n = \frac{h^{n+1}}{(n+1)!}f^{(n+1)}(\xi) \qquad (a < \xi < b)$$

THEOREM B-5. Weirstrass M test: Let $f_1(x) + f_2(x) + \cdots + f_n(x) + \cdots$ be a series of functions of x defined in the interval $a < x < b$. If there exists a convergent series of positive constants

$$M_1 + M_2 + \cdots + M_n + \cdots,$$

such that

$$|f_i(x)| \leq M_i \qquad \text{(for all } i\text{)}$$

for all x in the open interval $a < x < b$, then the series of functions is uniformly and absolutely convergent for $a < x < b$.

DEFINITION B-3. If in an interval $a \leq x \leq b$, the function $f(x)$ is continuous with the single exception of an infinite discontinuity at the point $x = x_k$ within the range of integration $(a < x_k < b)$, then the

$$\int_a^b f(x)\, dx$$

is said to exist provided each of the indicated limits exists, where by definition

$$\int_a^b f(x)\, dx = \lim_{\epsilon \to 0} \int_a^{x_k - \epsilon} f(x)\, dx + \lim_{\delta \to 0} \int_{x_k + \delta}^b f(x)\, dx$$

(Note that the point $x_k - \epsilon$ approaches the point of discontinuity x_k from the left, and $x_k + \delta$ approaches x_k independently from the right.)

THEOREM B-6. In the interval $a \leq x \leq b$, let the function $f(x)$ be continuous and positive everywhere except for the end point b where it has an infinite discontinuity. Then the integral

$$\int_a^b f(x)\, dx$$

exists provided both a positive number $\mu < 1$ and a fixed number M independent of x can be found such that everywhere in the interval $a \leq x < b$, the following inequality is true:

$$f(x) \leq \frac{M}{(b - x)^\mu}$$

On the other hand, the integral does not exist provided both a positive number $\nu \geq 1$ and a fixed number N can be found such that everywhere in the interval $a \leq x < b$, the following inequality is true:

$$f(x) \geq \frac{N}{(b - x)^\nu}$$

THEOREM B-7. Mean value theorem of differential calculus for multivariable functions: Let $f(x, y, z)$ be continuous and have continuous first partial derivatives in a domain D. Furthermore, let (x_0, y_0, z_0) and $(x_0 + h, y_0 + k, z_0 + l)$ be points in D such that the line segment joining these points lies in D. Then

$$f(x_0 + h, y_0 + k, z_0 + l) - f(x_0, y_0, z_0) = h\frac{\partial f}{\partial x} + k\frac{\partial f}{\partial y} + l\frac{\partial f}{\partial z}$$

where each partial derivative is evaluated at the point

$$(x_0 + \alpha h, y_0 + \alpha k, z_0 + \alpha l) \qquad 0 < \alpha < 1$$

Data C

TABLE C-1
CURVEFIT PARAMETERS FOR K VALUES* USED FOR THE PACKED DISTILLATION COLUMN
FOR THE TEMPERATURE RANGE OF 140°F TO 550°F AT A PRESSURE OF 165 PSIA

Component	a_{1i}	a_{2i}	a_{3i}	a_{4i}
Ethane	−0.392486	0.202187×10^{-2}	-0.219947×10^{-5}	0.789466×10^{-9}
Propane	−0.608797	0.241038×10^{-2}	-0.243232×10^{-5}	0.828446×10^{-9}
i-Butane	−0.542121	0.207116×10^{-2}	-0.206689×10^{-5}	0.719788×10^{-9}
n-Butane	−0.444393	0.164848×10^{-2}	-0.152905×10^{-5}	0.500920×10^{-9}
i-Pentane	−0.285285	0.944312×10^{-3}	-0.640445×10^{-6}	0.139459×10^{-9}
n-Pentane	−0.423656	0.146623×10^{-2}	-0.131675×10^{-5}	0.429443×10^{-9}
Hexane	−0.341693	0.109048×10^{-2}	-0.878400×10^{-6}	0.270849×10^{-9}
Heptane	0.109876	-0.702886×10^{-3}	0.137773×10^{-5}	-0.661598×10^{-9}
Octane	0.85093×10^{-1}	-0.557757×10^{-3}	0.106589×10^{-5}	-0.477751×10^{-9}
Nonane	0.182534	-0.892440×10^{-3}	0.139593×10^{-5}	-0.581714×10^{-9}
Decane	0.200104	-0.930235×10^{-3}	0.137897×10^{-5}	-0.553851×10^{-9}

*$\sqrt[3]{K_i/T} = a_{1i} + a_{2i}T + a_{3i}T^2 + a_{4i}T^3$ (T in °R); *Equilibrium Ratio Data Book*, Equation Ratio Committee of NGAA, Natural Gasoline Association of America, Tulsa, Oklahoma (1957).

407

TABLE C-2
CURVEFIT PARAMETERS FOR THE ENTHALPIES*
USED FOR THE PACKED DISTILLATION COLUMN

I. Curve fit Parameters† for the Liquid Enthalpies* for a Temperature Range of 50°F to 550°F at a Pressure of 10 atm

Component	b_{1i}	b_{2i}	b_{3i}
Ethane	-0.848570×10	0.162866	-0.194986×10^{-4}
Propane	-0.145001×10^2	0.198022	-0.290488×10^{-4}
i-Butane	-0.165534×10^2	0.216186	-0.314762×10^{-4}
n-Butane	-0.202981×10^2	0.230057	-0.386634×10^{-4}
i-Pentane	-0.233564×10^2	0.250175	-0.439179×10^{-4}
n-Pentane	-0.243715×10^2	0.256362	-0.464997×10^{-4}
Hexane	-0.238704×10^2	0.267680	-0.441978×10^{-4}
Heptane	0.375457×10	0.214683	-0.337505×10^{-5}
Octane	0.430269×10	0.233325	-0.130472×10^{-4}
Nonane	0.749894	0.257811	-0.231804×10^{-4}
Decane	0.375661×10	0.263258	-0.215909×10^{-4}

$\dagger\sqrt{h_i} = b_{1i} + b_{2i}T + b_{3i}T^2$ (T in °R).

II. Curve fit Parameters† for the Vapor Enthalpies* for a Temperature Range of 50°F to 550°F at a Pressure of 10 atm

Component	c_{1i}	c_{2i}	c_{3i}
Ethane	0.613345×10^2	0.588754×10^{-1}	0.119847×10^{-4}
Propane	0.817959×10^2	0.389819×10^{-1}	0.364709×10^{-4}
i-Butane	0.147654×10^3	-0.118529	0.152877×10^{-3}
n-Butane	0.152668×10^3	-0.115348	0.146641×10^{-3}
i-Pentane	0.130967×10^3	-0.197986×10^{-1}	0.825499×10^{-4}
n-Pentane	0.128901×10^3	-0.205096×10^{-3}	0.645015×10^{-4}
Hexane	0.858349×10^2	0.152339	-0.340186×10^{-4}
Heptane	0.128517×10^3	0.621876×10^{-1}	0.377143×10^{-4}
Octane	0.123936×10^3	0.998119×10^{-1}	0.188722×10^{-4}
Nonane	0.127235×10^3	0.115774	0.127238×10^{-4}
Decane	0.134026×10^3	0.117599	0.173987×10^{-4}

$\dagger\sqrt{H_i} = c_{1i} + c_{2i}T + c_{3i}T^2$ (T in °R).

*J. B. Maxwell, *Data on Hydrocarbons* (New York: D. Van Nostrand Company, Inc., 1955).

TABLE C-3
CURVEFIT PARAMETERS FOR K VALUES USED FOR ABSORBERS

I. K Values* Used for the Packed Absorber for the Temperature Range of $-25°F$ to $40°F$ and at a Pressure of 800 Psia

Component	a_{1i}	a_{2i}	a_{3i}	a_{4i}
Carbon dioxide	$-0.62822223 \times 10^{-1}$	$0.30688802 \times 10^{-3}$	$0.39996468 \times 10^{-6}$	$-0.5789983 \times 10^{-9}$
Nitrogen	0.50596821	$-0.43488364 \times 10^{-3}$	$-0.15009991 \times 10^{-5}$	$0.34494154 \times 10^{-8}$
Methane	0.15584934	$-0.15205775 \times 10^{-3}$	$0.50349212 \times 10^{-6}$	$-0.17713546 \times 10^{-9}$
Ethane	$0.91486037 \times 10^{-1}$	$-0.16355944 \times 10^{-3}$	$0.33741924 \times 10^{-6}$	$0.14797150 \times 10^{-9}$
Propane	$0.37769508 \times 10^{-1}$	$-0.64491702 \times 10^{-4}$	$0.29233627 \times 10^{-6}$	$-0.48597680 \times 10^{-11}$
i-Butane	$0.36708355 \times 10^{-1}$	$-0.94310963 \times 10^{-4}$	$0.28026648 \times 10^{-6}$	$0.10462797 \times 10^{-10}$
n-Butane	$0.37231278 \times 10^{-1}$	$-0.13635085 \times 10^{-3}$	$0.37584653 \times 10^{-6}$	$-0.69237741 \times 10^{-10}$
i-Pentane	$0.19747034 \times 10^{-1}$	$-0.40284984 \times 10^{-4}$	$0.14439195 \times 10^{-6}$	$0.56656790 \times 10^{-10}$
n-Pentane	$0.15414596 \times 10^{-1}$	$-0.34736106 \times 10^{-4}$	$0.12591028 \times 10^{-6}$	$0.73157133 \times 10^{-10}$
Hexane	$0.88765752 \times 10^{-3}$	$0.37082646 \times 10^{-4}$	$-0.40746951 \times 10^{-7}$	$0.15187203 \times 10^{-9}$
Heptane	$0.63677356 \times 10^{-2}$	$-0.64409760 \times 10^{-5}$	$0.31793974 \times 10^{-7}$	$0.78284379 \times 10^{-10}$
Octane	$0.99674799 \times 10^{-2}$	$-0.34673591 \times 10^{-4}$	$0.82305291 \times 10^{-7}$	$0.21022392 \times 10^{-10}$
Nonane	$0.78793392 \times 10^{-2}$	$-0.23886125 \times 10^{-4}$	$0.52435951 \times 10^{-7}$	$0.25793478 \times 10^{-10}$
Decane	$0.64146556 \times 10^{-2}$	$-0.16131104 \times 10^{-4}$	$0.30005250 \times 10^{-7}$	$0.30266026 \times 10^{-10}$

*$\sqrt[3]{K_i/T} = a_{1i} + a_{2i}T + a_{3i}T^2 + a_{4i}T^3$ (T in °R); based on data provided by the Exxon Company, Baytown, Texas.

TABLE C-4
CURVEFIT PARAMETERS FOR THE ENTHALPIES USED FOR ABSORBERS

I. Liquid Enthalpies* Used for Packed Absorbers for the Temperature Range of −25°F to 40°F at $P = 800$ Psia.

Component	b_{1i}	b_{2i}	b_{3i}	b_{4i}
Carbon dioxide	0.22524075×10^4	0.54462643×10^1	$0.27910080 \times 10^{-1}$	$-0.18765335 \times 10^{-4}$
Nitrogen	0.15837112×10^4	0.37315121×10^1	$0.17655857 \times 10^{-1}$	$-0.14662071 \times 10^{-4}$
Methane	0.81635181×10^3	0.72064600×10^1	$0.15354034 \times 10^{-1}$	$-0.84406456 \times 10^{-5}$
Ethane	0.97404712×10^3	0.11454294×10^2	$0.79399534 \times 10^{-2}$	$-0.42183183 \times 10^{-6}$
Propane	0.21237510×10^4	0.46383524×10^1	$0.31726830 \times 10^{-1}$	$-0.12580301 \times 10^{-4}$
i-Butane	0.17543628×10^4	0.92456856×10^1	$0.30206113 \times 10^{-1}$	$-0.89584664 \times 10^{-5}$
n-Butane	0.32309192×10^4	0.66175547×10^1	$0.38262386 \times 10^{-1}$	$-0.16110935 \times 10^{-4}$
i-Pentane	0.33611663×10^4	0.39552670×10^1	$0.54925647 \times 10^{-1}$	$-0.25869682 \times 10^{-4}$
n-Pentane	0.43454375×10^4	0.10596339×10^2	$0.43731511 \times 10^{-1}$	$-0.19637475 \times 10^{-4}$
Hexane	-0.44150469×10^4	0.70354599×10^2	$-0.67470074 \times 10^{-1}$	$0.60245657 \times 10^{-4}$
Heptane	0.66707016×10^2	0.18159073×10^2	$0.38164884 \times 10^{-1}$	$-0.42837073 \times 10^{-5}$
Octane	-0.10632578×10^2	0.19229950×10^2	$0.40186413 \times 10^{-1}$	$-0.70521889 \times 10^{-6}$
Nonane	-0.79141992×10^4	0.81615143×10^2	$-0.79501927 \times 10^{-1}$	$0.83943509 \times 10^{-4}$
Decane	-0.67810352×10^4	0.74108551×10^2	$-0.58315706 \times 10^{-1}$	$0.75087155 \times 10^{-4}$

*$h_i = b_{1i} + b_{2i}T + b_{3i}T^2 + b_{4i}T^3$ (T in °R); based on data provided by the Exxon Company, Baytown, Texas.

TABLE C-4 (CONTINUED)

II. Vapor Enthalpies* Used for Packed Absorbers for the Temperature Range of −25°F to 40°F at P = 800 Psia.

Component	c_{1i}	c_{2i}	c_{3i}	c_{4i}
Carbon dioxide	0.13978977×10^5	-0.96359463×10^1	$0.38228422 \times 10^{-1}$	$-0.26870170 \times 10^{-4}$
Nitrogen	0.48638672×10^4	-0.21227379×10^1	$0.17565668 \times 10^{-1}$	$-0.11367006 \times 10^{-4}$
Methane	0.63255430×10^4	-0.20747757×10^1	$0.18532634 + 10^{-1}$	$-0.10630416 \times 10^{-4}$
Ethane	0.10628934×10^5	-0.28718834×10^1	$0.24877094 \times 10^{-1}$	$-0.13233222 \times 10^{-4}$
Propane	0.13954383×10^5	-0.41930256×10^1	$0.32614145 \times 10^{-1}$	$-0.15483340 \times 10^{-4}$
i-Butane	0.94088984×10^4	0.39262680×10^2	$-0.55596594 \times 10^{-1}$	$0.51507392 \times 10^{-4}$
n-Butane	0.57302344×10^4	0.75117737×10^2	-0.13120884×10^0	$0.10517908 \times 10^{-3}$
i-Pentane	0.83081953×10^4	0.75267792×10^2	-0.12945843×10^0	$0.10845697 \times 10^{-3}$
n-Pentane	0.12804211×10^5	0.61654007×10^2	$-0.97365201 \times 10^{-1}$	$0.84398722 \times 10^{-4}$
Hexane	0.23001684×10^5	0.27744919×10^2	$0.31545494 \times 10^{-1}$	$0.49981289 \times 10^{-4}$
Heptane	0.14876816×10^5	0.59342438×10^2	$-0.81853271 \times 10^{-1}$	$0.81429855 \times 10^{-4}$
Octane	0.32793215×10^5	-0.35040283×10^2	0.11162955×10^0	$-0.42647429 \times 10^{-4}$
Nonane	0.47024656×10^5	-0.95395035×10^2	0.24547529×10^0	$-0.13209638 \times 10^{-3}$
Decane	0.55238211×10^5	-0.13195618×10^3	0.32518369×10^0	$-0.18188384 \times 10^{-3}$

*$H_i = c_{1i} + c_{2i}T + c_{3i}T^2 + c_{4i}T^3$ (T in °R); based on data provided by the Exxon Company, Baytown, Texas.

Table C-5

COEFFICIENTS AND PARAMETERS* FOR THE BLACK AND HARTWIG EQUATIONS
FOR THE PREDICTION OF THE PHASE EQUILIBRIA FOR KEROSENE-SULFUR DIOXIDE AT −10°C.

Coefficients and Parameters	Value of K_s								
	0.269	0.287	0.30	0.310	0.316	0.33	0.36	0.5	0.7
$B_{z_a a}$	0.1461	0.1429	0.1402	0.1379	0.1363	0.1332	0.1255	0.08918	0.04826
B_{ra}	0.0294	0.0274	0.0260	0.0250	0.0244	0.0231	0.0205	0.01184	0.00506
B_{aa}	0.2012	0.1917	0.18504	0.1805	0.1775	0.1712	0.1585	0.11014	0.05831
$B_{z_r r}$	0.0745	0.0722	0.0705	0.0695	0.0686	0.0670	0.0635	0.0480	0.0305
$B_{z_r a}$	0.0300	0.0292	0.0286	0.0280	0.0276	0.0265	0.0225	0.0	0.0
$B_{z_r H}$	0.0475	0.0415	0.0375	0.0358	0.0342	0.0313	0.0260	0.0110	0.0065
A	0.06085	0.05855	0.05702	0.05580	0.0551	0.05345	0.05022	0.03633	0.00851
$B \times 10^2$	0.1423	0.1370	0.1330	0.1306	0.1293	0.1256	0.1176	0.0850	0.0190
C	0.8780	0.8180	0.78336	0.7560	0.7415	0.7100	0.6500	0.44904	0.28441
D	0.01421	0.01345	0.0129	0.0125	0.01227	0.01178	0.0108	0.0074	0.0046
E	0.1037	0.0940	0.08793	0.0832	0.0808	0.0754	0.0652	0.03627	0.01522
$F \times 10^4$	0.01423	0.01279	0.0120	0.01116	0.01097	0.01023	0.00886	0.0050	0.0022
L	0.0578	0.06043	0.061267	0.06081	0.0606	0.06019	0.05658	0.03744	0.0096
$M \times 10^3$	0.9487	0.9133	0.8866	0.87065	0.8620	0.8373	0.78398	0.5666	0.1266
P	0.1026	0.1021	0.10017	0.0986	0.0974	0.0954	0.09105	0.06850	0.03304
$Q \times 10^2$	0.4404	0.4096	0.3922	0.3783	0.3710	0.3552	0.3251	0.2248	0.14267
$R \times 10^3$	0.7895	0.7472	0.7167	0.6945	0.6816	0.6544	0.6000	0.4111	0.2555
$G^\dagger$ (for $Z_a = -6$)	-0.0352	-0.0349	-0.0345	-0.0342	-0.0340	-0.0336	-0.0327	-0.0277	-0.0144
$G^\dagger$ (for $Z_a = -12$)	-0.0442	-0.0437	-0.0433	-0.0429	-0.0429	-0.0422	-0.0409	-0.0347	-0.0180

*Values of $B_{z_a a}$, B_{ra}, B_{aa}, A, B, C, D, E, F, and G were given by Black and Hartwig (1), and values of the remaining coefficients and parameters were deduced by Hutton (2) on the basis of the results given by Black and Hartwig (1).

1. Black, C. and G. M. Hartwig, "Phase Equilibria and Extraction Results for Petroleum Fractions with Sulfur Dioxide: Part 1, Binary and Ternary Equilibria; Part 2, Multicomponent Equilibria and Extraction Results for Kerosenes," *Chem. Engr. Progr. Sym. Ser.*, **64**, No. 88 (1968), 66.

2. Hutton, A. E., "Liquid-Liquid Extraction in Packed Columns at Steady State Operation," Ph.D. dissertation, Texas A&M University, College Station, Texas, 1971.

†Where G is a linear function of Z_a.

TABLE C-6

DATA PRESENTED BY BLACK AND HARTWIG* FOR TWO REFERENCE SYSTEMS AT −10°C

K_s	n-Dodecane(r)-Toluene(a)-SO$_2$(s) ($Z_a = -6$)			n-Dodecane(r)-1-Methylnophthalene(a)-SO$_2$(s) ($Z_a = -12$)		
	$\log_{10} K_r$	$\log_{10} K_a$	$\log_{10} K_H$	$\log_{10} K_r$	$\log_{10} K_a$	$\log_{10} K_H$
0.269	2.0188	0.0879	1.1359	1.9494	0.0161	1.0607
0.287	1.9017	0.0847	1.0157	1.8062	−0.0059	0.9243
0.30	1.8311	0.08337	0.9481	1.7235	−0.01773	0.8519
0.310†	1.7763	0.08167	0.9011	1.6660	−0.0251	0.8047
0.316	1.7467	0.0808	0.8760	1.6345	−0.0288	0.7789
0.33	1.6872	0.07922	0.8266	1.5700	−0.03518	0.7259
0.36	1.5576	0.0750	0.7409	1.4472	−0.0440	0.6335
0.50	1.0632	0.05875	0.4675	1.0043	−0.03574	0.3784
0.70	0.5643	0.03859	0.2398	0.5433	−0.01773	0.1892

*See Reference (1) of Table C-5.
†Values for $K_s = 0.310$ obtained by Lagrangian-parabolic interpolation.

413

TABLE C-7
PREDICTED PSEUDO TERNARY DISTRIBUTION COEFFICIENTS
FOR THE KEROSENE FEED USED IN THE FIELD TESTS*

K_s	$K_{r,P}$	$K_{a,P}$	K_H
0.20000	800.00000	5.15000	800.00000
0.20200	770.00000	5.10000	400.00000
0.20500	740.00000	5.00000	270.00000
0.21000	648.00000	4.82000	185.00000
0.22000	500.00000	4.53000	110.00000
0.23000	390.00000	4.29000	75.00000
0.24000	310.00000	4.02000	54.00000
0.25000	257.37420	3.87327	39.73058
0.26000	202.46710	3.65495	31.46449
0.27000	162.81100	3.46204	25.54285
0.28000	133.83630	3.29177	21.25528
0.29000	112.74670	3.14062	18.16028
0.20000	96.60904	3.00997	15.85352
0.31000	83.22389	2.90278	14.03456
0.32000	73.08731	2.79731	12.57523
0.33000	65.07506	2.70445	11.42850
0.34000	58.03027	3.62280	10.44734
0.35000	51.93427	2.55158	9.61976
0.36000	46.64510	2.49006	8.92209
0.37000	41.98181	2.42860	8.26628
0.38000	37.89198	2.37165	7.68234
0.39000	34.29739	2.31896	7.16174
0.40000	31.13193	2.27032	6.69706
0.41000	28.33858	2.22552	6.28190
0.42000	25.86911	2.18437	5.91072
0.43000	23.68173	2.14669	5.57865
0.44000	21.74094	2.11233	5.28152
0.35000	20.01559	2.08116	5.01570
0.46000	18.47925	2.05304	4.77801
0.47000	17.10945	2.02788	4.56560
0.48000	15.88614	2.00555	4.37616
0.49000	14.79196	1.98599	4.20755
0.50000	13.81222	1.96911	4.05800
0.51000	12.77535	1.93734	3.87216
0.52000	11.83026	1.90621	3.69949
0.53000	10.96810	1.87570	3.53896
0.54000	10.18087	1.84580	3.38964
0.55000	9.46132	1.81650	3.25069
0.56000	8.80307	1.78779	3.12135
0.57000	8.20031	1.75964	3.00092
0.58000	7.64791	1.73206	2.88876
0.59000	7.14115	1.70502	2.78427
0.60000	6.67589	1.67851	2.68693
0.61000	6.24835	1.65253	2.59625
0.62000	5.85512	1.62705	2.51177
0.63000	5.49313	1.60208	2.43310
0.64000	5.15965	1.57759	2.35984
0.65000	4.85215	1.55359	2.29167
0.66000	5.46840	1.53005	2.22825
0.67000	4.30634	1.50696	2.16931
0.68000	4.06413	1.48433	2.11457
0.69000	3.84009	1.46213	2.06380
0.70000	3.63270	1.44035	2.01678

*See Reference (2) of Table C-5.

Calculation of Partial Molar Enthalpies

D

If the expression for the enthalpy H of one mole of a mixture is given by

$$H = H^\circ + \Omega \tag{D-1}$$

then a suitable set of partial molar enthalpies (called the "virtual values") for computing the enthalpy of the mixture is given by

$$\overset{v}{\overline{H}}_i = H_i^\circ + \Omega \tag{D-2}$$

where

$H^\circ = \sum_{k=1}^{c} H_k^\circ N_k$, the enthalpy of one mole of the mixture at the standard state pressure and at the temperature of the mixture. (This expression assumes, of course, that the standard state pressure is selected such that the mixture forms an ideal solution at any temperature at this pressure.)

N_i = mole fraction of component i in the mixture of c components $(i = 1, 2, \ldots, c)$. (If the mixture under consideration is a liquid, $N_i = x_i$, and if it is a vapor, $N_i = y_i$.)

Ω = homogeneous function of order zero in the moles of each component i of the mixture at a fixed temperature and pressure.

Ω_i = a function of P and T for pure component i. For a pure component, Equations (D-1) and (D-2) reduce to $H = \overset{v}{\overline{H}}_i = H_i^\circ + \Omega_i$.

$\overset{v}{\overline{H}}$ = virtual value of the partial molar enthalpy of component i.

415

The quantities $\overset{\text{v}}{\overline{H}_i}$ are called the "virtual values" of the partial molar enthalpies because they may be used to compute the correct enthalpy of a mixture, but they may differ individually from the true values for the partial molar enthalpies.

In order to demonstrate the development of the expressions for the calculation of the partial molar enthalpies, two types of correlations are considered for illustrative purposes, namely, (1) correlations based on the law of corresponding states, and (2) correlations based on an equation of state such as the Benedict-Webb-Rubin equation of state.

The development of the formulas for the calculation of partial molar enthalpies makes use of the characteristics of homogeneous functions. A function $f(x_1, x_2, \ldots, x_c)$ is said to be homogeneous of order m if

$$f(\lambda x_1, \lambda x_2, \ldots, \lambda x_c) = \lambda^m f(x_1, x_2, \ldots, x_c) \tag{D-3}$$

where λ is any arbitrary, real number. By this definition, the total enthalpy of a mixture, H_T, is homogeneous of degree one because the enthalpy of two moles of a given mixture is twice that of one mole of the same mixture at a fixed temperature and pressure; that is,

$$H_T(\lambda n_1, \lambda n_2, \ldots, \lambda n_c) = \lambda H_T(n_1, n_2, \ldots, n_c) \tag{D-4}$$

$$H_T = n_T H \tag{D-5}$$

where n_i denotes the moles of each component i and

$$n_T = n_1 + n_2 + \cdots + n_c.$$

If the function $f(x_1, x_2, \ldots, x_c)$ is homogeneous of degree m in the x_i's and has continuous first partial derivatives, then by Euler's Theorem,

$$x_1 \frac{\partial f}{\partial x_1} + x_2 \frac{\partial f}{\partial x_2} + \cdots + x_c \frac{\partial f}{\partial x_c} = m f(x_1, x_2, \ldots, x_c) \tag{D-6}$$

Since the total enthalpy of a mixture is homogeneous of degree one in the n_i's at a fixed temperature and pressure, Euler's Theorem gives

$$n_1 \left(\frac{\partial H_T}{\partial n_1} \right)_{P,T,n_j} + n_2 \left(\frac{\partial H_T}{\partial n_2} \right)_{P,T,n_j} + \cdots + n_c \left(\frac{\partial H_T}{\partial n_c} \right)_{P,T,n_j} = H_T(n_1, n_2, \ldots, n_c) \tag{D-7}$$

The subscript n_j is used to denote the fact that the moles of all components except the particular one with which the partial derivative is to be taken are to be held fixed. The partial derivatives appearing in Equation (D-7) were given the name "partial molar enthalpies" by Gibbs, and they are commonly identified by the symbol $\overline{H}_i$; that is,

$$\overline{H}_i = \left(\frac{\partial H_T}{\partial n_i} \right)_{P,T,n_j} \tag{D-8}$$

Calculation of Partial Molar Enthalpies from Correlations Based on the Law of Corresponding States

At a given value of the critical compressibility factor Z_c for the mixture, Lydersen *et al.* (5) were able to correlate the enthalpies of mixtures as functions of the reduced temperatures and pressures. These correlations were of the general form:

$$\frac{H - H^\circ}{RT_c} = \frac{f_1(T_r, P_r)}{f_2(T_r, P_r)}, \quad \text{(at a given } Z_c\text{)} \tag{D-9}$$

where

H = enthalpy of one mole of mixture at a given temperature and pressure

$P_c = \sum\limits_{i=1}^{c} P_{ci}N_i$, pseudo critical pressure of the mixture. P_{ci} is the critical pressure of pure component i.

$T_c = \sum\limits_{i=1}^{c} T_{ci}N_i$, pseudo critical temperature of the mixture. T_{ci} is the critical temperature of pure component i.

$Z_c = \sum\limits_{i=1}^{c} Z_{ci}N_i$, pseudo critical compressibility factor for the mixture. Z_{ci} is the critical compressibility factor for pure component i.

$P_r = P/P_c$, reduced pressure.

$T_r = T/T_c$, reduced temperature.

The simplified formulas for calculating the enthalpy result from the fact that the reduced temperature and reduced pressure are homogeneous functions of degree zero in the n_i's, which is shown as follows. Since

$$T_r(n_1, n_2, \ldots, n_c) = \frac{T}{T_c} = \frac{n_T T}{\sum\limits_{i=1}^{c} T_{ci}n_i}$$

then

$$T_r(\lambda n_1, \lambda n_2, \ldots, \lambda n_c) = \frac{\lambda n_T T}{\sum\limits_{i=1}^{c} T_{ci}\lambda n_i} = T_r(n_1, n_2, \ldots, n_c)$$

Since T_r and P_r are homogeneous of degree zero, any function of these variables which has continuous first derivatives with respect to the n_i's at a fixed temperature and pressure is also homogeneous of degree zero. Also, it can be shown that any combinations of a set of homogeneous functions is also homogeneous.

Since T_c, f_1, and f_2 are all homogeneous of degree zero, then $T_c f_1/f_2$ is homogeneous of degree zero, and consequently, Equation (D-9) may be stated in the form given by Equation (D-1) by setting

$$\Omega = RT_c f_1/f_2 \tag{D-10}$$

Now let each member of Equation (D-1) be multiplied by the total number of moles to give

$$H_T = H_T^\circ + n_T \Omega \tag{D-11}$$

where

$$H_T^\circ = \sum_{i=1}^{c} H_i^\circ n_i; \ n_T H^\circ = H_T^\circ$$

Then at constant temperature and pressure, partial differentiation of each member of Equation (D-11) with respect to n_i with T and P and the moles of each component except i held fixed gives

$$\bar{H}_i = \sum_{k=1}^{c} H_k^\circ \left(\frac{\partial n_k}{\partial n_i}\right)_{P,T,n_j} + \left(\frac{\partial n_T}{\partial n_i}\right)_{P,T,n_j} \Omega + n_T \left(\frac{\partial \Omega}{\partial n_i}\right)_{P,T,n_j}$$

or

$$\bar{H}_i = H_i^\circ + \Omega + n_T \left(\frac{\partial \Omega}{\partial n_i}\right)_{P,T,n_j} \tag{D-12}$$

The difficulty of computing the partial molar enthalpies lies in the evaluation of the term $(\partial \Omega/\partial n_i)_{P,T,n_j}$. For example, page 478 of Reference (4) is required to present the formula obtained by Papadopoulos *et al.* (6) for the partial molar enthalpy when it is based on the Benedict-Webb-Rubin equation of state.

Fortunately, it can be shown as follows that the term $(\partial \Omega/\partial n_i)_{P,T,n_j}$ may be neglected in the calculation of the enthalpy of a mixture. Multiplication of each term of Equation (D-12) by n_i followed by the summation over all components yields

$$\sum_{i=1}^{c} \bar{H}_i n_i = \sum_{i=1}^{c} H_i^\circ n_i + \Omega n_T + n_T \sum_{i=1}^{c} n_i \left(\frac{\partial \Omega}{\partial n_i}\right)_{P,T,n_j} \tag{D-13}$$

By Equation (D-7), the left-hand side of Equation (D-13) is equal to H_T, and by Euler's Theorem [Equation (D-6)], it follows that

$$\sum_{i=1}^{c} n_i \left(\frac{\partial \Omega}{\partial n_i}\right)_{P,T,n_j} = 0 \tag{D-14}$$

since Ω is homogeneous of degree zero. Thus, Equation (D-13) reduces to Equation (D-11), and division of each term of this expression by n_T gives Equation (D-1). Since the last term of Equation (D-12) disappears in the calculation of the enthalpy of a mixture, it follows that this derivative may be omitted in the expression for $\bar{H}_i$ to give the expression for $\overset{\text{v}}{\bar{H}}_i$, Equation (D-2). Note that the virtual values of the partial molar enthalpies give the correct enthalpy for the mixture. For, multiplication of each term of Equation (D-2) by n_i followed by the summation overall components i gives

$$\sum_{i=1}^{c} \overset{\text{v}}{\bar{H}}_i n_i = n_T H^\circ + n_T \Omega \tag{D-15}$$

Comparison of Equations (D-11) and (D-15) shows that

$$H_T = \sum_{i=1}^{c} \overset{v}{H_i} n_i \qquad (D\text{-}16)$$

Calculation of Partial Molar Enthalpies By Use of the Benedict-Webb-Rubin Equations

The Benedict-Webb-Rubin equation of state (1) is given by

$$P = RT\rho + \left(B_0 RT - A_0 - \frac{C_0}{T^2}\right)\rho^2 + $$
$$(bRT - a)\rho^3 + a\alpha\rho^6 + \frac{c\rho^3}{T^2}(1 + \sigma\rho^2)e^{-\sigma\rho^2} \qquad (D\text{-}17)$$

The usual mixing rules for this equation of state are:

$$B_0 = \sum_{i=1}^{c} N_i B_{0i} \qquad a = \left[\sum_{i=1}^{c} N_i a_i^{1/3}\right]^3$$

$$A_0 = \left[\sum_{i=1}^{c} N_i A_{0i}^{1/2}\right]^2 \qquad c = \left[\sum_{i=1}^{c} N_i c_i^{1/3}\right]^3$$

$$C_0 = \left[\sum_{i=1}^{c} N_i C_{0i}^{1/2}\right]^2 \qquad \alpha = \left[\sum_{i=1}^{c} N_i \alpha_i^{1/3}\right]^3$$

$$b = \left[\sum_{i=1}^{c} N_i b_i^{1/3}\right]^3 \qquad \gamma = \left[\sum_{i=1}^{c} N_i \gamma_i^{1/2}\right]^2$$

where N_i is the mole fraction of component i in either phase. The Lorentz rule or a quadratic form of B_0 of the mixture is sometimes used:

$$B_0 = \sum_{ij} N_i N_j [(B_{0i})^{1/3} + (B_{0j})^{1/3}]/8 \quad \text{(Lorentz)}$$

$$B_0 = \left[\sum_{i=1}^{c} N_i B_{0i}^{1/2}\right]^{1/2} \quad \text{(Quadratic)}$$

Following others, Orye (7) suggested that the following mixing rule for A_0 be used instead of the one listed above

$$A_0 = \sum_{i=1}^{c} N_i^2 A_{0i} + \sum_i \sum_{\substack{j \\ j \neq i \\ j > i}} M_{ij} N_i N_j A_{0i}^{1/2} A_{0j}^{1/2}$$

An examination of all of the above expressions for the evaluation of the constants A_0, B_0, C_0, b, a, c, α, and γ shows that they are all homogeneous of degree zero.

Orye (7) presents the following expression for Ω of Equation (D-1).

$$\Omega = 0.18504\{(B_0 RT - 2A_0 - 4C_0/T^2)\rho$$
$$+ (2bRT - 3a)\rho^2/2 + 6a\alpha\rho^5/5$$

$$+ \frac{c\rho^2}{T^2}\left[3\left\{\frac{1 - \exp(-\gamma\rho^2)}{\gamma\rho^2}\right\} - \frac{\exp(-\gamma\rho^2)}{2}\right.$$

$$+ \gamma\rho^2(\exp(-\gamma\rho^2) + \frac{\rho}{T}\frac{dC_0}{dT} + \rho T \frac{dA_0}{dT}$$

$$\left. - \frac{RT^2\rho^2}{2}\frac{db}{dT} + \frac{c}{\gamma^2 T}\left[1 - \{\exp(-\gamma\rho^2)\}\left\{\gamma\rho^2 + 1 + \frac{\gamma^2\rho^4}{2}\right\}\right]\right]\frac{d\gamma}{dT} \quad \text{(D-18)}$$

Thus, the functional dependency of Ω may be represented as follows:

$$\Omega = \Omega(n_1, n_2, \ldots, n_c, \rho, T)$$

Then

$$\left(\frac{\partial H_T}{\partial n_i}\right)_{P,T,n_j} = H_i^\circ + \Omega + n_T\left(\frac{\partial\Omega}{\partial n_i}\right)_{P,T,n_j} \quad \text{(D-19)}$$

where

$$\left(\frac{\partial\Omega}{\partial n_i}\right)_{P,T,n_j} = \left(\frac{\partial\Omega}{\partial\rho}\right)_{T,n_i}\left(\frac{\partial\rho}{\partial n_i}\right)_{P,T,n_j} + \left(\frac{\partial\Omega}{\partial n_i}\right)_{\rho,T,n_j}$$

Since the density of one mole of a given mixture is equal to the density of two moles of the same mixture at a given temperature and pressure, the density ρ of a mixture is homogeneous of degree zero in the n_i's at a given temperature and pressure. Then

$$\left(\frac{\partial\Omega}{\partial\rho}\right)_{T,n_i}\sum_{i=1}^{c} n_i\left(\frac{\partial\rho}{\partial n_i}\right)_{P,T,n_j} = 0$$

Also,

$$\sum_{i=1}^{c} n_i\left(\frac{\partial\Omega}{\partial n_i}\right)_{\rho,T,n_j} = 0$$

Consequently,

$$\sum_{i=1}^{c} n_i\left(\frac{\partial\Omega}{\partial n_i}\right)_{P,T,n_j} = 0$$

Thus, it follows that the term involving the derivative $\partial\Omega/\partial n_i$ in Equation (D-19) may be omitted to give Equation (D-2), the expression for the virtual value partial molar enthalpy.

REFERENCES

1. Benedict, M., G. B. Webb, and L. C. Rubin, "An Empirical Equation for Thermodynamic Properties of Light Hydrocarbons and Their Mixtures," *J. Chem. Phys.*, **8**, (1940), 334, **10**, (1942), 747.

2. Burningham, W. and F. D. Otto, "Which Computer Design for Absorbers?" *Hydrocarbon Processing*, **46**, (1967), 163.

3. Chao, K. C. and J. D. Seader, "A General Correlation of Vapor-Liquid Equilibria in Hydrocarbon Mixtures," *AIChE Journal*, **10**, (1964), 698.

4. Holland, C. D., *Multicomponent Distillation*. Englewood Cliffs, N.J.: Prentice-Hall, Inc., 1963.

5. Lydersen, A. L., R. A. Greenkorn, and O. A. Hougen, *Wisc. Univ. Eng. Exp. Sta. Rept.*, 4, (October 1955).

6. Papadopoulos, A., R. L. Pigford, and L. Friend, "Partial Molal Enthalpies of the Lighter Hydrocarbons in Solution with Other Hydrocarbons," *Chemical Engineering Progress Symposium Series*, **49**, 7, (1953), 119.

7. Orye, R. V., "Prediction and Correlation of Phase Equilibria and Thermal Properties with the BWR Equation of State," *I & EC Process Design and Development*, **8**, (1969), 579.

8. Redlich, O. and J. N. S. Kwong, "On the Thermodynamics of Solutions V." An Equation of State. Fugacities of Gaseous Solutions, *Chem. Rev.*, **44**, (1949), 233.

9. Yen, L. C. and R. E. Alexander, "Estimation of Vapor and Liquid Enthalpies," *AIChE Journal*, **11**, (1965), 334.

Index

Index